AQA Chemistry

2nd Edition

A LEVEL YEAR 1 AND AS

Ted Lister
Janet Renshaw

OXFORD
UNIVERSITY PRESS

OXFORD
UNIVERSITY PRESS

Great Clarendon Street, Oxford, OX2 6DP, United Kingdom

Oxford University Press is a department of the University of Oxford.
It furthers the University's objective of excellence in research,
scholarship, and education by publishing worldwide. Oxford is a
registered trade mark of Oxford University Press in the UK and in
certain other countries

British Library Cataloguing in Publication Data
Data available

978-0-19-835181-8

10 9 8 7 6 5

Paper used in the production of this book is a natural, recyclable
product made from wood grown in sustainable forests.
The manufacturing process conforms to the environmental regulations
of the country of origin.

Printed in Great Britain by Bell and Bain Ltd, Glasgow

Approval message from AQA

This textbook has been approved by AQA for use with our qualification.
This means that we have checked that it broadly covers the specification and
we are satisfied with the overall quality. Full details of our approval process
can be found on our website.

We approve textbooks because we know how important it is for teachers
and students to have the right resources to support their teaching and
learning. However, the publisher is ultimately responsible for the editorial
control and quality of this book.

Please note that when teaching the *AQA AS or A-Level Chemistry* course, you
must refer to AQA's specification as your definitive source of information.
While this book has been written to match the specification, it does not
provide complete coverage of every aspect of the course.

A wide range of other useful resources can be found on the relevant subject
pages of our website: www.aqa.org.uk.

AS/A Level course structure

This book has been written to support students studying for AQA AS Chemistry and for students in their first year of studying for AQA A Level Chemistry. It covers the AS sections from the specification, the content of which will also be examined at A Level. The sections covered are shown in the contents list, which also shows you the page numbers for the main topics within each section. There is also an index at the back to help you find what you are looking for. If you are studying for AS Chemistry, you will only need to know the content in the blue box.

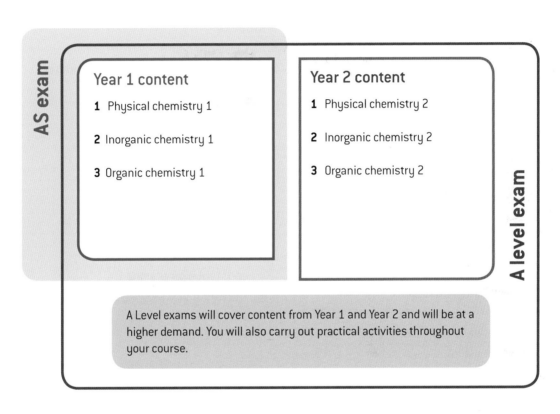

AS exam

Year 1 content

1 Physical chemistry 1

2 Inorganic chemistry 1

3 Organic chemistry 1

Year 2 content

1 Physical chemistry 2

2 Inorganic chemistry 2

3 Organic chemistry 2

A level exam

A Level exams will cover content from Year 1 and Year 2 and will be at a higher demand. You will also carry out practical activities throughout your course.

| How to use this book | vi |
| Kerboodle | ix |

Section 1 Physical chemistry 1   2

1 Atomic structure   4
1.1 Fundamental particles   4
1.2 Mass number, atomic number, and isotopes   6
1.3 The arrangement of the electrons   8
1.4 The mass spectrometer   10
1.5 More about electron arrangements in atoms   14
1.6 Electron arrangements and ionisation energy   17
    Practice questions   20

2 Amount of substance   22
2.1 Relative atomic and molecular masses, the Avogadro constant, and the mole   22
2.2 Moles in solution   25
2.3 The ideal gas equation   27
2.4 Empirical and molecular formulae   30
2.5 Balanced equations and related calculations   35
2.6 Balanced equations, atom economies, and percentage yields   39
    Practice questions   42

3 Bonding   44
3.1 The nature of ionic bonding   44
3.2 Covalent bonding   47
3.3 Metallic bonding   50
3.4 Electronegativity – bond polarity in covalent bonds   52
3.5 Forces acting between molecules   54
3.6 The shapes of molecules and ions   60
3.4 Bonding and physical properties   64
    Practice questions   71

4 Energetics   72
4.1 Exothermic and endothermic reactions   72
4.2 Enthalpy   75
4.3 Measuring enthalpy changes   77
4.4 Hess's law   82
4.5 Enthalpy changes of combustion   85
4.6 Representing thermochemical cycles   87
4.7 Bond enthalpies   91
    Practice questions   94

5 Kinetics   96
5.1 Collision theory   96
5.2 The Maxwell-Boltzmann distribution   98
5.3 Catalysts   100
    Practice questions   104

6 Equilibria   106
6.1 The idea of equilibrium   106
6.2 Changing the conditions of an equilibrium reaction   108
6.3 Equilibrium reactions in industry   111
6.4 The Equilibrium constant, K_c 114
6.5 Calculations using equilibrium constant expressions   116
6.6 The effect of changing conditions on equilibria   119
    Practice questions   122

7 Oxidation, reduction, and redox reactions   124
7.1 Oxidation and reduction   124
7.2 Oxidation states   127
7.3 Redox equations   130
    Practice questions   134
End of Section 1 questions   136
Section 1 summary   138

Section 2 Inorganic chemistry 1   140

8 Periodicity   142
8.1 The Periodic Table   142
8.2 Trends in the properties of elements Period 3   145
8.3 More trends in the properties of the elements of Period 3   147

8.4	A closer look at ionisation energies	150
	Practice questions	152
9	**Group 2, the Alkaline Earth Metals**	**154**
9.1	The physical and chemical properties of Group 2	154
	Practice questions	158
10	**Group 7(17), the Halogens**	**160**
10.1	The Halogens	160
10.2	The chemical reactions of the Halogens	162
10.3	Reactions of halide ions	164
10.4	Uses of chlorine	167
	Practice questions	168
	End of Section 2 questions	**170**
	Section 2 summary	**172**

Section 3 Organic chemistry 1   174

11	**Introduction to organic chemistry**	**176**
11.1	Carbon compounds	176
11.2	Nomenclature – naming organic compounds	181
11.3	Isomerism	186
	Practice questions	188
12	**Alkanes**	**190**
12.1	Alkanes	190
12.2	Fractional distillation of crude oil	193
12.3	Industrial cracking	196
12.4	Combustion of alkanes	198
12.5	The formation of halogenoalkanes	201
	Practice questions	203
13	**Halogenoalkanes**	**206**
13.1	Halogenoalkanes – introduction	206
13.2	Nucleophilic substitution in halogenoalkanes	208
13.3	Elimination reaction in halogenoalkanes	211
	Practice questions	214
14	**Alkenes**	**216**
14.1	Alkenes	216
14.2	Reactions of alkenes	220

14.3	Addition polymers	224
	Practice questions	228
15	**Alcohols**	**230**
15.1	Alcohols – introduction	230
15.2	Ethanol production	233
15.3	The reactions of alcohols	235
	Practice questions	240
16	**Organic analysis**	**242**
16.1	Test-tube reactions	242
16.2	Mass spectrometry	243
16.3	Infrared spectroscopy	245
	Practice questions	250
	End of Section 3 questions	**252**
	Section 3 summary	**254**

Section 4 Practical skills	**256**
Section 5 Mathematical skills	**260**

Reference	
Data table	268
Periodic Table	269
Glossary	270
Answers	274
Index	286
Acknowledgements	292

How to use this book

Learning objectives

→ At the beginning of each topic, there is a list of learning objectives.

→ These are matched to the specification and allow you to monitor your progress.

→ A specification reference is also included.
Specification reference: 3.1.1

This book contains many different features. Each feature is designed to foster and stimulate your interest in chemistry, as well as supporting and developing the skills you will need for your examinations. Terms that you will need to be able to define and understand are highlighted in **bold orange text**. You can look these words up in the glossary.

Sometimes a word appears in **bold**. These are words that are useful to know but are not used on the specification. They therefore do not have to be learnt for examination purposes.

Synoptic link

These highlight how the sections relate to each other. Linking different areas of chemistry together becomes increasingly important, and you will need to be able to do this.

There are also links to the mathematical skills on the specification. More detail can be found in the maths section.

 ## Application features

These features contain important and interesting applications of chemistry in order to emphasise how scientists and engineers have used their scientific knowledge and understanding to develop new applications and technologies. There are also application features to develop your maths skills, with the icon 🔲, and to develop your practical skills, with the icon 🔺.

 ## Extension features

These features contain material that is beyond the specification designed to stretch and provide you with a broader knowledge and understanding and lead the way into the types of thinking and areas you might study in further education. As such, neither the detail nor the depth of questioning will be required for the examinations. But this book is about more than getting through the examinations.

1 Extension and application features have questions that link the material with concepts that are covered in the specification. Short answers are inverted at the bottom of the feature, whilst longer answers can be found in the answers section at the back of the book.

Study tips

Study tips contain prompts to help you with your revision. They can also support the development of your practical skills (with the practical symbol 🔺) and your mathematical skills (with the math symbol 🔲).

Hint

Hint features give other information or ways of thinking about a concept to support your understanding. They can also relate to practical or mathematical skills and use the symbols 🔺 and 🔲.

Summary questions

1 These are short questions that test your understanding of the topic and allow you to apply the knowledge and skills you have aquired. The questions are ramped in order of difficulty.

2 Questions that will test and develop your mathematical and practical skills are labelled with the mathematical symbol (🔲) and the practial symbol (🔺).

Introduction at the opening of each section summarises what you need to know.

Section 1

about a century ago that scientists began to discover the ..., for example, that they were built up from smaller ... led to an understanding of how atoms are held together, ...gement of the Periodic Table makes sense, and how the ...elements and compounds can be explained. This unit ...standing of how atoms behave to explain some of the most important ideas in chemistry.

2 Amount of substance
3 Bonding
4 Energetics
5 Kinetics
6 Equilibria
7 Oxidation, reduction, and redox reactions

Atomic structure revises the idea of the atom, looking at some of the evidence for sub-atomic particles. It introduces the mass spectrometer, which is used to measure the masses of atoms. The evidence for the arrangement of electrons is studied and you will see how a more sophisticated model using atomic orbitals rather than circular orbits was developed.

Amount of substance is about quantitative chemistry, that is, how much product you get from a given amount of reactants. The idea of the mole is used as the unit of quantity to compare equal numbers of atoms and molecules of different substances, including gases and solutions. Balanced equations are used to describe and measure the efficiency of chemical processes.

Bonding revisits the three types of strong bonds that hold atoms together – ionic, covalent, and metallic. It introduces three weaker types of forces that act between molecules, the most significant of these being hydrogen bonding. It examines how the different types of forces are responsible for the solid, liquid, and gaseous states, and explores how the electrons contribute to the shapes of molecules and ions.

Energetics revisits exothermic and endothermic reactions and introduces the concept of enthalpy – heat energy measured under specific conditions. It looks at different ways of measuring enthalpy changes and then uses Hess's law to predict the energy changes of reactions. The idea of bond energies is explored to work out theoretical enthalpy changes by measuring the energy needed to make and break bonds.

Kinetics deals with the rate at which reactions take place, reinforcing the idea that reactions only happen when molecules of the reactants collide with enough energy to break bonds. The Maxwell–Boltzmann distribution shows us mathematically what fraction of the reactant molecules have enough collision energy at a given temperature. The role of catalysis is then explored.

Equilibria is about reactions that do not go to completion so that the end result is a mixture of reactants and products. It examines how to get the greatest proportion of desired products in the mixture by changing the conditions, and how to calculate the equilibrium composition. Some industrially important reversible reactions are then discussed.

Redox reactions expands the definition of oxidation as addition of oxygen to include reactions that involve electron transfers. It explains the idea of an oxidation state for elements and ions, and uses this to help balance complex redox (reduction–oxidation) equations.

The applications of science are found throughout the chapters, where they will provide you with an opportunity to apply your knowledge in a fresh context.

What you already know

The material in this section builds upon knowledge and understanding that you will have developed at GCSE, in particular the following:

☐ There are just over 100 elements, all made up of atoms.
☐ The atoms of any element are essentially the same as each other but they are different from the atoms of any other element.
☐ Atoms are tiny and cannot be weighed individually.
☐ Atoms are made of protons, neutrons, and electrons.
☐ Atoms bond together to obtain full outer shells of electrons.
☐ Atoms may lose or gain electrons to form ions with full outer electron shells.
☐ Chemical reactions may give out (exothermic) or take in (endothermic) heat.
☐ The rates of chemical reactions are affected by temperature, concentration of reactants, surface area of solids, and catalysts.
☐ Some chemical reactions are reversible – they do not go to completion.
☐ Reactions can be classified as oxidation (addition of oxygen) or reduction (removal of oxygen).

A checklist to help you assess your knowledge from KS4, before starting work on the section.

2

Visual summaries of each section show how some of the key concepts of that section interlink.

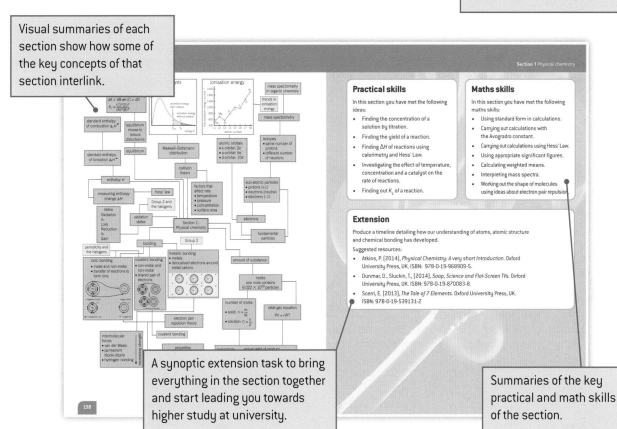

Section 1 Physical chemistry

Practical skills

In this section you have met the following ideas:

- Finding the concentration of a solution by titration.
- Finding the yield of a reaction.
- Finding ΔH of reactions using calorimetry and Hess' Law.
- Investigating the effect of temperature, concentration and a catalyst on the rate of reactions.
- Finding out K_c of a reaction.

Maths skills

In this section you have met the following maths skills:

- Using standard form in calculations.
- Carrying out calculations with the Avogadro constant.
- Carrying out calculations using Hess' Law.
- Using appropriate significant figures.
- Calculating weighted means.
- Interpreting mass spectra.
- Working out the shape of molecules using ideas about electron pair repulsion.

Extension

Produce a timeline detailing how our understanding of atoms, atomic structure and chemical bonding has developed.

Suggested resources:

- Atkins, P. (2014), *Physical Chemistry: A very short Introduction*. Oxford University Press, UK. ISBN: 978-0-19-968909-5.
- Dunmar, D., Sluckin, T., (2014), *Soap, Science and Flat-Screen TVs*. Oxford University Press, UK. ISBN: 978-0-19-870083-8.
- Scerri, E. (2013), *The Tale of 7 Elements*. Oxford University Press, UK. ISBN: 978-0-19-539131-2

138

A synoptic extension task to bring everything in the section together and start leading you towards higher study at university.

Summaries of the key practical and math skills of the section.

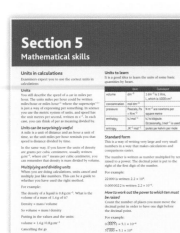

Section 5
Mathematical skills

Mathematical section to support and develop your mathematical skills required for your course. Remember, at least 20% of your exam will involve mathematical skills.

Section 4
Practical skills

Practical skills section with questions for each suggested practical on the specification. Remember, at least 15% of your exam will be based on practical skills.

Practice questions

Practice questions at the end of each chapter and each section, including questions that cover practical and maths skills. There are also additional practice questions at the end of the book.

Kerboodle

This book is supported by next generation Kerboodle, offering unrivalled digital support for independent study, differentiation, assessment, and the new practical endorsement.

If your school subscribes to Kerboodle, you will also find a wealth of additional resources to help you with your studies and with revision:

- Study guides
- Maths skills boosters and calculation worksheets
- On your marks activities to help you achieve your best
- Practicals and follow up activities to support the practical endorsement
- Interactive objective tests that give question-by-question feedback
- Animations and revision podcasts
- Self-assessment checklists.

Revise with ease using the study guides to guide you through each chapter and direct you towards the resources you need.

Oxford A Level Sciences

AQA Chemistry

15.3 The reactions of alcohols
Method sheet

Safety

- Concentrated phosphoric acid is CORROSIVE
- Cyclohexanol is HARMFUL
- Anhydrous calcium chloride is an IRRITANT
- Bromine water is HARMFUL
- Saturated sodium chloride solution is LOW HAZARD
- Eye protection should be worn
- Wear gloves when handling concentrated phosphoric acid.

Equipment and materials

- concentrated phosphoric acid
- 10 cm³ measuring cylinder
- round bottom flasks
- cork ring
- Liebig condenser
- separating funnel and bung
- balance (accurate to at least **two** decimal places)
- anti-bumping granules
- spatula
- protective gloves

- saturated sodium chloride solution
- anhydrous calcium chloride
- cyclohexanol
- bromine water
- 0–100 °C thermometer and thermometer pocket
- 50 cm³ conical flask
- test tube and bung
- Bunsen burner
- heat proof mat
- clamp and clamp stand.

Method

Measure out 10 cm³ of cyclohexanol in a measuring cylinder and place into a round bottom flask.

Slowly, add 4 cm³ of concentrated phosphoric acid and swirl to ensure the contents are thoroughly mixed. Add a few anti-bumping granules to the flask. Sit your round bottom flask in a cork ring if you need to put it down.

Set up your equipment as shown in Figure 2, holding all of the equipment in place using a clamp and clamp stand.

Make sure that the cold water enters the *bottom* of the condenser to ensure complete cooling of the vapours.

Using a Bunsen burner, gently heat the flask and distil slowly, collecting the distillate formed between 70 °C and 90 °C.

Pour the distillate into a separating funnel and add an equal volume of saturated sodium chloride solution. Place the bung on the separating funnel and invert to mix the contents thoroughly.

On standing, two layers will form in the separating funnel. The upper layer will contain the crude alkene; discard the lower aqueous layer, which will contain impurities, and then place the upper layer into a small conical flask.

Oxford University Press 2015   www.oxfordsecondary.co.uk/acknowledgements

This source sheet may have been changed from the original   2

1.4 The mass spectrometer

Explain how a mass spectrometer works and what it measures

A mass spectrometer is used by scientists to accurately determine the relative atomic mass of an element. See how mass spectrometry works in the animation '1.4 Animation: Mass spectrometry'.

You can calculate relative atomic mass from the spectrum produced. Practise calculating isotopic abundance using '1.4 Maths skills: Mass spectrometry and calculating relative atomic mass' – the activity will take you through the process step-by-step.

2.2 Mass spectrometry

2.2 Maths skill: Mass spectrometry and calculation of relative atomic mass

If you are a teacher reading this, Kerboodle also has plenty of further assessment resources, answers to the questions in the book, and a digital markbook along with full teacher support for practicals and the worksheets, which include suggestions on how to support and stretch your students. All of the resources are pulled together into teacher guides that suggest a route through each chapter.

Section 1
Physical chemistry 1

Chapters in this section

1 Atomic structure

2 Amount of substance

3 Bonding

4 Energetics

5 Kinetics

6 Equilibria

7 Oxidation, reduction, and redox reactions

It was only about a century ago that scientists began to discover the nature of atoms, for example, that they were built up from smaller particles. This led to an understanding of how atoms are held together, why the arrangement of the Periodic Table makes sense, and how the properties of elements and compounds can be explained. This unit uses the understanding of how atoms behave to explain some of the most important ideas in chemistry.

Atomic structure revises the idea of the atom, looking at some of the evidence for sub-atomic particles. It introduces the mass spectrometer, which is used to measure the masses of atoms. The evidence for the arrangement of electrons is studied and you will see how a more sophisticated model using atomic orbitals rather than circular orbits was developed.

Amount of substance is about quantitative chemistry, that is, how much product you get from a given amount of reactants. The idea of the mole is used as the unit of quantity to compare equal numbers of atoms and molecules of different substances, including gases and solutions. Balanced equations are used to describe and measure the efficiency of chemical processes.

Bonding revisits the three types of strong bonds that hold atoms together – ionic, covalent, and metallic. It introduces three weaker types of forces that act between molecules, the most significant of these being hydrogen bonding. It examines how the various types of forces are responsible for the solid, liquid, and gaseous states, and explores how the electrons contribute to the shapes of molecules and ions.

Energetics revisits exothermic and endothermic reactions and introduces the concept of enthalpy – heat energy measured under specific conditions. It looks at different ways of measuring enthalpy changes and then uses Hess's law to predict the energy changes of reactions. The idea of bond energies is explored to work out theoretical enthalpy changes by measuring the energy needed to make and break bonds.

Kinetics deals with the rate at which reactions take place, reinforcing the idea that reactions only happen when molecules of the reactants collide with enough energy to break bonds. The Maxwell–Boltzmann distribution shows us mathematically what fraction of the reactant molecules have enough collision energy at a given temperature. The role of catalysts is then explored.

Equilibria is about reactions that do not go to completion so that the end result is a mixture of reactants and products. It examines how to get the greatest proportion of desired products in the mixture by changing the conditions, and how to calculate the equilibrium composition. Some industrially important reversible reactions are then discussed.

Redox reactions expands the definition of oxidation as addition of oxygen to include reactions that involve electron transfers. It explains the idea of an oxidation state for elements and ions, and uses this to help balance complex redox (reduction–oxidation) equations.

The applications of science are found throughout the chapters, where they will provide you with an opportunity to apply your knowledge in a fresh context.

What you already know

The material in this section builds upon knowledge and understanding that you will have developed at GCSE, in particular the following:

- ☐ There are just over 100 elements, all made up of atoms.
- ☐ The atoms of any element are essentially the same as each other but they are different from the atoms of any other element.
- ☐ Atoms are tiny and cannot be weighed individually.
- ☐ Atoms are made of protons, neutrons, and electrons.
- ☐ Atoms bond together to obtain full outer shells of electrons.
- ☐ Atoms may lose or gain electrons to form ions with full outer electron shells.
- ☐ Chemical reactions may give out (exothermic) or take in (endothermic) heat.
- ☐ The rates of chemical reactions are affected by temperature, concentration of reactants, surface area of solids, and catalysts.
- ☐ Some chemical reactions are reversible – they do not go to completion.
- ☐ Reactions can be classified as oxidation (addition of oxygen) or reduction (removal of oxygen).

Learning objectives:

→ State the relative masses of protons, neutrons, and electrons.

→ State the relative charges of protons, neutrons, and electrons.

→ Explain how these particles are arranged in an atom.

Specification reference: 3.1.1

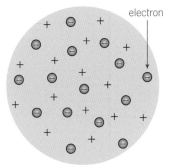

electron

sphere of positive charge

▲ **Figure 1** *The plum pudding model of the atom – electrons located in circular arrays within a sphere of positive charge*

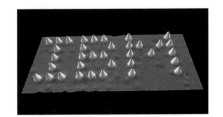

▲ **Figure 2** *Atoms can only be seen indirectly. This photograph of xenon atoms was taken by an instrument called a scanning tunnelling electron microscope*

Developing ideas of the atom

The Greek philosophers had a model in which matter was made up of a single continuous substance that produced the four elements – earth, fire, water, and air. The idea that matter was made of individual atoms was not taken seriously for another 2000 years. During this time alchemists built up a lot of evidence about how substances behave and combine. Their aim was to change other metals into gold. Here are a few of the steps that led to our present model.

1661 Robert Boyle proposed that there were some substances that could not be made simpler. These were the chemical elements, as we now know them.

1803 John Dalton suggested that elements were composed of indivisible atoms. All the atoms of a particular element had the same mass and atoms of different elements had different masses. Atoms could not be broken down.

1896 Henri Becquerel discovered radioactivity. This showed that particles could come from inside the atom. Therefore the atom was not indivisible. The following year, J J Thomson discovered the electron. This was the first sub-atomic particle to be discovered. He showed that electrons were negatively charged and electrons from all elements were the same.

As electrons had a negative charge, there had to be some source of positive charge inside the atom too. Also, as electrons were much lighter than whole atoms, there had to be something to account for the rest of the mass of the atom. Thompson suggested that the electrons were located within the atom in circular arrays, like plums in a pudding of positive charge, see Figure 1.

1911 Ernest Rutherford and his team found that most of the mass and all the positive charge of the atom was in a tiny central nucleus.

So, for many years, it has been known that atoms themselves are made up of smaller particles, called sub-atomic particles. The complete picture is still being built up in 'atom smashers' such as the one at CERN, near Geneva.

The sub-atomic particles

Atoms are made of three fundamental particles – **protons**, **neutrons**, and **electrons**.

The protons and neutrons form the **nucleus**, in the centre of the atom.

• Protons and neutrons are sometimes called **nucleons** because they are found in the nucleus.

• The electrons surround the nucleus.

The properties of the sub-atomic particles are shown in Table 1.

▼ **Table 1** *The properties of the sub-atomic particles*

Property	Proton *p*	Neutron *n*	Electron *e*
Mass / kg	1.673×10^{-27}	1.675×10^{-27}	0.911×10^{-30} (very nearly 0)
Charge / C	$+1.602 \times 10^{-19}$	0	-1.602×10^{-19}
Position	in the nucleus	in the nucleus	around the nucleus

These numbers are extremely small. In practice, *relative* values for mass and charge are used. The relative charge on a proton is taken to be +1, so the charge on an electron is −1. Neutrons have no charge, see Table 2.

▼ **Table 2** *The relative masses and charges of the sub-atomic particles*

	Proton *p*	Neutron *n*	Electron *e*
Relative mass	1	1	$\dfrac{1}{1840}$
Relative charge	+1	0	−1

Study tip

You must remember the relative masses and charges of a proton, neutron, and an electron as given in Table 2.

In a neutral atom, the number of electrons must be the same as the number of protons because their charge is equal in size and opposite in sign.

The arrangement of the sub-atomic particles

The sub-atomic particles (protons, neutrons, and electrons) are arranged in the atom as shown in Figure 3.

The protons and neutrons are in the centre of the atom, held together by a force called the **strong nuclear force**. This is much stronger than the **electrostatic forces** of attraction that hold electrons and protons together in the atom, so it overcomes the repulsion between the protons in the nucleus. It acts only over very short distances, that is, within the nucleus.

The nucleus is surrounded by electrons. Electrons are found in a series of levels, sometimes referred to as orbits or shells, which get further and further away from the nucleus. This is a simplified picture that will develop in Topic 1.5.

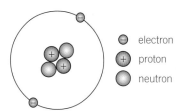

▲ **Figure 3** *The sub-atomic particles in a helium atom (not to scale)*

Summary questions

1 **a** Identify which of the following – protons, neutrons, or electrons:

　　i are nucleons

　　ii have the same relative mass

　　iii have opposite charges

　　iv have no charge

　　v are found outside the nucleus

   b Explain why we assume that there are the same number of protons and electrons in an atom.

Extension

The diameter of the nucleus of a hydrogen atom is about 2×10^{-15} m, while the diameter of the atom itself is about 1×10^{-10} m, about 50 000 times larger. This means that if the nucleus were the size of a fly, the whole atom would be roughly the size of a cathedral.

St Paul's Cathedral is roughly 200 m long. Estimate the length of a fly and, without using a calculator, check that the analogy is realistic.

Mass number, atomic number, and isotopes

Learning objectives:

→ Define the terms mass number, atomic number, and isotope.

→ Explain why isotopes of the same element have identical chemical properties.

Specification reference: 3.1.1

Mass number and atomic number

Atomic number Z

As you have seen in Topic 1.1, atoms consist of a tiny nucleus made up of protons and neutrons that is surrounded by electrons. The number of protons in the nucleus is called the atomic number or the **proton number** Z.

The number of electrons in the atom is equal to the proton number, so atoms are electrically neutral. The number of electrons in the outer shell of an atom determines the chemical properties of an element (how it reacts) and what sort of element it is. The atomic number defines the chemical identity of an element.

atomic number (proton number) Z = number of protons

All atoms of the same element have the same atomic number. Atoms of different elements have different atomic numbers.

Mass number A

The total number of protons plus neutrons in the nucleus (the total number of nucleons) is called the mass number A. It is the nucleons that are responsible for almost all of the mass of an atom because electrons weigh virtually nothing.

mass number A = number of protons + number of neutrons

Isotopes

Every single atom of any particular element has the same number of protons in its nucleus and therefore the same number of electrons. But the number of neutrons may vary.

- Atoms with the same number of protons but different numbers of neutrons are called isotopes.
- Different isotopes of the same element react chemically in exactly the same way as they have the same electron configuration.
- Atoms of different isotopes of the same element vary in mass number because of the different number of neutrons in their nuclei.

All atoms of the element carbon, for example, have atomic number 6. That is what makes them carbon rather than any other element. However, carbon has three isotopes with mass numbers 12, 13, and 14 respectively (Table 1). All three isotopes will react in the same way, for example, burning in oxygen to form carbon dioxide.

Isotopes are often written like this $^{13}_{6}C$. The superscript 13 is the mass number of the isotope and the subscript 6 the atomic number.

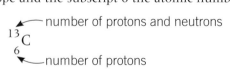

▼ **Table 1** *Isotopes of carbon*

Name of isotope	carbon-12	carbon-13	carbon-14
symbol	$^{12}_{6}C$	$^{13}_{6}C$	$^{14}_{6}C$
number of protons	6	6	6
number of neutrons	6	7	8
abundance	98.89%	1.11%	trace

Carbon dating

Isotopes of an element have different numbers of neutrons in their nuclei and most elements have some isotopes. Sometimes these isotopes are unstable and the nucleus of the atom itself breaks down giving off bits of the nucleus or energetic rays. This is the cause of radioactivity. Radioactive isotopes have many uses. Each radioactive isotope decays at a rate measured by its half-life. This is the time taken for half of its radioactivity to decay.

One well-known radioactive isotope is carbon-14. It has a half-life of 5730 years and is produced by cosmic-ray activity in the atmosphere. It is used to date organic matter. Radiocarbon dating can find the age of carbon-based material up to 60 000 years old, though it is most accurate for materials up to 2000 years old.

There is always a tiny fixed proportion of carbon-14 in all living matter. All living matter takes in and gives out carbon in the form of food and carbon dioxide, respectively. As a result, the level of carbon-14 stays the same. Once the living material dies, this stops happening. The radioactive carbon breaks down and the level of radioactivity slowly falls. So, knowing the half-life of carbon-14, scientists work backwards. They work out how long it has taken for the level of radioactivity to fall from what it is in a living organism to what it is in the sample. So, a sample with half the level of radioactivity expected in a living organism would have been dead for 5730 years, while one with a quarter of the expected level would have been dead for twice as long.

The radioactivity in a wooden bowl was found to be $\frac{1}{8}$ of that found in a sample of living wood.

1 How old is the wood from the bowl?
2 Does this tell us the age of the bowl? Explain your answer.

Summary questions

1 Isotopes are usually identified by the name of the element and the mass number of the isotope, as in carbon-13. However, Isotopes of hydrogen have their own names. Hydrogen-2 is often called deuterium and hydrogen-3 tritium. However, both these isotopes behave chemically just like the most common isotope, hydrogen-1. State how many protons, neutrons, and electrons the atoms of the following have.

   a deuterium

   b tritium

2 $^{31}_{15}W$, $^{14}_{7}X$, $^{16}_{8}Y$, $^{15}_{7}Z$ Identify which of these atoms (not their real symbols) is a pair of isotopes.

3 For each element in question **2**, state:

   a the number of protons

   b the mass number

   c the number of neutrons

Carbon-14

Radiocarbon dating was introduced in 1949 by the American Willard Libby who won the Nobel Prize for the technique. Carbon-14 is produced in the atmosphere by a nuclear reaction in which a neutron (from a cosmic ray) hits a nitrogen atom and ejects a proton:

$$^{14}_{7}N + ^{1}_{0}n \rightarrow ^{14}_{6}C + ^{1}_{1}p$$

If the half life of ^{14}C is taken to be 6000 years, 24 000 years is four half lives so the remaining radioactivity will be

$$\frac{1}{2} \times \frac{1}{2} \times \frac{1}{2} \times \frac{1}{2} = \frac{1}{16}$$

of the original activity.

Suggest why 60 000 years is the practical limit for ^{14}C dating.

Learning objectives:

→ Describe how electrons are arranged in an atom.

→ Recognise that the electron can behave as a particle, a wave, or a cloud of charge.

→ Describe how the structure of an atom developed from Dalton to Schrödinger.

Specification reference: 3.1.1

Synoptic link

This topic revises your knowledge of electron arrangements from GCSE. This will be useful when you study electron arrangements in Topic 1.5, More about electron arrangements in atoms.

Quantum theory in practice

Quantum Theory makes predictions that seem to contradict our everyday experience such as the fact that an electron can pass through two different holes at once! However, it is an extremely successful theory and underlies electronic gadgets such as computers, mobile phones, and DVD players.

The atom and electrons

During the early years of the twentieth century, physicists made great strides in understanding the structure of the atom. These are some of the landmarks.

1913 Niels Bohr put forward the idea that the atom consisted of a tiny positive nucleus orbited by negatively-charged electrons to form an atom like a tiny solar system. The electrons orbited in shells of fixed size and the movement of electrons from one shell to the next explained how atoms absorbed and gave out light. This was the beginning of what is called quantum theory.

1926 Erwin Schrödinger, a mathematical physicist, worked out an equation that used the idea that electrons had some of the properties of waves as well as those of particles. This led to a theory called quantum mechanics which can be used to predict the behaviour of sub-atomic particles.

1932 James Chadwick discovered the neutron.

At the same time, chemists were developing their ideas about how electrons allowed atoms to bond together. One important contributor was the American, Gilbert Lewis. He put forward the ideas that:

• the inertness of the noble gases was related to their having full outer shells of electrons

• ions were formed by atoms losing or gaining electrons to attain full outer shells

• atoms could also bond by sharing electrons to form full outer shells.

Lewis' theories are the basis of modern ideas of chemical bonding and explain the formulae of many simple compounds using the idea that atoms tend to gain the stable electronic structure of the nearest noble gas.

Evolving ideas

Early theories model the electron as a minute solid particle. Later theories suggest you can also think of electrons as smeared out clouds of charge, so you can never say exactly where an electron is at any moment. You can merely state the probability that it can be found in a particular volume of space that has a particular shape. However, chemists still use different models of the atom for different purposes.

• Dalton's model can still be used to explain the geometries of crystals.

• Bohr's model can be used for a simple model of ionic and covalent bonding.

• The charge cloud idea is used for a more sophisticated explanation of bonding and the shapes of molecules.

• The simple model of electrons orbiting in shells is useful for many purposes, particularly for working out bonding between atoms.

You will be familiar with the electron diagrams in this section from GCSE. They lead on to the more sophisticated models of electron structure described in Topic 1.5. However, they can still be useful, for example, in predicting and explaining the formulae of simple compounds and the shapes of molecules.

Electron shells

The first shell, which is closest to the nucleus, fills first, then the second, and so on. The number of electrons in each shell = $2n^2$, where n is the number of the shell, so:

- the first shell holds up to two electrons
- the second shell holds up to eight electrons
- the third shell holds up to 18 electrons.

Electron diagrams

If you know the number of protons in an atom, you also know the number of electrons it has. This is because the atom is neutral. You can therefore draw an electron diagram for any element. For example, carbon has six electrons. The four electrons in the outer shell are usually drawn spaced out around the atom (Figure 1).

Sulfur has 16 electrons. It has six electrons in its outer shell. It helps when drawing bonding diagrams to space out the first four (as in carbon), and then add the next two electrons to form pairs (Figure 2).

You can also draw electron diagrams of ions, as long as you know the number of electrons. For example, a sodium *atom*, Na, has 11 electrons, but its *ion* has 10, so it has a positive charge, Na^+ (Figure 3).

An oxygen *atom* has eight electrons, but its *ion* has 10, so it has a negative charge, O^{2-} (Figure 4).

You can write electron diagrams in shorthand:

- write the number of electrons in each shell, starting with the inner shell and working outwards
- separate each number by a comma.

For carbon you write 2,4; for sulfur 2,8,6; for Na^+ 2,8.

Study tip

You must remember the number of electrons in each shell.

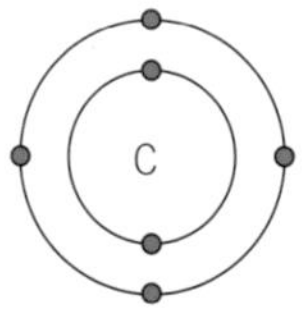

carbon (2,4)

▲ **Figure 1** *Electron diagram of carbon*

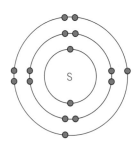

sulfur (2,8,6)

▲ **Figure 2** *Electron diagram of sulfur*

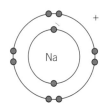

Na^+ sodium ion
11 protons, 10 electrons
(2,8)

▲ **Figure 3** *Electron diagram of a sodium ion*

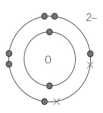

O^{2-} oxygen ion
8 protons,
10 electrons (2,8)

▲ **Figure 4** *Electron diagram of an oxygen ion*

Summary questions

1 Draw the electron arrangement diagrams of atoms that have the following numbers of electrons:

a 3 **b** 9 **c** 14

2 State, in shorthand, the electron arrangements of atoms with:

a 4 electrons **b** 13 electrons **c** 18 electrons

3 Identify which of the following are atoms, positive ions, or negative ions. Give the size of the charge on each ion, including its sign. Use the Periodic Table to identify the elements A–E.

	Number of protons	Number of electrons
A	12	10
B	2	2
C	17	18
D	10	10
E	3	2

Learning objectives:

→ Explain how a mass spectrometer works and what it measures.

Specification reference: 3.1.1

Synoptic link

Mass spectrometry can also be used to measure relative molecular masses and much more, as you will see in Topic 16.2, Mass spectrometry.

Synoptic link

You will find out more about relative atomic mass in Topic 2.1, Relative atomic and molecular masses, the Avogadro constant, and the mole.

The mass spectrometer

The mass spectrometer is the most useful instrument for the accurate determination of **relative atomic masses** A_r. Relative atomic masses are measured on a scale on which the mass of an atom of ^{12}C is defined as *exactly* 12. No other isotope has a relative atomic mass that is exactly a whole number. This is because neither the proton nor the neutron has a mass of exactly 1.

$$\text{relative atomic mass } A_r = \frac{\text{average mass of 1 atom}}{\frac{1}{12} \text{ mass of 1 atom of } ^{12}C}$$

$$\text{relative molecular mass } M_r = \frac{\text{average mass of molecule}}{\frac{1}{12} \text{ mass of 1 atom of } ^{12}C}$$

The mass spectrometer determines the mass of separate atoms (or molecules). Mass spectrometers are an essential part of a chemist's toolkit of equipment. For example, they are used by forensic scientists to help identify substances such as illegal drugs.

There are several types of mass spectrometer but all work on the principle of forming ions from the sample and then separating the ions according to the ratio of their charge to their mass. The type described here is called an electro spray ionisation time of flight (TOF) instrument. The layout of this type of mass spectrometer is shown in Figure 2.

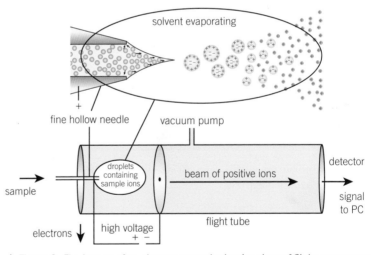

▲ **Figure 2** *The layout of an electron spray ionisation time of flight mass spectrometer*

What happens in a time of flight mass spectrometer?

In outline, the substance(s) in the sample are converted to positive ions, accelerated to high speeds (which depend on their mass to charge ratio), and arrive at a detector. The steps are described in more detail below.

- **Vacuum** The whole apparatus is kept under a high vacuum to prevent the ions that are produced colliding with molecules from the air.

- **Ionisation** The sample to be investigated is dissolved in a volatile solvent and forced through a fine hollow needle that is connected

▲ **Figure 1** *A modern mass spectrometer*

to the positive terminal of a high voltage supply. This produces tiny positively charged droplets which have lost electrons to the positive charge of the supply. The solvent evaporates from the droplets into the vacuum and the droplets get smaller and smaller until they may contain no more than a single positively charged ion.

- **Acceleration** The positive ions are attracted towards a negatively charged plate and accelerate towards it. Lighter ions and more highly charged ions achieve a higher speed.

- **Ion drift** The ions pass through a hole in the negatively charged plate, forming a beam and travel along a tube, called the flight tube, to a detector.

- **Detection** When ions with the same charge arrive at the detector, the lighter ones are first as they have higher velocities. The flight times are recorded. The positive ions pick up an electron from the detector, which causes a current to flow.

- **Data analysis** The signal from the detector is passed to a computer which generates a mass spectrum like those in Figures 3 and 4.

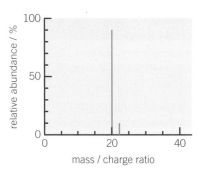

▲ **Figure 3** *The mass spectrum of neon. There is no peak at 20.2 because no neon atoms actually have this mass*

Mass spectra of elements

The mass spectrometer can be used to identify the different isotopes that make up an element. It detects individual ions, so different isotopes are detected separately because they have different masses. This is how the data for the neon, germanium, and chlorine isotopes in Figures 3, 4, and 5 were obtained. The peak height gives the relative abundance of each isotope and the horizontal scale gives the *m/z* which, for a singly charged ion is numerically the same as the mass number *A*.

Mass spectrometers can measure relative atomic masses to five decimal places of an atomic mass unit – this is called high resolution mass spectrometry. However most work is done to one decimal point – this is called low resolution mass spectrometry.

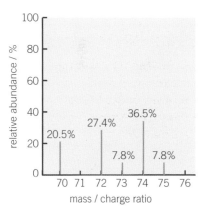

▲ **Figure 4** *The mass spectrum of germanium (the percentage abundance of each peak is given)*

Low resolution mass spectrometry

The low resolution mass spectrum of neon is shown in Figure 3. This shows that neon has two isotopes, with mass numbers 20 and 22, and abundances to the nearest whole number of 90% and 10%, respectively. From this we can say that neon has an average relative atomic mass of:

$$\frac{(90 \times 20) + (10 \times 22)}{100} = 20.2$$

When calculating the relative atomic mass of an element, you must take account of the relative abundances of the isotopes. The relative atomic mass of neon is not 21 because there are far more atoms of the lighter isotope.

Another example is the mass spectrum of the element germanium, which is shown in Figure 4.

Isotopes of chlorine

Chlorine has two isotopes. They are $^{35}_{17}\text{Cl}$, with a mass number of 35, and $^{37}_{17}\text{Cl}$, with a mass number of 37. They occur in the ratio of almost exactly 3 : 1.

^{35}Cl	^{35}Cl	^{35}Cl	^{37}Cl
three of these			to every one of this

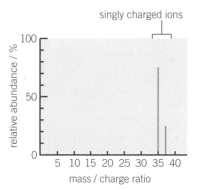

▲ **Figure 5** *The mass spectrum of chlorine*

Relative atomic masses are weighted averages of the mass numbers of the isotopes of the element, taking account both the masses and their abundances, relative to the ^{12}C isotope, which is exactly 12. Chlorine has isotopes of mass number 35 and 37 but the relative atomic mass of chlorine is *not* 36, it is 35.5.

So there is 75% ^{35}Cl and 25% ^{37}Cl atoms in naturally occurring chlorine gas.

The average mass of these is 35.5, as shown below.

Mass of 100 atoms = $(35 \times 75) + (37 \times 25) = 3550$

$$\text{Average mass} = \frac{3550}{100} = 35.5$$

This explains why the relative atomic mass of chlorine is *approximately* 35.5.

Identifying elements

All elements have a characteristic pattern that shows the relative abundances of their isotopes. This can be used to help identify any particular element. Chlorine, for example, shows two peaks at mass 35 and mass 37. The peak of mass 35 is three times the height of the peak of mass 37 because there are three times as many ^{35}Cl atoms in chlorine.

The spectrum will also show peaks caused by ionised Cl_2 molecules. These are called molecular ions. There will be three of these:

- at m/z 70, due to $^{35}Cl^{35}Cl$
- at m/z 72, due to $^{35}Cl^{37}Cl$
- at m/z 74, due to $^{37}Cl^{37}Cl$

High resolution mass spectrometers can measure the masses of atoms to several decimal places. This allows us to identify elements by the exact masses of their atoms that (apart from carbon-12 whose relative atomic mass is exactly 12) are not exactly whole numbers.

√x̄ What will be the relative abundances of the three Cl_2^+ ions of m/z 70, 72, and 74 respectively? The relative abundances of the atoms are ^{35}Cl : $^{37}Cl = 3 : 1$. i.e., $\frac{3}{4} : \frac{1}{4}$

Mass spectrometers in space

Space probes such as the Mars Rover Curiosity carry mass spectrometers. They are used to identify the elements in rock samples. The Huygens spacecraft that landed on Titan, one of the moons of Saturn, in January 2005 carried a mass spectrometer used to identify and measure the amounts of the gases in Titan's atmosphere. After landing, it also analysed vaporised samples of the surface.

▲ **Figure 6** *The Mars Rover Curiosity carries a mass spectrometer to look for compounds of carbon that may suggest that there was once life on Mars.*

Mini-mass spectrometer

The latest development in mass spectrometry is a unit small enough to carry as a back pack. The unit, including rechargeable batteries, weighs 10 kg, light enough to be carried by scene of crime officers looking for drugs, explosive, or chemical weapons. Other uses include investigating chemical spills.

▲ **Figure 7** *The mini-mass spectrometer*

The mass spectrometer includes software to match spectra of samples investigated with a library of spectra and so identify them. The instrument can be used by operators with little or no chemical knowledge.

Summary questions

1 Explain why the ions formed in a mass spectrometer have a positive charge.

2 Explain what causes the ions to accelerate through the mass spectrometer.

3 Describe what forms the ions into a beam.

4 State which ions will arrive at the detector first.

5 Use the information about germanium in Figure 4 to Calculate out its relative atomic mass.

6 Figure 8 shows the mass spectrum of copper. Calculate the relative atomic mass of copper.

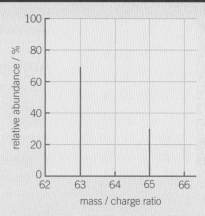

▲ **Figure 8** *The mass spectrum of copper*

1.5 More about electron arrangements in atoms

→ Illustrate how the electron configurations of atoms and ions are written in terms of s, p, and d electrons.

Specification reference: 3.1.1

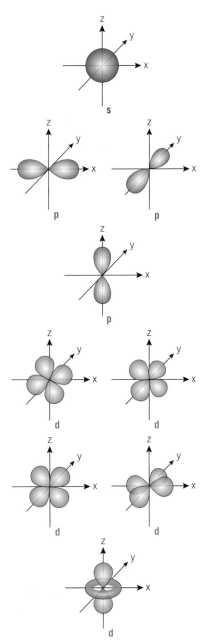

▲ **Figure 3** *The shapes of s-, p-, and d-orbitals*

As you have seen in Topic 1.3, in a simple model of the atom the electrons are thought of as being arranged in shells around the nucleus. The shells can hold increasing numbers of electrons as they get further from the nucleus – the pattern is 2, 8, 18, and so on.

Energy levels

Electrons in different shells have differing amounts of energy. They can therefore be represented on an energy level diagram. The shells are called main energy levels and they are labelled 1, 2, 3, and so on (Figure 1). Each main energy level can hold up to a maximum number of electrons given by the formula $2n^2$, where n is the number of the main level. So, you can have two electrons in the first main level, eight in the next, 18 in the next, and so on.

Apart from the first level, which has only an s-sub-level, these main energy levels are divided into sub-levels, called s, p, d, and f, which have slightly different energies (Figure 2). Level 2 has an s-sub-level and a p-sub-level. Level 3 an s-sub-level, a p-sub-level, and a d-sub-level.

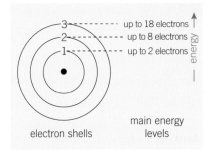

▲ **Figure 1** *Electron shells and energy levels*

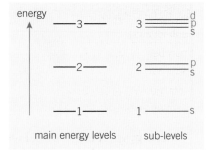

▲ **Figure 2** *Energy levels and sub-levels*

Quantum mechanics

For a more complete description of the electrons in atoms a theory called quantum mechanics is used, which was developed during the 1920s. This describes the atom mathematically with an equation (the Schrödinger equation). The solutions to this equation give the *probability* of finding an electron in a given *volume* of space called an atomic orbital.

Atomic orbitals

The electron is no longer considered to be a particle but a cloud of negative charge. An electron fills a volume in space called its **atomic orbital**. The concept of the main levels and the sub-levels is then included in the following way.

- Different atomic orbitals have different energies. Each orbital has a number that tells us the main energy level that it corresponds to: 1, 2, 3, and so on.

- The atomic orbitals of each main level have different shapes, which in turn have slightly different energies. These are the sub-levels. They are described by the letters s, p, d, and f. The shapes of the s-, p-, and d-orbitals are shown in Figure 3. The shapes of f-orbitals are even more complicated.

- These shapes represent a volume of space in which there is a 95% probability of finding an electron and they influence the shapes of molecules.
- The first main energy level consists of a single s-orbital. The second main level has a single s-orbital and three p-orbitals of a slightly higher energy, the third main level has a single s-orbital, three p-orbitals of slightly higher energy, and five d-orbitals of slightly higher energy still, and so on, see Figure 4.
- Any single atomic orbital can hold a maximum of two electrons.
- s-orbitals can hold up to two electrons.
- p-orbitals can hold up to two electrons each, but always come in groups of three of the same energy, to give a total of up to six electrons in the p-sub-level.
- d-orbitals can hold up to two electrons each, but come in groups of five of the same energy to give a total of up to 10 electrons in the d-sub-level.

Table 1 summarises the number of electrons in the different levels and sub-levels.

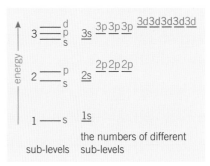

▲ **Figure 4** *The subdivisions of orbitals*

▼ **Table 1** *The number of electrons in the different levels and sub-levels*

Main energy level (shell)	1	2		3			4			
sub-level(s)	s	s	p	s	p	d	s	p	d	f
number of orbitals in sub-level	1 (2 electrons)	1 (2e⁻)	3 (6e⁻)	1 (2e⁻)	3 (6e⁻)	5 (10e⁻)	1 (2e⁻)	3 (6e⁻)	5 (10e⁻)	7 (14e⁻)
total number of electrons in main energy level	2	8		18			32			

The energy level diagram in Figure 5 shows the energies of the orbitals for the first few elements of the Periodic Table. Notice that the first main energy level has only an s-orbital. The second main level has an s- and p-sub-level and the p-sub-level is composed of three p-orbitals of equal energy. The third main level has an s-, p-, and d-sub-level, and the d-sub-level is composed of five atomic orbitals of equal energy.

- Each 'box' in Figure 5 represents an orbital of the appropriate shape that can hold up to two electrons.
- Notice that 4s is actually of slightly lower energy than 3d for neutral atoms, though this can change when ions are formed.

Spin

Electrons also have the property called spin.

- Two electrons in the same orbital must have opposite spins.
- The electrons are usually represented by arrows pointing up or down to show the different directions of spin.

Putting electrons into atomic orbitals

Remember that the label of an atomic orbital tells us about the energy (and shape) of an electron cloud. For example, the atomic orbital 3s means the main energy level is 3 and the sub-level (and therefore the shape) is spherical.

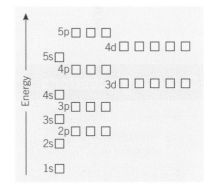

▲ **Figure 5** *The energy levels of the first few atomic orbitals*

Study tip

Make sure that you can work out the shorthand electronic structure of all the elements at least up to krypton (atomic number 36).

There are three rules for allocating electrons to atomic orbitals.

1 Atomic orbitals of lower energy are filled first – so the lower main level is filled first and, within this level, sub-levels of lower energy are filled first.

2 Atomic orbitals of the same energy fill singly before pairing starts. This is because electrons repel each other.

3 No atomic orbital can hold more than two electrons.

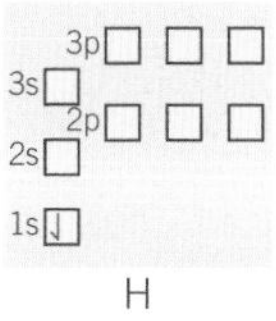

The electron diagrams for the elements hydrogen to sodium are shown in Figure 6.

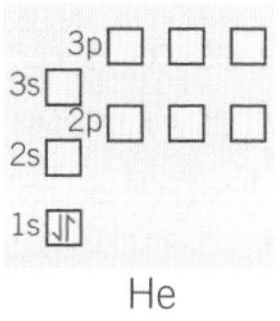

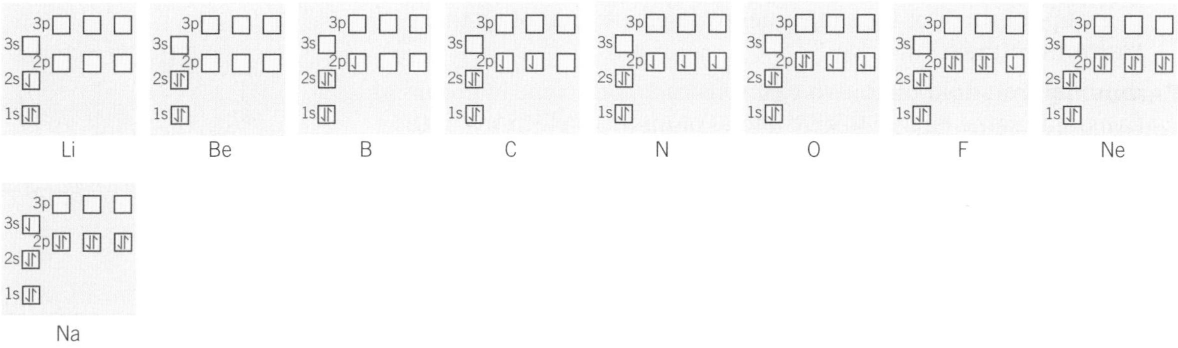

▲ **Figure 6** *The electron arrangements for the elements hydrogen to sodium – note how they obey the rule above*

Writing electronic structures

A shorthand way of writing electronic structures is as follows, for example, for sodium which has 11 electrons:

$1s^2$	$2s^2\ 2p^6$	$3s^1$
2	8	1

Note how this matches the simpler 2,8,1 you used at GCSE.

Calcium, with 20 electrons would be:

$1s^2\ \ 2s^2\ 2p^6\ \ 3s^2\ 3p^6\ \ 4s^2$ which matches 2,8,8,2

Notice how the 4s orbital is filled before the 3d orbital because it is of lower energy.

After calcium, electrons begin to fill the 3d orbitals, so vanadium with 23 electrons is: $1s^2\ 2s^2\ 2p^6\ 3s^2\ 3p^6\ 3d^3\ 4s^2$

Krypton with 36 electrons is: $1s^2\ 2s^2\ 2p^6\ 3s^2\ 3p^6\ 3d^{10}\ 4s^2\ 4p^6$

Sometimes it simplifies things to use the previous noble gas symbol. So the electron arrangement of calcium, Ca, could be written [Ar] $4s^2$ as a shorthand for [$1s^2\ 2s^2\ 2p^6\ 3s^2\ 3p^6$] $4s^2$ because $1s^2\ 2s^2\ 2p^6\ 3s^2\ 3p^6$ is the electron arrangement of argon.

You can use the same notation for ions. So a sodium ion, Na^+, would have the electron arrangement $1s^2\ 2s^2\ 2p^6$, one less than a sodium atom, $1s^2\ 2s^2\ 2p^6\ 3s^1$.

Summary questions

1 a Give the full electron arrangement for phosphorus.

b Give the electron arrangement for phosphorus using an inert gas symbol as a shorthand.

2 a Give the full electron arrangements of:
i Ca^{2+} and ii F^-

b Give their electron arrangements using an inert gas symbol as a shorthand.

1.6 Electron arrangements and ionisation energy

The patterns in first ionisation energies across a period provide evidence for electron energy sub-levels.

Ionisation energy

Electrons can be removed from atoms and the energy it takes to remove them can be measured. This is called **ionisation energy** because as the electrons are removed, the atoms become positive ions.

- Ionisation energy is the energy required to remove a mole of electrons from a mole of atoms in the gaseous state and is measured in kJ mol^{-1}.
- Ionisation energy has the abbreviation IE.

Removing the electrons one by one

You can measure the energies required to remove the electrons one by one from an atom, starting from the outer electrons and working inwards.

- The first electron needs the least energy to remove it because it is being removed from a neutral atom. This is the first IE.
- The second electron needs more energy than the first because it is being removed from a +1 ion. This is the second IE.
- The third electron needs even more energy to remove it because it is being removed from a +2 ion. This is the third IE.
- The fourth needs yet more, and so on.

These are called **successive ionisation energies**.

For example, sodium:

$$\text{Na(g)} \rightarrow \text{Na}^+\text{(g)} + e^- \quad \text{first IE} \quad = +496 \text{ kJ mol}^{-1}$$
$$\text{Na}^+\text{(g)} \rightarrow \text{Na}^{2+}\text{(g)} + e^- \quad \text{second IE} \quad = +4563 \text{ kJ mol}^{-1}$$
$$\text{Na}^{2+}\text{(g)} \rightarrow \text{Na}^{3+}\text{(g)} + e^- \quad \text{third IE} \quad = +6913 \text{ kJ mol}^{-1}$$

and so on, see Table 1.

▼ Table 1 Successive ionisation energies of sodium

Electron removed	1st	2nd	3rd	4th	5th	6th	7th	8th	9th	10th	11th
Ionisation energy / kJ mol^{-1}	496	4 563	6 913	9 544	13 352	16 611	20 115	25 491	28 934	141 367	159 079

Notice that the second IE is *not* the energy change for

$$\text{Na(g)} \rightarrow \text{Na}^{2+}\text{(g)} + 2e^-$$

The energy for this process would be (first IE + second IE).

If you plot a graph of the values shown in Table 1 you get Figure 1.

Notice that one electron is relatively easy to remove, then comes a group of eight that are more difficult to remove, and finally two that are very difficult to remove.

Learning objectives:

→ State the definition of ionisation energy.

→ Describe the trend in ionisation energies a) down a group and b) across a period in terms of electron configurations.

→ Explain how trends in ionisation energies provide evidence for the existence of electron energy levels and sub-levels.

Specification reference: 3.1.1

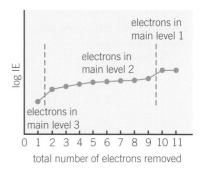

▲ Figure 1 *The successive ionisation energies of sodium against number of electrons removed. Note that the log of the ionisation energy has been plotted in order to fit the large range of values on the scale*

Study tip

The shape of the graph in Figure 1 has to be thought about carefully. The first electron removed is in the outer main level and the 10th and 11th electrons removed are in the innermost main level.

This suggests that sodium has:

- *one* electron furthest away from the positive nucleus (easy to remove)
- *eight* electrons nearer in to the nucleus (harder to remove)
- *two* electrons very close to the nucleus (very difficult to remove because they are nearest to the positive charge of the nucleus).

This tells you about the number of electrons in each main level or orbit: 2,8,1. The eight electrons in shell 2 are in fact sub-divided into two further groups that correspond to the $2s^2$, $2p^6$ electrons in the second main level, but this is not visible on the scale of Figure 1. It is just visible in figure 2.

You can find the number of electrons in each main level of *any* element by looking at the jumps in successive ionisation energies.

Trends in ionisation energies across a period in the Periodic Table

The trends in first ionisation energies moving across a period in the Periodic Table can also give information about the energies of electrons in main levels and sub-levels. Ionisation energies generally increase across a period because the nuclear charge is increasing and this makes it more difficult to remove an electron.

The data for Period 3 are shown in Table 2.

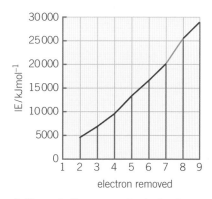

▲ **Figure 2** *The successive ionisation energies of the electrons in shell 2 in sodium. You can just see the jump between electron 7 and 8*

▼ **Table 2** *The first ionisation energies of the elements in Period 3 in kJ mol⁻¹*

Na	Mg	Al	Si	P	S	Cl	Ar
496	738	578	789	1012	1000	1251	1521

nuclear charge increasing →

Plotting a graph of these values shows that the increase is not regular (Figure 3). In going from magnesium ($1s^2$, $2s^2$, $2p^6$, $3s^2$) to aluminium ($1s^2$, $2s^2$, $2p^6$, $3s^2$, $3p^1$), the ionisation energy actually goes down, despite the increase in nuclear charge. This is because the outer electron in aluminium is in a 3p orbital which is of a slightly higher energy than the 3s orbital. It therefore needs less energy to remove it, see Figure 4.

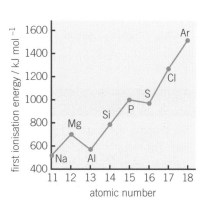

▲ **Figure 3** *Trends in first ionisation energies across Period 3*

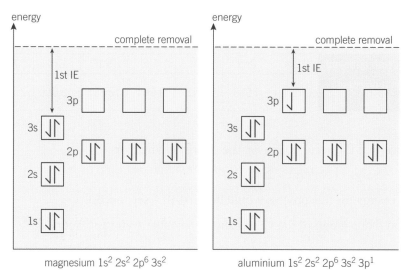

▲ **Figure 4** *The first ionisation energy of aluminium is less than that of magnesium*

In Figure 3, notice the small drop between phosphorus ($1s^2$, $2s^2$, $2p^6$, $3s^2$, $3p^3$) and sulfur ($1s^2$, $2s^2$, $2p^6$, $3s^2$, $3p^4$). In phosphorus, each of the three 3p orbitals contains just one electron, while in sulfur, one of the 3p orbitals must contain two electrons. The repulsion between these paired electrons makes it easier to remove one of them, despite the increase in nuclear charge, see Figure 5.

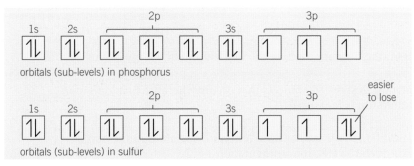

▲ **Figure 5** *Electron arrangements of phosphorus and sulfur*

Both these cases, which go against the expected trend, are evidence that confirms the existence of s- and p-sub-levels. These were predicted by quantum theory and the Schrödinger equation.

Trends in ionisation energies down a group in the Periodic Table

Figure 6 shows that there is a general decrease in first ionisation energy going down Group 2 and the same pattern is seen in other groups. This is because the outer electron is in a main level that gets further from the nucleus in each case.

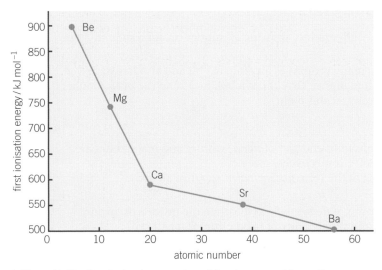

▲ **Figure 6** *The first ionisation energies of the elements of Group 2*

Going down a group, the nuclear charge increases. At first sight you might expect that this would make it *more* difficult to remove an electron. However, the actual positive charge 'felt' by an electron in the outer shell is less than the full nuclear charge. This is because of the effect of the inner electrons shielding the nuclear charge.

1 State why the second ionisation energy of any atom is larger than the first ionisation energy.

2 Sketch a graph similar to Figure 1 of the successive ionisation energies of aluminium (electron arrangement 2,8,3).

3 An element X has the following values (in kJ mol^{-1}) for successive ionisation energies: 1093, 2359, 4627, 6229, 37 838, 47 285.

   a Identify which group in the Periodic Table it is in.

   b Explain your answer to **a**.

Practice questions

1   The following diagram shows the first ionisation energies of some Period 3 elements.

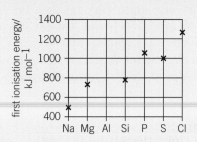

(a) Draw a cross on the diagram to show the first ionisation energy of aluminium.

(1 mark)

(b) Write an equation to show the process that occurs when the first ionisation energy of aluminium is measured.

(2 marks)

(c) State which of the first, second, or third ionisations of aluminium would produce an ion with the electron configuration $1s^2\ 2s^2\ 2p^6\ 3s^1$.

(1 mark)

(d) Explain why the value of the first ionisation energy of sulfur is less than the value of the first ionisation energy of phosphorus.

(2 marks)

(e) Identify the element in Period 2 that has the highest first ionisation energy and give its electron configuration.

(2 marks)

(f) State the trend in first ionisation energies in Group 2 from beryllium to barium. Explain your answer in terms of a suitable model of atomic structure.

(3 marks)
AQA, 2010

2   (a) One isotope of sodium has a relative mass of 23.
(i) Define, in terms of the fundamental particles present, the meaning of the term *isotopes*.
(ii) Explain why isotopes of the same element have the same chemical properties.

(3 marks)

(b) Give the electronic configuration, showing all sub-levels, for a sodium atom.

(1 mark)

(c) An atom has half as many protons as an atom of ^{28}Si and also has six fewer neutrons than an atom of ^{28}Si. Give the symbol, including the mass number and the atomic number, of this atom.

(2 marks)
AQA, 2004

3   The values of the first ionisation energies of neon, sodium and magnesium are 2080, 494 and 736 kJ mol^{-1}, respectively.
(a) Explain the meaning of the term *first ionisation energy* of an atom.

(2 marks)

(b) Write an equation using state symbols to illustrate the process occurring when the **second** ionisation energy of magnesium is measured.

(2 marks)

(c) Explain why the value of the first ionisation energy of magnesium is higher than that of sodium.

(2 marks)

(d) Explain why the value of the first ionisation energy of neon is higher than that of sodium.

(2 marks)
AQA, 2004

4 A sample of iron from a meteorite was found to contain the isotopes ^{54}Fe, ^{56}Fe, and ^{57}Fe.
   (a) The relative abundances of these isotopes can be determined using a mass
       spectrometer. In the mass spectrometer, the sample is first vaporised and then ionised.
       (i) State what is meant by the term *isotopes*.
       (ii) Explain how, in a mass spectrometer, ions are detected and how their
             abundance is measured.

(5 marks)

   (b) **(i)** Define the term *relative atomic mass* of an element.
       (ii) The relative abundances of the isotopes in this sample of iron were
             found to be as follows.

m/z	54	56	57
Relative abundance %	5.80	91.60	2.60

Use the data above to calculate the relative atomic mass of iron in this sample.
Give your answer to the appropriate number of significant figures.

(2 marks)
AQA, 2005

5 The diagram shows the layout of a time of flight mass spectrometer.

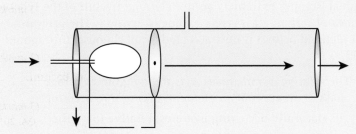

   (a) Explain how positive ions are formed from the sample.

(1 mark)

   (b) Explain why the instrument is kept under vacuum.

(1 mark)

   (c) Explain how the ions are accelerated and separated by mass in the instrument.

(3 marks)

   (d) Explain how an electric current is produced when an ion arrives at the detector.

(1 mark)

   (e) The low resolution mass spectrum of magnesium shows three peaks

Mass / charge	Relative abundance / %
24	79.0%
25	10.0%
26	11.0%

   (i) Give the numbers of protons and neutrons in the nuclei of each isotope.

(1 mark)

   (ii) Calculate the relative atomic mass of a sample of magnesium.
        Give your answer to the appropriate number of significant figures.

(2 marks)

2 Amount of substance

2.1 Relative atomic and molecular masses, the Avogadro constant, and the mole

Learning objectives:

- → State the definition of relative atomic mass.
- → State the definition of relative molecular mass.
- → State the meaning of the Avogadro constant.
- → State what the same numbers of moles of different substances have in common.
- → Calculate the number of moles present in a given mass of an element or compound.

Specification reference: 3.1.2

Relative atomic mass A_r

The actual mass in grams of any atom or molecule is too tiny to find by weighing. Instead, the masses of atoms are compared and *relative* masses are used.

This was done in the past by defining the relative atomic mass of hydrogen, the lightest element, as 1. The average mass of an atom of oxygen (for example) is 16 times heavier, to the nearest whole number, so oxygen has a relative atomic mass of 16. Scientists now use the isotope carbon-12 as the baseline for relative atomic mass because the mass spectrometer has allowed us to measure the masses of individual isotopes extremely accurately. One twelfth of the relative atomic mass of carbon-12 is given a value of *exactly* 1. The carbon-12 standard (defined below) is now accepted by all chemists throughout the world.

The relative atomic mass A_r is the weighted average mass of an atom of an element, taking into account its naturally occurring isotopes, relative to $\frac{1}{12}$ the relative atomic mass of an atom of carbon-12.

$$\text{relative atomic mass } A_r = \frac{\text{average mass of one atom of an element}}{\frac{1}{12}\text{ mass of one atom of }^{12}\text{C}}$$

$$= \frac{\text{average mass of one atom of an element} \times 12}{\text{mass of one atom of }^{12}\text{C}}$$

Study tip

The weighted average mass must be used to allow for the presence of isotopes, using their percentage abundances in calculations.

Relative molecular mass M_r

Molecules can be handled in the same way, by comparing the mass of a molecule with that of an atom of carbon-12.

The relative molecular mass, M_r, of a molecule is the mass of that molecule compared to $\frac{1}{12}$ the relative atomic mass of an atom of carbon-12.

$$\text{relative molecular mass } M_r = \frac{\text{average mass of one molecule}}{\frac{1}{12}\text{ mass of one atom of }^{12}\text{C}}$$

$$= \frac{\text{average mass of one molecule} \times 12}{\text{mass of one atom of }^{12}\text{C}}$$

Study tip

You need to learn the exact definitions of A_r and M_r.

You find the **relative molecular mass** by adding up the relative atomic masses of all the atoms present in the molecule and you find this from the formula.

▼ **Table 1** *Examples of relative molecular mass*

Molecule	Formula	A_r of atoms	M_r
water	H_2O	$(2 \times 1.0) + 16.0$	18.0
carbon dioxide	CO_2	$12.0 + (2 \times 16.0)$	44.0
methane	CH_4	$12.0 + (4 \times 1.0)$	16.0

Relative formula mass

The term **relative formula mass** is used for ionic compounds because they don't exist as molecules. However, this has the same symbol M_r.

▼ **Table 2** *Some examples of the relative formula masses of ionic compounds*

Ionic compound	Formula	A_r of atoms	M_r
calcium fluoride	CaF_2	$40.1 + (2 \times 19.0)$	78.1
sodium sulfate	Na_2SO_4	$(2 \times 23.0) + 32.1 + (4 \times 16.0)$	142.1
magnesium nitrate	$Mg(NO_3)_2$	$24.3 + (2 \times (14.0 + (16.0 \times 3)))$	148.3

The Avogadro constant and the mole

One atom of any element is too small to see with an optical microscope and impossible to weigh individually. So, to count atoms, chemists must weigh large numbers of them. This is how cashiers count money in a bank (Figure 1).

Working to the nearest whole number, a helium atom ($A_r = 4$) is four times heavier than an atom of hydrogen. A lithium atom ($A_r = 7$) is seven times heavier than an atom of hydrogen. To get the same number of atoms in a sample of helium or lithium, as the number of atoms in 1 g of hydrogen, you must take 4 g of helium or 7 g of lithium.

In fact if you weigh out the relative atomic mass of *any* element, this amount will also contain this same number of atoms.

The same logic applies to molecules. Water H_2O, has a relative molecular mass M_r of 18. So, one molecule of water is 18 times heavier than one atom of hydrogen. Therefore, 18 g of water contain the same number of *molecules* as there are *atoms* in 1 g of hydrogen. A molecule of carbon dioxide is 44 times heavier than an atom of hydrogen, so 44 g of carbon dioxide contain this same number of molecules.

If you weigh out the relative or formula mass M_r of a compound in grams you have *this same number* of **entities**.

The Avogadro constant

The actual number of atoms in 1 g of hydrogen atoms is unimaginably huge:

602 200 000 000 000 000 000 000 usually written 6.022×10^{23}.

The difference between this scale, based on H = 1 and the scale used today based on ^{12}C, is negligible, for most purposes.

> The **Avogadro constant** or **Avogadro number** is the number of atoms in 12 g of carbon-12.

The mole

The amount of substance that contains 6.022×10^{23} particles is called a **mole**.

▲ **Figure 1** *Large numbers of coins or bank notes are counted by weighing them*

The relative atomic mass of any element in grams contains one mole of atoms. The relative molecular mass (or relative formula mass) of a substance in grams contains one mole of entities. You can also have a mole of ions or electrons.

It is easy to confuse moles of *atoms* and moles of *molecules*, so always give the formula when working out the mass of a mole of entities. For example, 10 moles of hydrogen could mean 10 moles of hydrogen atoms or 10 moles of hydrogen molecules, H_2, which contains twice the number of atoms. Using the mole, you can compare the *numbers* of different particles that take part in chemical reactions.

▼ **Table 3** *Examples of moles*

Entities	Formula	Relative mass to nearest whole number	Mass of a mole / g = molar mass
oxygen atoms	O	16.0	16.0
oxygen molecules	O_2	32.0	32.0
sodium ions	Na^+	23.0	23.0
sodium fluoride	NaF	42.0	42.0

Number of moles

If you want to find out how many moles are present in a particular mass of a substance you need to know the substance's formula. From the formula you can then work out the mass of one mole of the substance.

You use:

$$\text{number of moles } n = \frac{\text{mass } m \text{ (g)}}{\text{mass of 1 mole } M \text{ (g)}}$$

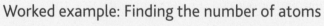

Worked example: Finding the number of moles

How many moles are there in 0.53 g of sodium carbonate, Na_2CO_3?

A_r Na = 23.0, A_r C = 12.0, A_r O = 16.0,

so M_r of Na_2CO_3 = (23.0 × 2) + 12.0 + (16.0 × 3) = 106.0,

so 1 mole of calcium carbonate has a mass of 106.0 g.

Number of moles = $\frac{0.53}{106.0}$ = 0.0050 mol

Worked example: Finding the number of atoms

You have 3.94 g of gold, Au, and 2.70 g of aluminium, Al. Which contains the greater number of atoms? (A_r Au = 197.0, A_r Al = 27.0)

Number of moles of gold atoms = $\frac{3.94}{197.0}$ = 0.020 mol

Number of moles of aluminium atoms = $\frac{2.70}{27.0}$ = 0.100 mol

There are more atoms of aluminium.

 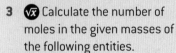

Solutions

A solution consists of a solvent with a solute dissolved in it, (Figure 1).

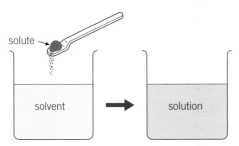

▲ **Figure 1** *A solution contains a solute and a solvent*

The units of concentration

The concentration of a solution tells us how much solute is present in a known volume of solution.

Concentrations of solutions are measured in $mol\,dm^{-3}$. $1\ mol\ dm^{-3}$ means there is 1 mole of solute per cubic decimetre of solution; 2 $mol\,dm^{-3}$ means there are 2 moles of solute per cubic decimetre of solution, and so on.

Worked example: Finding the concentration in $mol\ dm^{-3}$

1.17 g of sodium chloride was dissolved in water to make $500\ cm^3$ of solution. What is the concentration of the solution in $mol\,dm^{-3}$? A_r Na = 23.0, A_r Cl = 35.5

The mass of 1 mole of sodium chloride, NaCl, is 23.0 + 35.5 = 58.5 g.

$$\text{number of moles } n = \frac{\text{mass } m \text{ (g)}}{\text{mass of 1 mole } M \text{ (g)}}$$

So 1.17 g of NaCl contains $\frac{1.17}{58.5}$ mol = 0.020 mol to 3 s.f.

This is dissolved in $500\ cm^3$, so $1000\ cm^3$ ($1\ dm^3$) would contain 0.040 mol of NaCl. This means that the concentration of the solution is $0.040\ mol\,dm^{-3}$.

The general way of finding a concentration is to remember the relationship:

$$\text{concentration } c \text{ (mol dm}^{-3}) = \frac{\text{number of moles } n}{\text{volume } V \text{ (dm}^3)}$$

Substituting into this gives:

$$\text{concentration} = \frac{0.020}{0.500} = 0.040\ mol\,dm^{-3}$$

Learning objectives:

→ Calculate the number of moles of substance from the volume of a solution and its concentration.

Specification reference: 3.1.2

Study tip 🧪

To get a solution with a concentration of $1\ mol\ dm^{-3}$ you have to add the solvent to the solute until you have $1\ dm^3$ of solution. You *do not* add 1 mol of solute to $1\ dm^3$ of solvent. This would give more than $1\ dm^3$ of solution.

Study tip √x̄

1 decimetre = 10 cm, so one cubic decimetre, $1\ dm^3$, is $10\ cm \times 10\ cm \times 10\ cm = 1000\ cm^3$. This is the same as 1 litre (1 l or 1 L). If you are not confident about conversion factors and writing units, see the Maths Appendix.

The small negative in $mol\ dm^{-3}$ means per and is sometimes written as a slash, mol/dm^3.

Error in measurements

Every measurement has an inherent uncertainty (also known as error). In general, the uncertainty in a single measurement from an instrument is *half the value of the smallest division*. The uncertainty of a measurement may also be expressed by ± sign at the end, For example the mass of an electron is given as $9.109\,382\,92 \times 10^{-31}$ kg $\pm\,0.000\,000\,40 \times 10^{-31}$ kg, that is, it is between $9.109\,383\,31$ and $9.109\,382\,51 \times 10^{-31}$ kg

For example a 100 cm³ measuring cylinder has 1 cm³ as its smallest division so the measuring error can be taken as 0.5 cm³. So if you measure 50 cm³, the percentage error is $\left(\dfrac{0.5}{50}\right) \times 100\% = 1\%$

> What is the percentage error if you use a measuring cylinder to measure
> **a** 10 cm³
> **b** 100 cm³

a 5% b 0.5%

The number of moles in a given volume of solution

You often have to work out how many moles are present in a particular volume of a solution of known concentration. The general formula for the number of moles in a solution of concentration c (mol dm⁻³) and volume V (cm³) is:

number of moles in solution $n =$

$$\frac{\text{concentration } c \text{ (mol dm}^{-3}) \times \text{volume } V \text{ (cm}^3)}{1000}$$

Here is an example of how you reach this formula in steps.

> **Worked example: Moles in a solution**
>
> How many moles are present in 25.0 cm³ of a solution of concentration 0.10 mol dm⁻³?
>
> From the definition,
>
> 1000 cm³ of a solution of 1.00 mol dm⁻³ contains 1 mol
>
> So 1000 cm³ of a solution of 0.100 mol dm⁻³ contains 0.100 mol
>
> So 1.0 cm³ of a solution of 0.100 mol dm⁻³ contains
> $\dfrac{0.10}{1000}$ mol = 0.000 10 mol
>
> So 25.0 cm³ of a solution of 0.10 mol dm⁻³ contains
> $25.0 \times 0.000\,10 = 0.0025$ mol
>
> Using the formula gives the same answer:
> $$n = \frac{c \times V}{1000}$$
> $$= \frac{0.10 \times 25.0}{1000} = 0.0025 \text{ mol}$$

Summary questions

1 $\sqrt{x}$ Calculate the concentration in mol dm⁻³ of the following.

   a 0.500 mol acid in 500 cm³ of solution

   b 0.250 mol acid in 2000 cm³ of solution

   c 0.200 mol solute in 20 cm³ of solution

2 $\sqrt{x}$ Calculate how many moles of solute there are in the following.

   a 20.0 cm³ of a 0.100 mol dm⁻³ solution

   b 50.0 cm³ of a 0.500 mol dm⁻³ solution

   c 25.0 cm³ of a 2.00 mol dm⁻³ solution

3 $\sqrt{x}$ 0.234 g of sodium chloride was dissolved in water to make 250 cm³ of solution.

   a State the M_r for NaCl.
      A_r Na = 23.0, A_r Cl = 35.5

   b Calculate how many moles of NaCl is in 0.234 g.

   c Calculate the concentration in mol dm⁻³.

The Hindenburg airship (Figure 1) was originally designed in the 1930s to use helium as its lifting gas, rather than hydrogen, but the only source of large volumes of helium was the USA and they refused to sell it to Germany because of Hitler's aggressive policies. The airship was therefore made to use hydrogen. It held about 210 000 m³ of hydrogen gas but this volume varied with temperature and pressure.

The volume of a given mass of any gas is not fixed. It changes with pressure and temperature. However, there are a number of simple relationships for a given mass of gas that connect the pressure, temperature, and volume of a gas.

Boyle's law

The product of pressure and volume is a constant as long as the temperature remains constant.

$$\text{pressure } P \times \text{volume } V = \text{constant}$$

Charles' law

The volume is proportional to the temperature as long as the pressure remains constant.

$$\text{volume } V \propto \text{temperature } T \quad \text{and} \quad \frac{\text{volume } V}{\text{temperature } T} = \text{constant}$$

Gay-Lussac's law (also called the constant volume law)

The pressure is proportional to the temperature as long as the volume remains constant.

$$\text{pressure } P \propto \text{temperature } T \quad \text{and} \quad \frac{\text{pressure } P}{\text{temperature } T} = \text{constant}$$

Combining these relationships gives us the equation:

$$\frac{\text{pressure } P \times \text{volume } V}{\text{temperature } T} = \text{constant for a fixed mass of gas}$$

The ideal gas equation

In one mole of gas, the constant is given the symbol R and is called the gas constant. For n moles of gas:

$$\begin{array}{ccccccc}
\text{pressure} & \times & \text{volume} & = & \text{number of} & \times & \text{gas constant} & \times & \text{temperature} \\
P \text{ (Pa)} & & V \text{ (m}^3\text{)} & & \text{moles } n & & R \text{ (J K}^{-1}\text{ mol}^{-1}\text{)} & & T \text{ (K)}
\end{array}$$

$$PV = nRT$$

The value of R is $8.31 \text{ J K}^{-1}\text{ mol}^{-1}$.

This is the ideal gas equation. No gases obey it exactly, but at room temperature and pressure it holds quite well for many gases. It is often useful to imagine a gas which obeys the equation perfectly – an ideal gas.

Notes on units

When using the ideal gas equation, consistent units must be used. If you want to calculate n, the number of moles:

P must be in Pa (N m^{-2}) T must be in K

V must be in m³ R must be in $\text{J K}^{-1}\text{ mol}^{-1}$

Learning objectives:

→ State the ideal gas equation.

→ Describe how it is used to calculate the number of moles of a gas at a given volume, temperature and pressure.

Specification reference: 3.1.2

▲ **Figure 1** *The German airship Hindenburg held about 210 000 m³ of hydrogen gas*

Maths link

If you are not sure about proportionality and changing the subject of an equation, see Section 5, Mathematical skills.

Study tip √x

The units used here are part of the Système Internationale (SI) of units. This is a system of units for measurements used by scientists throughout the world. The basic units used by chemists are: metre *m*, second *s*, Kelvin *K*, and kilogram *kg*.

Using the ideal gas equation

Using the ideal gas equation, you can calculate the volume of one mole of gas at any temperature and pressure. Since none of the terms in the equation refers to a particular gas, this volume will be the same for any gas.

This may seem very unlikely at first sight, but it is the space between the gas molecules that accounts for the volume of a gas. Even the largest gas particle is extremely small compared with the space in between the particles.

Rearranging the ideal gas equation to find a volume gives:

$$V = \frac{nRT}{P}$$

The worked example tells you that the volume of a mole of *any* gas at room temperature and pressure is approximately 24 000 cm^3 (24 dm^3). For example, one mole of sulfur dioxide gas, SO_2 (mass 64.1 g) has the same volume as one mole of hydrogen gas, H_2 (mass 2.0 g).

In a similar way, pressure can be found using $P = \dfrac{nRT}{V}$

Finding the number of moles *n* of a gas

If you rearrange the equation $PV = nRT$ so that n is on the left-hand side, you get:

$$n = \frac{PV}{RT}$$

If T, P, and V are known, then you can find n.

Worked example: Volume from the ideal gas equation

If temperature = 20.0 °C (293.0 K), pressure = 100 000 Pa, and $n = 1$ for one mole of gas

$V = \dfrac{8.31\ \text{J K}^{-1}\,\text{mol}^{-1} \times 293\ \text{K}}{100\,000\ \text{Pa}}$

$= 0.024\,3\ \text{m}^3$

$= 0.024\,3 \times 10^6\ \text{cm}^3$

$= 24\,300\ \text{cm}^3$

Study tip $\sqrt{x}$

Remember to convert to SI units and to cancel the units.

Study tip $\sqrt{x}$

To convert °C to K add 273.

Worked example: Finding the number of moles

How many moles of hydrogen molecules are present in a volume of 100 cm^3 at a temperature of 20.0 °C and a pressure of 100 kPa? $R = 8.31\ \text{J K}^{-1}\,\text{mol}^{-1}$

First, convert to the base units:

P must be in Pa, and 100 kPa = 100 000 Pa

V must be in m^3, and 100 cm^3 = 100 × 10^{-6} m^3

T must be in K, and 20 °C = 293 K (add 273 to the temperature in °C)

Substituting into the ideal gas equation:

$n = \dfrac{PV}{RT}$

$= \dfrac{100\,000 \times 100 \times 10^{-6}}{8.31 \times 293}$

$= 0.004\,11$ moles

Finding the relative molecular mass of a gas

If you know the number of moles present in a given mass of gas, you can find the mass of one mole of gas and this tells us the relative molecular mass.

Study tip

Using 24 000 cm^3 as the volume of a mole of any gas is not precise and it is always necessary to apply the ideal gas equation in calculations.

Finding the relative molecular mass of lighter fuel

The apparatus used to find the relative molecular mass of lighter fuel is shown in Figure 2.

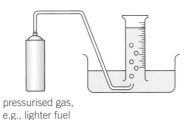

pressurised gas,
e.g., lighter fuel

▲ **Figure 2** *Measuring the relative molecular mass of lighter fuel*

The lighter fuel canister was weighed.

1000 cm³ of gas was dispensed into the measuring cylinder, until the levels of the water inside and outside the measuring cylinder were the same, so that the pressure of the collected gas was the same as atmospheric pressure.

The canister was reweighed.

Atmospheric pressure and temperature were noted.

The results were

$$\text{loss of mass of the can} = 2.29 \text{ g}$$

$$\text{temperature} = 14\,^{\circ}\text{C} = 287 \text{ K}$$

$$\text{atmospheric pressure} = 100\,000 \text{ Pa}$$

$$\text{Volume of gas} = 1000 \text{ cm}^3 = 1000 \times 10^{-6} \text{ m}^3$$

$$n = \frac{PV}{RT}$$

$$= \frac{100\,000 \times 1000 \times 10^{-6}}{8.31 \times 287}$$

$$= 0.042 \text{ mol}$$

0.042 mol has a mass of 2.29 g

So, 1 mol has a mass of $\dfrac{2.29}{0.042 \text{ g}} = 54.5$ g

So, $M_r = 54.5$

Summary questions

1 √x̄ a   Calculate approximately how many moles of H_2 molecules were contained in the Hindenburg airship at 298 K.

  b   The original design used helium. State how many moles of helium atoms it would have contained.

2 √x̄ a   Calculate the volume of 2 moles of a gas if the temperature is 30 °C, and the pressure is 100 000 Pa.

  b   Calculate the pressure of 0.5 moles of a gas if the volume is 11 000 cm³, and the temperature is 25 °C.

3 √x̄   Calculate how many moles of hydrogen molecules are present in a volume of 48 000 cm³, at 100 000 Pa and 25 °C.

4   State how many moles of carbon dioxide molecules would be present in **3**? Explain your answer.

Study tip

The mass of 1 mole in grams is the same as the relative atomic mass of the element.

The empirical formula

The **empirical formula** is the formula that represents the simplest whole number ratio of the atoms of each element present in a compound. For example, the empirical formula of carbon dioxide, CO_2, tells us that for every carbon atom there are two oxygen atoms.

To find an empirical formula:

1 Find the masses of each of the elements present in a compound (by experiment).
2 Work out the number of moles of atoms of each element.

$$\text{number of moles} = \frac{\text{mass of element}}{\text{mass of 1 mol of element}}$$

3 Convert the number of moles of each element into a whole number ratio.

Worked example: Finding empirical formula of calcium carbonate

10.01 g of a white solid contains 4.01 g of calcium, 1.20 g of carbon, and 4.80 g of oxygen. What is its empirical formula?
(A_r Ca = 40.1, A_r C = 12.0, A_r O = 16.0)

Step 1 Find the masses of each element.

Mass of calcium = 4.01 g

Mass of carbon = 1.20 g

Mass of oxygen = 4.80 g

Step 2 Find the number of moles of atoms of each element.

A_r Ca = 40.1
Number of moles of calcium = $\frac{4.01}{40.1}$ = 0.10 mol
A_r C = 12.0
Number of moles of carbon = $\frac{1.2}{12.0}$ = 0.10 mol
A_r O = 16
Number of moles of oxygen = $\frac{4.8}{16.0}$ = 0.30 mol

Step 3 Find the simplest ratio.

Ratio in moles of	calcium	:	carbon	:	oxygen
	0.10	:	0.10	:	0.30

So the simplest whole number ratio is: 1 : 1 : 3

The formula is therefore $CaCO_3$.

Worked example: Finding the empirical formula of copper oxide

0.795 g of black copper oxide is reduced to 0.635 g of copper when heated in a stream of hydrogen (Figure 1). What is the formula of copper oxide? A_r Cu = 63.5, A_r O =16.0

Step 1 Find the masses of each element.

Mass of copper = 0.635 g

Started with 0.795 g of copper oxide and 0.635 g of copper were left, so:

Mass of oxygen = 0.795 − 0.635 = 0.160 g

Step 2 Find the number of moles of atoms of each element.

A_r Cu = 63.5

Number of moles of copper = $\dfrac{0.635}{63.5}$ = 0.01

A_r O =16.0

Number of moles of oxygen = $\dfrac{0.16}{16.0}$ = 0.01

Step 3 Find the simplest ratio.

The ratio of moles of copper to moles of oxygen is:

copper : oxygen

0.01 : 0.01

So the simplest whole number ratio is 1 : 1

The simplest formula of black copper oxide is therefore one Cu to one O, CuO. You may find it easier to make a table.

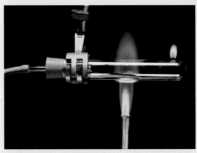

▲ **Figure 1** *Finding the empirical formula of copper oxide*

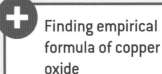

Finding empirical formula of copper oxide

In this experiment explain why:

1 there is a flame at the end of the tube
2 this flame goes green
3 droplets of water form near the end of the tube
4 the flame at the end of the tube is kept alight until the apparatus is cool.

	Copper Cu	Oxygen O
mass of element	0.635 g	0.160 g
A_r of element	63.5	16.0
number of moles = $\dfrac{\text{mass of element}}{A_r}$	$\dfrac{0.635}{63.5}$ = 0.01	$\dfrac{0.160}{16.0}$ = 0.01
ratio of elements	1	1

Erroneous results

One student carried out the experiment to find the formula of black copper oxide with the following results:
0.735 g of the oxide was reduced to 0.635 g after reduction.

1 Confirm that these results lead to a ratio of 0.01 mol, copper to 0.006 25 mol of oxygen, which is incorrect.
2 Suggest what the student might have done wrong to lead to this apparently low value for the amount of oxygen.

Another oxide of copper

There is another oxide of copper, which is red. In a reduction experiment similar to that for finding the formula of black copper oxide, 1.43 g of red copper oxide was reduced with a stream of hydrogen and 1.27 g of copper were formed. Use the same steps as for black copper oxide to find the formula of the red oxide.

1   Find the masses of each element.
2   Find the number of moles of atoms of each element.
3   Find the simplest ratio.

Finding the simplest ratio of elements

Sometimes you will end up with ratios of moles of atoms of elements that are not easy to convert to whole numbers. If you divide each number by the smallest number you will end up with whole numbers (or ratios you can recognise more easily). Here is an example.

Worked example: Empirical formula

Compound X contains 50.2 g sulfur and 50.0 g oxygen. What is its empirical formula? A_r S = 32.1, A_r O = 16.0

Step 1 Find the number of moles of atoms of each element.

A_r S = 32.1

Number of moles of sulfur $= \dfrac{50.2}{32.1} = 1.564$

A_r O = 16

Number of moles of oxygen $= \dfrac{50.0}{16.0} = 3.125$

Step 2 Find the simplest ratio.

Ratio of sulfur : oxygen : 1.564 : 3.125

Now divide each of the numbers by the smaller number.

Ratio of sulfur : oxygen

$\dfrac{1.564}{1.564} : \dfrac{3.125}{1.564} = 1:2$

The empirical formula is therefore SO_2. Sometimes you may end up with a ratio of moles of atoms, such as 1:1.5. In these cases you must find a whole number ratio, in this case 2:3.

Study tip

• When calculating empirical formulae from percentages, check that all the percentages of the compositions by mass add up to 100%. (Don't forget any oxygen that may be present.)

• Remember to use relative atomic masses from the Periodic Table, *not* the atomic number.

Finding the molecular formula

The **molecular formula** gives the actual number of atoms of each element in one molecule of the compound. (It applies only to substances that exist as molecules.)

The empirical formula is not always the same as the molecular formula. There may be several units of the empirical formula in the molecular formula.

For example, ethane (molecular formula C_2H_6) would have an empirical formula of CH_3.

To find the number of units of the empirical formula in the molecular formula, divide the relative molecular mass by the relative mass of the empirical formula.

For example, ethene is found to have a relative molecular mass of 28.0 but its empirical formula, CH_2, has a relative mass of 14.0.

$$\frac{\text{Relative molecular mass of ethene}}{\text{Relative mass of empirical formula of ethene}} = \frac{28.0}{14.0} = 2$$

So there must be two units of the empirical formula in the molecule of ethene. So ethene is $(CH_2)_2$ or C_2H_4.

Synoptic link

Once we know the formula of a compound we can use techniques such as infra-red spectroscopy and mass spectrometry to help work out its structure, see Chapter 16, Organic analysis.

Combustion analysis

Organic compounds are based on carbon and hydrogen. One method of finding empirical formulae of new compounds is called combustion analysis (Figure 2). It is used routinely in the pharmaceutical industry. It involves burning the unknown compound in excess oxygen and measuring the amounts of water, carbon dioxide, and other oxides that are produced. The gases are carried through the instrument by a stream of helium.

The basic method measures carbon, hydrogen, sulfur, and nitrogen. It is assumed that oxygen makes up the difference after the other four elements have been measured. Once the sample has been weighed and placed in the instrument, the process is automatic and controlled by computer.

The sample is burnt completely in a stream of oxygen. The final combustion products are water, carbon dioxide, and sulfur dioxide. The instrument measures the amounts of these by infrared absorption. They are removed from the gas stream leaving the unreacted nitrogen which is measured by thermal conductivity. The measurements are used to calculate the masses of each gas present and hence the masses of hydrogen, sulfur, carbon, and nitrogen in the original sample. Oxygen is found by difference.

Traditionally, the amounts of water and carbon dioxide were measured by absorbing them in suitable chemicals and measuring the increase in mass of the absorbents. This is how the composition data for the worked example below were measured. The molecular formula can then be found, if the relative molecular mass has been found using a mass spectrometer.

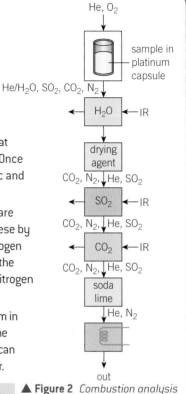

▲ **Figure 2** *Combustion analysis*

Soda lime is a mixture containing mostly calcium hydroxide, $Ca(OH)_2$. Construct a balanced symbol equation for the reaction of calcium hydroxide with carbon dioxide.

$$Ca(OH)_2 + CO_2 \rightarrow CaCO_3 + H_2O$$

Worked example: Molecular formula

An organic compound containing only carbon, hydrogen, and oxygen was found to have 52.17% carbon and 13.04% hydrogen. What is its molecular formula if $M_r = 46.0$?

100.00 g of this compound would contain 52.17 g carbon, 13.04 g hydrogen and (the rest) 34.79 g oxygen.

Step 1 Find the empirical formula.

	Carbon	Hydrogen	Oxygen
mass of element/g	52.17	13.04	34.79
A_r of element	12.0	1.0	16.0
number of moles $= \dfrac{\text{mass of element}}{A_r}$	$\dfrac{52.17}{12.0} = 4.348$	$\dfrac{13.04}{10.0} = 13.04$	$\dfrac{34.79}{16.0} = 2.174$
divide through by the smallest	$\dfrac{4.348}{2.174} = 2$	$\dfrac{13.04}{2.174} = 6$	$\dfrac{2.174}{2.174} = 1$
ratio of elements	2	6	1

So the empirical formula is C_2H_6O.

Step 2 Find M_r of the empirical formula.

$$(2 \times 12.0) + (6 \times 1.0) + (1 \times 16.0) = 46.0$$
$$\quad C \qquad\quad H \qquad\qquad O$$

So, the molecular formula is the same as the empirical formula, C_2H_6O.

Worked example: Molecular formula by combustion analysis

0.53 g of a compound X containing only carbon, hydrogen, and oxygen, gave 1.32 g of carbon dioxide and 0.54 g of water on complete combustion in oxygen. What is its empirical formula? What is its molecular formula if its relative molecular mass is 58.0?

To calculate the empirical formula:

carbon   1.32 g of CO_2 (M_r = 44.0) is $\dfrac{1.32}{44.0}$ = 0.03 mol CO_2

As each mole of CO_2 has 1 mole of C, the sample contained 0.03 mol of C atoms.

hydrogen   0.54 g of H_2O (M_r = 18.0) is $\dfrac{0.54}{18.0}$ = 0.03 mol H_2O

As each mole of H_2O has 2 moles of H, the sample contained 0.06 mol of H atoms.

oxygen   0.03 mol of carbon atoms (A_r = 12.0) has a mass of 0.36 g

               0.06 mol of hydrogen atoms (A_r = 1.0) has a mass of 0.06 g

               Total mass of carbon and hydrogen is 0.42 g

The rest (0.58 − 0.42) must be oxygen, so the sample contained 0.16 g of oxygen.

0.16 g of oxygen (A_r = 16.0) is $\dfrac{0.16}{16.0}$ = 0.01 mol oxygen atoms

So the sample contains   0.03 mol C, 0.06 mol H, and 0.01 mol O

Dividing by the smallest number 0.06 gives the ratio: C   H   O

so the **empirical formula** is C_3H_6O.   3   6   1

M_r of this unit is 58, so the molecular formula is also C_3H_6O.

Summary questions

1   √x Calculate the empirical formula of each of the following compounds? (You could try to name them too.)

   a   A liquid containing 2.0 g of hydrogen, 32.1 g sulfur, and 64.0 g oxygen.

   b   A white solid containing 4.0 g calcium, 3.2 g oxygen, and 0.2 g hydrogen.

   c   A white solid containing 0.243 g magnesium and 0.710 g chlorine.

2   3.888 g magnesium ribbon was burnt completely in air and 6.448 g of magnesium oxide was produced.

   √x **a**   Calculate how many moles of magnesium and oxygen are present in 6.488 g of magnesium oxide.

   b   State the empirical formula of magnesium oxide.

3   State the empirical formula of each of the following molecules?

   a   cyclohexane, C_6H_{12} **b**   dichloroethene, $C_2H_2Cl_2$ **c**   benzene, C_6H_6

4   M_r for ethane-1,2-diol is 62.0. It is composed of carbon, hydrogen, and oxygen in the ratio by moles of 1 : 3 : 1. Identify its molecular formula.

5   An organic compound containing only carbon, hydrogen, and oxygen was found to have 62.07% carbon and 10.33% hydrogen. Identify the molecular formula if M_r = 58.0.

6   A sample of benzene of mass 7.8 g contains 7.2 g of carbon and 0.6 g of hydrogen. If M_r is 78.0, identify:

   a   the empirical formula   **b**   the molecular formula.

Equations represent what happens when chemical reactions take place. They are based on experimental evidence. The starting materials are reactants. After these have reacted you end up with products.

reactants → products

Word equations only give the names of the reactants and products, for example:

hydrogen + oxygen → water

Once the idea of atoms had been established, chemists realised that atoms react together in simple whole number ratios. For example, two hydrogen molecules react with one oxygen molecule to give two water molecules.

2 hydrogen molecules + 1 oxygen molecule → 2 water molecules

| 2 | : | 1 | : | 2 |

The ratio in which the reactants react and the products are produced, in simple whole numbers, is called the **stoichiometry** of the reaction.

You can build up a stoichiometric relationship from experimental data by working out the number of moles that react together. This leads us to a balanced symbol equation.

Balanced symbol equations

Balanced symbol equations use the formulae of reactants and products. There are the same number of atoms of each element on both sides of the arrow. (This is because atoms are never created or destroyed in chemical reactions.) Balanced equations tell us about the amounts of substances that react together and are produced.

State symbols can also be added. These are letters, in brackets, which can be added to the formulae in equations to say what state the reactants and products are in – (s) means solid, (l) means liquid, (g) means gas, and (aq) means aqueous solution (dissolved in water).

Writing balanced equations

When aluminium burns in oxygen it forms solid aluminium oxide. You can build up a balanced symbol equation from this and the formulae of the reactants and product – Al, O_2, and Al_2O_3.

1   Write the word equation

aluminium   +   oxygen   →   aluminium oxide

2   Write in the correct formulae

Al + O_2 → Al_2O_3

This is not balanced because:

- there is one aluminium atom on the reactants side (left-hand side) but two on the products side (right-hand side)
- there are two oxygen atoms on the reactants side (left-hand side) but three on the products side (right-hand side).

Learning objectives:

→ Demonstrate how an equation can be balanced if the reactants and products are known.

→ Calculate the amount of a product using experimental data and a balanced equation.

Specification reference: 3.1.2

Study tip

Make sure you learn these four state symbols.

3 To get two aluminium atoms on the left-hand side put a 2 in front of the Al:

$$2Al \quad + \quad O_2 \quad \rightarrow \quad Al_2O_3$$

Now the aluminium is correct but not the oxygen.

4 If you multiply the oxygen on the left-hand side by 3, and the aluminium oxide by 2, you have six O on each side:

$$2Al \quad + \quad 3O_2 \quad \rightarrow \quad 2Al_2O_3$$

5 Now you return to the aluminium. You need four Al on the left-hand side:

$$4Al \quad + \quad 3O_2 \quad \rightarrow \quad 2Al_2O_3$$

The equation is balanced because there are the same numbers of atoms of each element on both sides of the equation.

The numbers in front of the formulae (4, 3, and 2) are called coefficients.

6 You can add state symbols.

The equation tells you the numbers of moles of each of the substances that are involved. From this you can work out the masses that will react together: (using $Al = 27.0$, $O = 16.0$)

$4Al(s)$	$+$	$3O_2(g)$	$\rightarrow$	$2Al_2O_3(s)$
4 moles		3 moles		2 moles
108.0 g		96.0 g		204.0 g

The total mass is the same on both sides of the equation. This is another good way of checking whether the equation is balanced.

Ionic equations

In some reactions you can simplify the equation by considering the ions present. Sometimes there are ions that do not take part in the overall reaction. For example, when any acid reacts with an alkali in solution, you end up with a salt (also in solution) and water. Look at the reaction between hydrochloric acid and sodium hydroxide:

$$HCl(aq) \quad + \quad NaOH(aq) \quad \rightarrow \quad NaCl(aq) \quad + \quad H_2O(l)$$

hydrochloric acid + sodium hydroxide → sodium chloride + water

The ions present are:

$HCl(aq)$	$H^+(aq)$ and $Cl^-(aq)$
$NaOH(aq)$	$Na^+(aq)$ and $OH^-(aq)$
$NaCl(aq)$	$Na^+(aq)$ and $Cl^-(aq)$

If you write the equation using these ions and then strike out the ions that appear on each side we have:

$H^+(aq) + \cancel{Cl^-(aq)} + \cancel{Na^+(aq)} + OH^-(aq) \rightarrow \cancel{Na^+(aq)} + \cancel{Cl^-(aq)} + H_2O(l)$

Overall, the equation is

$$H^+(aq) + OH^-(aq) \rightarrow H_2O(l)$$

$Na^+(aq)$ and $Cl^-(aq)$ are called **spectator ions** – they do not take part in the reaction.

Study tip

The charges balance as well as the elements. On the left +1 and −1 (no overall charge) and no charge on the right.

Whenever an acid reacts with an alkali, the overall reaction will be the same as the one above.

Useful tips for balancing equations

- You *must* use the correct formulae – you cannot change them to make the equation balance.
- You can only change the numbers of atoms by putting a number, called a coefficient, in front of formulae.
- The coefficient in front of the symbol tells you how many moles of that substance are reacting.
- It often takes more than one step to balance an equation, but too many steps suggests that you may have an incorrect formula.
- When dealing with ionic equations the total of the charges on each side must also be the same.

Working out amounts

You can use a balanced symbol equation to work out how much product is produced from a reaction.

Worked example: Calculating the mass of product

How much magnesium chloride is produced by 0.120 g of magnesium ribbon and excess hydrochloric acid? A_r Mg = 24.3, A_r H = 1.0, A_r Cl = 35.5

(The word excess means there is more than enough acid to react with all the magnesium.)

Step 1 Write the correct formulae equation.

$$Mg(s) + HCl(aq) \rightarrow MgCl_2(aq) + H_2(g)$$
magnesium + hydrochloric acid → magnesium chloride + hydrogen

Step 2 Balance the equation. The number of Mg atoms is correct. There are two Cl atoms and two H atoms on the right-hand side so you need to add a 2 in front of the HCl.

$$Mg(s) + 2HCl(aq) \rightarrow MgCl_2(aq) + H_2(g)$$

Now find the numbers of moles that react.

$$Mg(s) + 2HCl(aq) \rightarrow MgCl_2(aq) + H_2(g)$$
1 mol   2 mol → 1 mol   1 mol

1 mol of Mg has a mass of 24.3 g because its A_r = 24.3.

So, 0.12 g of Mg is $\frac{0.12}{24.3}$ = 0.0049 mol.

From the equation, you can see that one mole of magnesium reacts to give one mole of magnesium chloride. Therefore, 0.0049 mol of magnesium produces 0.0049 mol of magnesium chloride.

M_r MgCl$_2$ = 24.3 + (2 × 35.5) = 95.3

So the mass of MgCl$_2$ = 0.0049 × 95.3 = 0.48 g to 2 s.f.

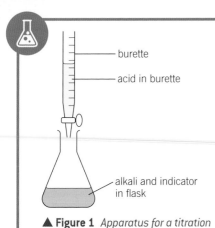

▲ **Figure 1** *Apparatus for a titration*

Finding concentrations using titrations

Titrations can be used to find the concentration of a solution, for example, an alkali by reacting the acid with an alkali using a suitable indicator.

You need to know the concentration of the acid and the equation for the reaction between the acid and alkali.

The apparatus is shown in Figure 1.

The steps in a titration are:

1 Fill a burette with the acid of known concentration.
2 Accurately measure an amount of the alkali using a calibrated pipette and pipette filler.
3 Add the alkali to a conical flask with a few drops of a suitable indicator.
4 Run in acid from the burette until the colour just changes, showing that the solution in the conical flask is now neutral.
5 Repeat the procedure, adding the acid dropwise as you approach the end point, until two values of the volume of acid used at neutralisation are the same, within experimental error.

Summary questions

1 Balance the following equations.

  a $Mg + O_2 \rightarrow MgO$

  b $Ca(OH)_2 + HCl$
$\rightarrow CaCl_2 + H_2O$

  c $Na_2O + HNO_3$
$\rightarrow NaNO_3 + H_2O$

2 State the concentration of hydrochloric acid if $20.0\ cm^3$ is neutralised by $25.0\ cm^3$ of sodium hydroxide of concentration $0.200\ mol\ dm^{-3}$.

3 In the reaction
$Mg(s) + 2HCl(aq)$
$\rightarrow MgCl_2(aq) + H_2$
$2.60\ g$ of magnesium was added to $100\ cm^3$ of $1.00\ mol\ dm^{-3}$ hydrochloric acid.

  a State is there be any magnesium left when the reaction finished. Explain your answer.

  b Calculate the volume of hydrogen produced at $25\ °C$ and $100\ kPa$.

4 **a** Write the balanced equation for the reaction between sulfuric acid and sodium hydroxide

     i in full

    ii in terms of ions.

  b Identify the spectator ions in this reaction.

Worked example: Finding concentration

$25.00\ cm^3$ of a solution of sodium hydroxide, NaOH, of unknown concentration was neutralised by $22.65\ cm^3$ of a $0.100\ mol\ dm^{-3}$ solution of hydrochloric acid, HCl. What is the concentration of the alkali?

First write a balanced symbol equation and then the numbers of moles that react:

$NaOH(aq)$	+	$HCl(aq)$	$\rightarrow$	$NaCl(aq)$	+	$H_2O(l)$
sodium hydroxide		hydrochloric acid		sodium chloride		water
1 mol		1 mol		1 mol		1 mol

1 mol of sodium hydroxide reacts with 1 mol of hydrochloric acid.

$$\text{number of moles of HCl} = \frac{c \times V}{1000} = \frac{22.65 \times 0.100}{1000}$$

From the equation, there must be an equal number of moles of sodium hydroxide and hydrochloric acid for neutralisation:

number of moles of NaOH = number of moles of HCl

So you must have $\dfrac{22.65 \times 0.100}{1000}$ mol of NaOH in the $25.00\ cm^3$ of sodium hydroxide solution.

The concentration of a solution is the number of moles in $1000\ cm^3$.

Therefore the concentration of the alkali

$$= \frac{22.65 \times 0.100}{1000} \times \frac{1000}{25.00}\ mol\ dm^{-3} = 0.0906\ mol\ dm^{-3}$$

A note on significant figures

22.65 and 25.00 both have 4 s.f. but 0.100 has only 3 s.f. So we can quote the answer to 3 s.f only. So rounding up the final digit gives the concentration of the alkali as $0.091\ mol\ dm^{-3}$ to 3 s.f.

Once you know the balanced equation for a chemical reaction, you can calculate the theoretical amount that you should be able to make of any of the products. Most chemical reactions produce two (or more) products but often only one of them is required. This means that some of the products will be wasted. In a world of scarce resources, this is obviously not a good idea. One technique that chemists use to assess a given process is to determine the percentage atom economy.

Atom economy

The **atom economy** of a reaction is found directly from the balanced equation. It is theoretical rather than practical. It is defined as:

$$\% \text{ atom economy} = \frac{\text{mass of desired product}}{\text{total mass of reactants}} \times 100$$

You can see what atom economy means by considering the following real reaction.

Chlorine, Cl_2, reacts with sodium hydroxide, NaOH, to form sodium chloride, NaCl, water, H_2O, and sodium chlorate, NaOCl. Sodium chlorate is used as household bleach – this is the useful product.

From the equation you can work out the mass of each reactant and product involved.

2NaOH	+	Cl_2	$\rightarrow$	NaCl	+	H_2O	+	NaOCl
2 mol		1 mol	$\rightarrow$	1 mol		1 mol		1 mol
80.0 g		71.0 g	$\rightarrow$	58.5 g		18.0 g		74.5 g
Total	151.0 g			Total		151.0 g		

$$\% \text{ Atom economy} = \frac{\text{mass of desired product}}{\text{total mass of reactants}} \times 100$$
$$= \frac{74.5}{151} \times 100$$
$$= 49.3\%$$

So only 49.3% of the starting materials are included in the desired product, the rest is wasted.

It may be easier to see what has happened if you colour the atoms involved. Those coloured in green are included in the final product and those in red are wasted – one atom of sodium, one of chlorine, two of hydrogen, and one of oxygen.

NaOH + NaOH + ClCl → NaCl + H₂O + NaOCl

Another example is the reaction where ethanol breaks down to ethene, the product wanted, and water, which is wasted.

C_2H_5OH	$\rightarrow$	CH_2=CH_2	+	H_2O
46.0 g	$\rightarrow$	28.0 g		18.0 g

$$\% \text{ Atom economy} = \frac{28.0}{46.0} \times 100 = 60.9\%$$

Learning objectives:
→ Describe the atom economy of a chemical reaction.
→ State how an equation is used to calculate an atom economy.
→ Describe the percentage yield of a chemical reaction.
→ Calculate percentage yields.

Specification reference: 3.1.2

Some reactions, in theory at least, have no wasted atoms.

For example, ethene reacts with bromine to form 1,2-dibromoethane

$$CH_2{=\!\!=}CH_2 \quad + \quad Br_2 \quad \rightarrow \quad CH_2BrCH_2Br$$

$$28.0\,g \qquad\qquad 160.0\,g \quad \rightarrow \qquad 188.0\,g$$

$$\text{Total } 188.0\,g \qquad\qquad \text{Total } 188.0\,g$$

$$\text{\% Atom economy} = \frac{188.0}{(28.0 + 160.0)} \times 100 = 100\%$$

Atom economies

There are clear advantages for industry and society to develop chemical processes with high atom economies. A good example is the manufacture of the over-the-counter painkiller and anti-inflammatory drug ibuprofen. the original manufacturing process had an atom economy of only 44%, but a newly-developed process has improved this to 77%.

Atom economy – a dangerous fuel

Hydrogen can be made by passing steam over heated coal, which is largely carbon.

$$C(s) + 2H_2O(g) \rightarrow 2H_2(g) + CO_2(g)$$

$$12.0 + (2 \times 18.0) \rightarrow (2 \times 2.0) + 44.0$$

As the only useful product is hydrogen, the atom economy of this reaction is $\left(\frac{4.0}{48.0}\right) \times 100\% = 8.3\%$ – not a very efficient reaction! The reason that it is so inefficient is that all of the carbon is discarded as useless carbon dioxide.

However, under different conditions a mixture of hydrogen and carbon monoxide can be formed (this was called water gas or town gas).

$$C(s) + H_2O(g) \rightarrow H_2(g) + CO(g)$$

Both hydrogen and carbon monoxide are useful fuels, so nothing is discarded and the atom economy is 100%. You can check this with a calculation if you like.

Carbon monoxide is highly toxic. However, almost incredibly to modern eyes, town gas was supplied as a fuel to homes in the days before the country converted to natural gas (methane, CH_4) from the North Sea.

Even methane is not without its problems. When it burns in a poor supply of oxygen, carbon monoxide is formed and this can happen in gas fires in poorly-ventilated rooms. This has sometimes happened in student flats, for example, where windows and doors have been sealed to reduce draughts and cut energy bills resulting in a lack of oxygen for the gas fire. Landlords are now recommended to fit a carbon monoxide alarm.

Write a balanced formula equation for the formation of carbon monoxide by the combustion of methane in a limited supply of oxygen.

$$2CH_4 + 3O_2 \rightarrow 2CO + 4H_2O$$

The percentage yield of a chemical reaction

The yield of a reaction is different from the atom economy.

- The atom economy tells us *in theory* how many atoms *must* be wasted in a reaction.
- The yield tells us about the practical efficiency of the process, how much is lost by:
  a the *practical* process of obtaining a product and
  b as a result of reactions that do not go to completion.

As you have seen, once you know the balanced symbol equation for a chemical reaction, you can calculate the amount of any product that you should be able to get from given amounts of starting materials if the reaction goes to completion. For example:

$$2KI(aq) \quad + Pb(NO_3)_2(aq) \rightarrow \quad PbI_2(s) \quad + \quad 2KNO_3(aq)$$

potassium iodide	lead nitrate	lead iodide	potassium nitrate
2 mol	1 mol	1 mol	2 mol
332 g	331 g	461 g	202 g

So starting from 3.32 g $\left(\frac{2}{100}\text{mol}\right)$ of potassium iodide in solution and adding 3.31 g $\left(\frac{1}{100}\text{mol}\right)$ of lead nitrate in aqueous solution should produce 4.61 g $\left(\frac{1}{100}\text{mol}\right)$ of a precipitate of lead iodide which can be filtered off and dried.

However, this is in theory only. When you pour one solution into another, some droplets will be left in the beaker. When you remove the precipitate from the filter paper, some will be left on the paper. This sort of problem means that in practice you never get as much product as the equation predicts. Much of the skill of the chemist, both in the laboratory and in industry, lies in minimising these sorts of losses.

$$\frac{\text{The yield of a}}{\text{chemical reaction}} = \frac{\text{the number of moles of a specified product}}{\text{theoretical maximum number of moles}} \times 100\%$$
$$\text{of the product}$$

It can equally well be defined as:

$$\frac{\text{the number of grams of a specified}}{\text{theoretical maximum number of}} \times 100\%$$
$$\text{grams of the product}$$

If you had obtained 4.00 g of lead iodide in the above reaction, the yield would have been:

$$\frac{4.00}{4.61} \times 100\% = 86.8\%$$

A further problem arises with reactions that are reversible and do not go to completion. This is not uncommon. One example is the Haber process in which ammonia is made from hydrogen and nitrogen. Here is it impossible to get a yield of 100% even with the best practical skills. However, chemists can improve the yield by changing the conditions.

Percentage yields

Yields of multi-step reactions can be surprisingly low because the overall yield is the yield of each step multiplied together. So a four step reaction in which each step had an 80% yield would be
$80\% \times 80\% \times 80\% \times 80\% = 41\%$

What would be the overall yield of a three step process if the yield of each separate step were 80%, 60%, and 75% respectively?

36%

1 √x̄ Lime (calcium oxide, CaO) is made by heating limestone (calcium carbonate, $CaCO_3$) to drive off carbon dioxide gas, CO_2.

$$CaCO_3 \rightarrow CaO + CO_2$$

Calculate the atom economy of the reaction.

2 Sodium sulfate can be made from sulfuric acid and sodium hydroxide.

$$H_2SO_4 + 2NaOH$$
$$\rightarrow Na_2SO_4 + H_2O$$

If sodium sulfate is the required product, calculate the atom economy of the reaction.

3 Ethanol, C_2H_6O, can be made by reacting ethene, C_2H_4, with water, H_2O.

$$C_2H_4 + H_2O \rightarrow C_2H_6O$$

Without doing a calculation, state the atom economy of the reaction. Explain your answer.

4 √x̄ Consider the reaction $CaCO_3 \rightarrow CO_2 + CaO$

a Calculate the theoretical maximum number of moles of calcium oxide, CaO, that can be obtained from 1 mole of calcium carbonate, $CaCO_3$.

b Starting from 10 g calcium carbonate, calculate the theoretical maximum number of grams of calcium oxide that can be obtained.

c If 3.6 g of calcium oxide was obtained, calculate the yield of the reaction.

1 Potassium nitrate, KNO_3, decomposes on strong heating, forming oxygen and solid **Y** as the only products.

   (a) A 1.00 g sample of KNO_3 (M_r = 101.1) was heated strongly until fully decomposed into **Y**.

     √x (i) Calculate the number of moles of KNO_3 in the 1.00 g sample.

       (ii) At 298 K and 100 kPa, the oxygen gas produced in this decomposition occupied a volume of 1.22×10^{-4} m^3.
State the ideal gas equation and use it to calculate the number of moles of oxygen produced in this decomposition.
(The gas constant R = 8.31 $JK^{-1}mol^{-1}$)

(5 marks)

   (b) Compound **Y** contains 45.9% of potassium and 16.5% of nitrogen by mass, the remainder being oxygen.

     (i) State what is meant by the term *empirical formula*.

     (ii) Use the data above to calculate the empirical formula of **Y**.

(4 marks)

   (c) Deduce an equation for the decomposition of KNO_3 into **Y** and oxygen.

(1 mark)

AQA, 2006

2 Ammonia is used to make nitric acid, HNO_3, by the Ostwald Process.
Three reactions occur in this process.

Reaction 1 $4NH_3(g) + 5O_2(g) \rightarrow 4NO(g) + 6H_2O(g)$

Reaction 2 $2NO(g) + O_2(g) \rightarrow 2NO_2(g)$

Reaction 3 $3NO_2(g) + H_2O(l) \rightarrow 2HNO_3(aq) + NO(g)$

   √x (a) In one production run, the gases formed in Reaction 1 occupied a total volume of 4.31 m^3 at 25 °C and 100 kPa.
Calculate the amount, in moles, of NO produced.
Give your answer to the appropriate number of significant figures.
(The gas constant R = 8.31 $JK^{-1}mol^{-1}$)

(4 marks)

   √x (b) In another production run, 3.00 kg of ammonia gas were used in Reaction 1 and all of the NO gas produced was used to make NO_2 gas in Reaction 2.
Calculate the mass of NO_2 formed from 3.00 kg of ammonia in Reaction 2 assuming an 80.0% yield.
Give your answer in kilograms.

(5 marks)

   √x (c) Consider Reaction 3 in this process.

       $3NO_2(g) + H_2O(l) \rightarrow 2HNO_3(aq) + NO(g)$

Calculate the concentration of nitric acid produced when 0.543 mol of NO_2 is reacted with water and the solution is made up to 250 cm^3.

(2 marks)

   (d) Suggest why a leak of NO_2 gas from the Ostwald Process will cause atmospheric pollution.

(1 mark)

   (e) Give one reason why excess air is used in the Ostwald Process.

(1 mark)

   (f) Ammonia reacts with nitric acid as shown in this equation.

       $NH_3 + HNO_3 \rightarrow NH_4NO_3$

Deduce the type of reaction occurring.

(1 mark)

AQA, 2013

3 Zinc forms many different salts including zinc sulfate, zinc chloride, and zinc fluoride.

(a) People who have a zinc deficiency can take hydrated zinc sulfate, $ZnSO_4.xH_2O$, as a dietary supplement.

A student heated 4.38 g of hydrated zinc sulfate and obtained 2.46 g of anhydrous zinc sulfate.

Use these data to calculate the value of the integer x in $ZnSO_4.xH_2O$.

Show your working.

(3 marks)

(b) Zinc chloride can be prepared in the laboratory by the reaction between zinc oxide and hydrochloric acid.

The equation for the reaction is:

$ZnO + 2HCl \rightarrow ZnCl_2 + H_2O$

A 0.0830 mol sample of pure zinc oxide was added to 100 cm^3 of 1.20 mol dm^{-3} hydrochloric acid.

Calculate the maximum mass of anhydrous zinc chloride that could be obtained from the products of this reaction. Give your answer to the appropriate number of significant figures.

(4 marks)

(c) Zinc chloride can also be prepared in the laboratory by the reaction between zinc and hydrogen chloride gas.

$Zn + 2HCl \rightarrow ZnCl_2 + H_2$

An impure sample of zinc powder with a mass of 5.68 g was reacted with hydrogen chloride gas until the reaction was complete. The zinc chloride produced had a mass of 10.7 g.

Calculate the percentage purity of the zinc metal. Give your answer to 3 significant figures.

(4 marks)
AQA, 2013

4 In this question give all your answers to the appropriate number of significant figures.

Magnesium nitrate decomposes on heating to form magnesium oxide, nitrogen dioxide, and oxygen as shown in the following equation.

$2Mg(NO_3)_2(s) \rightarrow 2MgO(s) + 4NO_2(g) + O_2(g)$

(a) Thermal decomposition of a sample of magnesium nitrate produced 0.741 g of magnesium oxide.

(i) Calculate the amount, in moles, of MgO in 0.741 g of magnesium oxide.

(2 marks)

(ii) Calculate the total amount, in moles, of gas produced from this sample of magnesium nitrate.

(1 mark)

(b) In another experiment, a different sample of magnesium nitrate decomposed to produce 0.402 mol of gas. Calculate the volume, in dm^3, that this gas would occupy at 333 K and 1.00×10^5 Pa.

(The gas constant $R = 8.31$ J K^{-1} mol^{-1})

(3 marks)

(c) A 0.0152 mol sample of magnesium oxide, produced from the decomposition of magnesium nitrate, was reacted with hydrochloric acid.

$MgO + 2HCl \rightarrow MgCl_2 + H_2O$

This 0.0152 mol sample of magnesium oxide required 32.4 cm^3 of hydrochloric acid for complete reaction. Use this information to calculate the concentration, in mol dm^{-3}, of the hydrochloric acid.

(2 marks)
AQA, 2010

Learning objectives:

→ State how ions form and why they attract each other.

→ State the properties of ionically bonded compounds.

→ Describe the structure of ionically bonded compounds.

Specification reference: 3.1.3

Noble gas compounds

The noble gases do form a few compounds although they are mostly unstable. The first, Xe PtF$_6$, was made in 1961 by Neil Bartlett. Here is how he describes the moment.

> I was not ready to carry [the experiment] out until about 7 pm on that Friday. When I broke the seal between the red PtF$_6$ gas and the colorless xenon gas, there was an immediate interaction, causing an orange-yellow solid to precipitate. At once I tried to find someone with whom to share the exciting finding, but it appeared that everyone had left for dinner!

There are as yet no compounds of helium or neon. Xenon has the largest number of known compounds. In most of them xenon forms a positive ion by losing an electron.

Suggest why it is easier for xenon to form a positive ion than for helium or neon.

Why do chemical bonds form?

The bonds between atoms always involve their outer electrons.

- Noble gases have full outer main levels of electrons (Figure 1) and are very unreactive.
- When atoms bond together they share or transfer electrons to achieve a more stable electron arrangement, often a full outer main level of electrons, like the noble gases.
- There are three types of strong chemical bonds – **ionic**, **covalent**, and **metallic**.

Ionic bonding

Metals have one, two, or three electrons in their outer main levels, so the easiest way for them to attain the electron structure of a noble gas is to lose their outer electrons. Non-metals have spaces in their outer main levels, so that the easiest way for them to attain the electron structure of a noble gas is to gain electrons.

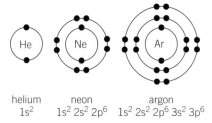

helium
$1s^2$

neon
$1s^2\ 2s^2\ 2p^6$

argon
$1s^2\ 2s^2\ 2p^6\ 3s^2\ 3p^6$

▲ **Figure 1** *Noble gases*

- Ionic bonding occurs between metals and non-metals.
- Electrons are transferred from metal atoms to non-metal atoms.
- Positive and negative ions are formed.

Sodium chloride (Figure 2) has ionic bonding.

- Sodium, Na, has 11 electrons (and 11 protons). The electron arrangement is $1s^2\ 2s^2\ 2p^6\ 3s^1$.
- Chlorine, Cl, has 17 electrons (and 17 protons). The electron arrangement is $1s^2\ 2s^2\ 2p^6\ 3s^2\ 3p^5$.
- An electron is transferred. The single outer electron of the sodium atom moves into the outer main level of the chlorine atom.
- Each outer main level is now full.

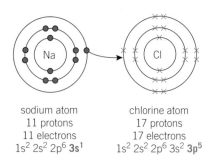

sodium atom
11 protons
11 electrons
$1s^2\ 2s^2\ 2p^6\ \mathbf{3s^1}$

chlorine atom
17 protons
17 electrons
$1s^2\ 2s^2\ 2p^6\ 3s^2\ \mathbf{3p^5}$

▲ **Figure 2** *A dot-and-cross diagram to show the transfer of the 3s^1 electron from the sodium atom to the 3p orbital on a chlorine atom. Remember that electrons are all identical whether shown by a dot or a cross*

• Both sodium and chlorine now have a noble gas electron arrangement. Sodium has the neon noble gas arrangement whereas chlorine has the argon noble gas arrangement (compare the ions in Figure 3 with the noble gas atoms in Figure 1).

The two charged particles that result from the transfer of an electron are called ions.

• The sodium ion is positively charged because it has *lost* a negative electron.

• The chloride ion is negatively charged because it has *gained* a negative electron.

• The two ions are attracted to each other and to other oppositely charged ions in the sodium chloride compound by **electrostatic forces**.

Therefore ionic bonding is the result of electrostatic attraction between oppositely charged ions. The attraction extends throughout the compound. Every positive ion attracts every negative ion and vice versa. Ionic compounds always exist in a structure called a **lattice**. Figure 4 shows the three-dimensional lattice for sodium chloride with its singly charged ions.

The formula of sodium chloride is NaCl because for every one sodium ion there is one chloride ion.

Example: magnesium oxide

Magnesium, Mg, has 12 electrons. The electron arrangement is $1s^2\ 2s^2\ 2p^6\ 3s^2$.

Oxygen, O, has eight electrons. The electron arrangement is $1s^2\ 2s^2\ 2p^4$.

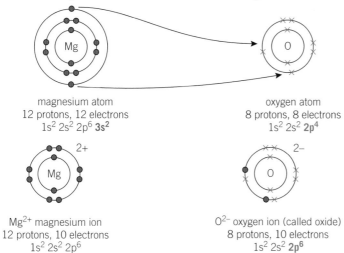

magnesium atom
12 protons, 12 electrons
$1s^2\ 2s^2\ 2p^6\ \mathbf{3s^2}$

oxygen atom
8 protons, 8 electrons
$1s^2\ 2s^2\ \mathbf{2p^4}$

Mg^{2+} magnesium ion
12 protons, 10 electrons
$1s^2\ 2s^2\ 2p^6$

O^{2-} oxygen ion (called oxide)
8 protons, 10 electrons
$1s^2\ 2s^2\ \mathbf{2p^6}$

▲ **Figure 5** *Ionic bonding in magnesium oxide, MgO*

This time, two electrons are transferred from the 3s orbitals on each magnesium atom. Each oxygen atom receives two electrons into its 2p orbital.

• The magnesium ion, Mg^{2+}, is positively charged because it has lost two negative electrons.

• The oxide ion, O^{2-}, is negatively charged because it has gained two negative electrons.

• The formula of magnesium oxide is MgO.

Na^+ sodium ion
11 protons, 10 electrons
$1s^2\ 2s^2\ 2p^6$

Cl^- chlorine ion (called chloride)
17 protons, 18 electrons
$1s^2\ 2s^2\ 2p^6\ 3s^2\ \mathbf{3p^6}$

▲ **Figure 3** *The ions that result from electron transfer*

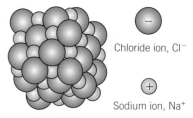

Chloride ion, Cl^-

Sodium ion, Na^+

▲ **Figure 4** *The sodium chloride structure. This is an example of a giant ionic structure. The strong bonding extends throughout the compound and because of this it will be difficult to melt.*

Study tip

Dot-and-cross diagrams can help you to understand the principles of bonding and to predict the shapes of molecules.

Hint

A current of electricity is a flow of charge. In metals, negative electrons move. In ionic compounds, charged ions move.

Properties of ionically bonded compounds

Ionic compounds are always solids at room temperature. They have giant structures and therefore high melting temperatures. This is because in order to melt an ionic compound, energy must be supplied to break up the lattice of ions.

Ionic compounds conduct electricity when molten or dissolved in water (aqueous) but not when solid. This is because the ions that carry the current are free to move in the liquid state but are not free in the solid state (Figure 6).

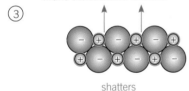

① a small displacement causes contact between ions with the same charge...

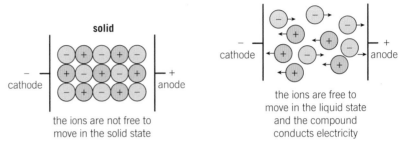

▲ **Figure 6** *Ionic liquids conduct electricity, ionic solids do not*

Ionic compounds are *brittle* and shatter easily when given a sharp blow. This is because they form a lattice of alternating positive and negative ions, see Figure 7. A blow in the direction shown may move the ions and produce contact between ions with like charges.

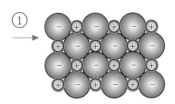

...and the structure shatters

shatters

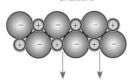

▲ **Figure 7** *The brittleness of ionic compounds*

Summary questions

1   Identify which of the following are ionic compounds and explain why.

    a CO   **b** KF   **c** CaO   **d** HF

2   Explain why ionic compounds have high melting temperatures.

3   Describe the conditions where ionic compounds conduct electricity.

4   Draw dot-and-cross diagrams to show the following:

    a the ions being formed when magnesium and fluorine react

    b the ions being formed when sodium and oxygen react.

5   Give the formulae of the compounds formed in question **4**.

6   Look at the electron arrangements of the Mg^{2+} and O^{2-} ions. State the noble gas they correspond to.

Non-metal atoms need to *receive* electrons to fill the spaces in their outer shells.

- A covalent bond forms between a pair of non-metal atoms.
- The atoms *share* some of their outer electrons so that each atom has a stable noble gas arrangement.
- A covalent bond is a shared pair of electrons.

Forming molecules by covalent bonding

A small group of covalently bonded atoms is called a molecule. For example, chlorine exists as a gas that is made of molecules, Cl_2, see Figure 1.

Chlorine has 17 electrons and an electron arrangement $1s^2\ 2s^2\ 2p^6\ 3s^2\ 3p^5$. Two chlorine atoms make a chlorine molecule:

- The two atoms share one pair of electrons.
- Each atom now has a stable noble gas arrangement.
- The formula is Cl_2.
- Molecules are neutral because no electrons have been transferred from one atom to another.

You can represent one pair of shared electrons in a covalent bond by a line, Cl—Cl.

Example: methane

Methane gas is a covalently bonded compound of carbon and hydrogen. Carbon, C, has six electrons with electron arrangement $1s^2\ 2s^2\ 2p^2$ and hydrogen, H, has just one electron $1s^1$.

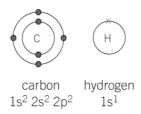

carbon   hydrogen
$1s^2\ 2s^2\ 2p^2$ $1s^1$

In order for carbon to attain a stable noble gas arrangement, there are four hydrogen atoms to every carbon atom.

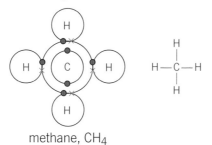

methane, CH_4

The formula of methane is CH_4. The four 2p electrons from carbon and the $1s^1$ electron from the four hydrogen atoms are shared.

Learning objectives:

→ Describe a covalent bond.

→ Describe a co-ordinate bond.

→ Describe the properties of covalently bonded molecules.

Specification reference: 3.1.3

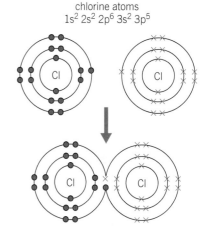

chlorine atoms
$1s^2\ 2s^2\ 2p^6\ 3s^2\ 3p^5$

a chlorine molecule

▲ **Figure 1** *Formation of a chlorine molecule – the two atoms share a 3p electron from each atom*

Hint

Another way of picturing covalent bonds is to think of electron orbitals on each atom merging to form a molecular orbital that holds the shared electrons.

Hint

The hydrogen has a filled outer main level with only two electrons ($1s^2$). It fills the first shell to get the structure of the noble gas helium. The carbon atoms have an electron arrangement $1s^2\ 2s^2\ 2p^6$.

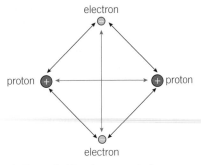

▲ **Figure 2** *The electrostatic forces within a hydrogen molecule*

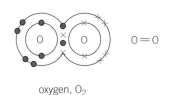

oxygen, O_2

▲ **Figure 3** *An oxygen molecule has a double bond which shares two 2p electrons from each atom*

How does sharing electrons hold atoms together?

Atoms with covalent bonds are held together by the electrostatic attraction between the nuclei and the shared electrons. This takes place within the molecule. The simplest example is hydrogen. The hydrogen molecule consists of two protons held together by a pair of electrons. The electrostatic forces are shown in Figure 2. The attractive forces are in black and the repulsive forces in red. These forces just balance when the nuclei are a particular distance apart.

Double covalent bonds

In a double bond, four electrons are shared. The two atoms in an oxygen molecule share two pairs of electrons so that the oxygen atoms have a double bond between them (Figure 3). You can represent the two pairs of shared electrons in a covalent bond by a double line, O=O.

When you are drawing covalent bonding diagrams you may leave out the inner main levels because the inner shells are not involved at all. Other examples of molecules with covalent bonds are shown in Table 1.

All the examples in Table 1 are neutral molecules. The atoms within the molecules are strongly bonded together with covalent bonds within the molecule. However, the molecules are *not* strongly attracted to each other.

▼ **Table 1** *Examples of covalent molecules. Only the outer shells are shown.*

Formula	Name	Formula	Name
H_2	hydrogen Each hydrogen atom has a full outer main level with just two electrons	NH_3	ammonia
HCl	hydrogen chloride	C_2H_4	ethene There is a carbon–carbon double bond in this molecule
H_2O	water	CO_2	carbon dioxide There are two carbon–oxygen double bonds in this molecule

Properties of substances with molecular structures

Substances composed of molecules are gases, liquids, or solids with low melting temperatures. This is because the strong covalent bonds are only *between the atoms* within the molecules. There is only weak attraction between the molecules so the molecules do not need much energy to move apart from each other.

They are poor conductors of electricity because the molecules are neutral overall. This means that there are no charged particles to carry the current.

If they dissolve in water, and remain as molecules, the solutions do not conduct electricity. Again, this is because there are no charged particles.

Co-ordinate bonding

A single covalent bond consists of a pair of electrons shared between two atoms. In most covalent bonds, each atom provides one of the electrons. But, in some bonds, one atom provides both the electrons. This is called **co-ordinate bonding**. It is also called **dative covalent bonding**.

In a co-ordinate or dative covalent bond:

- the atom that *accepts* the electron pair is an atom that does not have a filled outer main level of electrons – the atom is electron-deficient
- the atom that is *donating* the electrons has a pair of electrons that is not being used in a bond, called a **lone pair**.

Example: the ammonium ion

For example, ammonia, NH_3, has a lone pair of electrons. In the ammonium ion, NH_4^+, the nitrogen uses its lone pair of electrons to form a coordinate bond with an H^+ ion (a bare proton with no electrons at all and therefore electron-deficient).

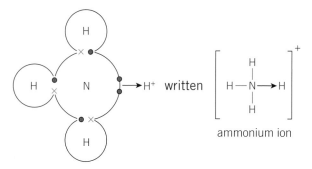

ammonium ion

Coordinate covalent bonds are represented by an arrow. The arrow points towards the atom that is accepting the electron pair. However, this is only to show how the bond was made. The ammonium ion is completely symmetrical and all the bonds have exactly the same strength and length.

- Coordinate bonds have exactly the same strength and length as ordinary covalent bonds between the same pair of atoms.

The ammonium ion has *covalently* bonded atoms but is a charged particle.

> **Hint**
>
> Some covalent compounds react with water to form ions. In such cases the resulting solution *will* conduct electricity. Hydrogen chloride is an example of this; $HCl(g) + aq \rightarrow H^+(aq) + Cl^-(aq)$

Summary questions

1 State what a covalent bond is.

2 Identify which of the following have covalent bonding and explain your answer.

   a Na_2O

   b CF_4

   c $MgCl_2$

   d C_2H_4

3 Draw a dot-and-cross diagram for hydrogen sulfide, a compound of hydrogen and sulfur.

4 Draw a dot-and-cross diagram to show a water molecule forming a coordinate bond with an H^+ ion.

3.3 Metallic bonding

Metals are shiny elements made up of atoms that can easily lose up to three outer electrons, leaving positive metal ions. For example, sodium, Na, 2,8,1 ($1s^2\ 2s^2\ 2p^6\ 3s^1$) loses its one outer electron, aluminium, Al, 2,8,3 ($1s^2\ 2s^2\ 2p^6\ 3s^2\ 3p^1$) loses its three outer electrons.

Metallic bonding

The atoms in a metal element cannot transfer electrons (as happens in ionic bonding) unless there is a non-metal atom present to receive them. In a metal element, the outer main levels of the atoms merge. The outer electrons are no longer associated with any one particular atom. A simple picture of **metallic bonding** is that metals consist of a lattice of positive ions existing in a 'sea' of outer electrons. These electrons are **delocalised**. This means that they are not tied to a particular atom. Magnesium metal is shown in Figure 1. The positive ions tend to repel one another and this is balanced by the electrostatic attraction of these positive ions for the negatively charged 'sea' of delocalised electrons.

> **Hint**
>
> In Figure 1 the metal ions are shown spaced apart for clarity. In fact metal atoms are more closely packed, and so metals tend to have high densities.

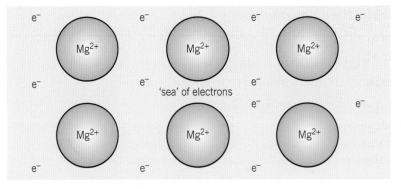

▲ **Figure 1** *The delocalised 'sea' of electrons in magnesium*

- The number of delocalised electrons depends on how many electrons have been lost by each metal atom.
- The metallic bonding spreads throughout so metals have giant structures.

Properties of metals

Metals are good conductors of electricity and heat
The delocalised electrons that can move throughout the structure explain why metals are such good conductors of electricity. An electron from the negative terminal of the supply joins the electron sea at one end of a metal wire while *at the same time* a different electron leaves the wire at the positive terminal, as shown in Figure 2.

Metals are also good conductors of heat – they have high thermal conductivities. The sea of electrons is partly responsible for this property, with energy also spread by increasingly vigorous vibrations of the closely packed ions.

> **Hint**
>
> The word delocalised is often used to describe electron clouds that are spread over more than two atoms.

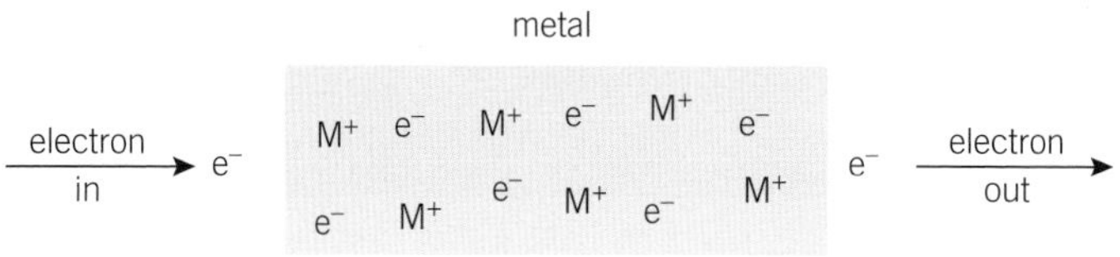

metal

$$\xrightarrow[\text{in}]{\text{electron}} e^- \quad \boxed{\begin{array}{cccccc} M^+ & e^- & M^+ & e^- & M^+ & e^- \\ e^- & M^+ & e^- & M^+ & e^- & M^+ \end{array}} \quad e^- \xrightarrow[\text{out}]{\text{electron}}$$

▲ **Figure 2** *The conduction of electricity by a metal*

The strength of metals

In general, the strength of any metallic bond depends on the following:

- the charge on the ion – the greater the charge on the ion, the greater the number of delocalised electrons and the stronger the electrostatic attraction between the positive ions and the electrons.

- the size of ion – the smaller the ion, the closer the electrons are to the positive nucleus and the stronger the bond.

Metals tend to be strong. The delocalised electrons also explain this. These extend throughout the solid so there are no individual bonds to break.

Metals are malleable and ductile

Metals are malleable (they can be beaten into shape) and ductile (they can be pulled into thin wires). After a small distortion, each metal ion is still in exactly the same environment as before so the new shape is retained, see Figure 3.

Contrast this with the brittleness of ionic compounds in Topic 3.1.

Metals have high melting points

Metals generally have high melting and boiling points because they have giant structures. There is strong attraction between metal ions and the delocalised sea of electrons. This makes the atoms difficult to separate.

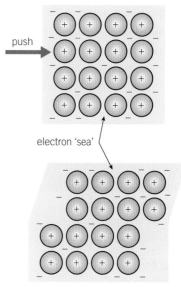

▲ **Figure 3** *The malleability and ductility of metals*

Summary questions

1. Give three differences in physical properties between metals and non-metals.

2. Write the electron arrangement of a calcium atom, Ca.

3. Which electrons will a calcium atom lose to gain a stable noble gas configuration.

4. State how many electrons each calcium atom will contribute to the delocalised sea of electrons that holds the metal atoms together.

5. Sodium forms +1 ions with a metallic radius of 0.191 nm. Magnesium forms +2 ions with a metallic radius of 0.160 nm. How would you expect the following properties of the two metals to compare? Explain your answers.

   a. The melting point

   b. The strength of the metals

Learning objectives:

→ State what is meant by the term electronegativity.

→ State what makes one atom more electronegative than another.

→ State what the symbols $\delta+$ and $\delta-$ mean when placed above atoms in a covalent bond.

Specification reference: 3.1.3

▼ **Table 1** *Some values for Pauling electronegativity*

H 2.1							He
Li 1.0	Be 1.5	B 2.0	C 2.5	N 3.0	O 3.5	F 4.0	Ne
Na 0.9	Mg 1.2	Al 1.5	Si 1.8	P 2.1	S 2.5	Cl 3.0	Ar
						Br 2.8	Kr

The forces that hold atoms together are all about the attraction of positive charges to negative charges. In ionic bonding there is complete transfer of electrons from one atom to another. But, even in covalent bonds, the electrons shared by the atoms will not be evenly spread if one of the atoms is better at attracting electrons than the other. This atom is more **electronegative** than the other.

Electronegativity

Flourine is better at attracting electrons than hydrogen – fluorine is said to be more electronegative than hydrogen.

Electronegativity is the power of an atom to attract the electron density in a covalent bond towards itself.

When chemists consider the electrons as charge clouds, the term **electron density** is often used to describe the way the negative charge is distributed in a molecule.

The Pauling scale is used as a measure of electronegativity. It runs from 0 to 4. The greater the number, the more electronegative the atom, see Table 1. The noble gases have no number because they do not, in general, form covalent bonds.

Electronegativity depends on:

1 the nuclear charge

2 the distance between the nucleus and the outer shell electrons

3 the shielding of the nuclear charge by electrons in inner shells.

Note the following:

• The smaller the atom, the closer the nucleus is to the shared outer main level electrons and the greater its electronegativity.

• The larger the nuclear charge (for a given shielding effect), the greater the electronegativity.

Trends in electronegativity

Going up a group in the Periodic Table, electronegativity increases (the atoms get smaller) and there is less shielding by electrons in inner shells.

Going across a period in the Periodic Table, the electronegativity increases. The nuclear charge increases, the number of inner main levels remain the same and the atoms become smaller.

So, the most electronegative atoms are found at the top right-hand corner of the Periodic Table (ignoring the noble gases which form few compounds). The most electronegative atoms are fluorine, oxygen, and nitrogen followed by chlorine.

Hint

Think of electronegative atoms as having more 'electron-pulling power'.

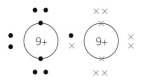
▲ **Figure 1** *Electron diagram of fluorine molecule*

▲ **Figure 2** *Electron cloud around fluorine molecule*

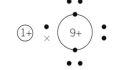
▲ **Figure 3** *Electron diagram of hydrogen fluoride molecule*

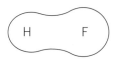
▲ **Figure 4** *Electron cloud around hydrogen fluoride molecule*

▼ **Table 2** *Trends in electronegativity*

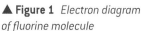

		Increasing electronegativity →					
Li	Be	B	C	N	O	F	
1.0	1.5	2.0	2.5	3.0	3.5	4.0	
						Cl	Increasing electronegativity ↑
						3.0	
						Br	
						2.8	

Polarity of covalent bonds

Polarity is about the unequal sharing of the electrons between atoms that are bonded together covalently. It is a property of the *bond*.

Covalent bonds between two atoms that are the same

When both atoms are the same, for example, in fluorine, F_2, the electrons in the bond *must* be shared equally between the atoms (Figure 1) – both atoms have exactly the same electronegativity and the bond is completely non-polar.

If you think of the electrons as being in a cloud of charge, then the cloud is uniformly spread between the two atoms, as shown in Figure 2.

Covalent bonds between two atoms that are different

In a covalent bond between two atoms of *different* electronegativity, the electrons in the bond will not be shared equally between the atoms. For example, the molecule hydrogen fluoride, HF, shown in Figure 3.

Hydrogen has an electronegativity of 2.1 and fluorine of 4.0. This means that the electrons in the covalent bond will be attracted more by the fluorine than the hydrogen. The electron cloud is distorted towards the fluorine, as shown in Figure 4.

The fluorine end of the molecule is therefore relatively negative and the hydrogen end relatively positive, that is, electron deficient. You show this by adding partial charges to the formula:

$$^{\delta+}H—F^{\delta-}$$

Covalent bonds like this are said to be **polar**. The greater the difference in electronegativity, the more polar is the covalent bond.

You could say that although the H—F bond is covalent, it has some ionic character. It is going some way towards the separation of the atoms into charged ions. It is also possible to have ionic bonds with some covalent character.

Hint

$\delta+$ and $\delta-$ are pronounced 'delta plus' and 'delta minus'.

The + and – signs represent one 'electron's worth' of charge.

$\delta+$ and $\delta-$ represent a small charge of less than one 'electron's worth'.

Summary questions

1 Explain why fluorine is more electronegative than chlorine.

2 Write $\delta+$ and $\delta-$ signs to show the polarity of the bonds in a hydrogen chloride molecule.

3 Identify of these covalent bonds is/are non-polar, and explain your answer.

   a H—H

   b F —F

   c H—F

4 a Arrange the following covalent bonds in order of increasing polarity:
   H—O,   H—F,   H—N

   b Explain your answer.

Specification reference: 3.1.3

Learning objectives:

→ State the three types of intermolecular force.

→ Describe how dipole–dipole and van der Waals forces arise.

→ Describe how van der Waals forces affect boiling temperatures.

→ State what is needed for hydrogen bonding to occur.

→ Explain why NH_3, H_2O, and HF have higher boiling temperatures than might be expected.

Hint

van der Waals is spelt with a small v, even at the beginning of a sentence.

Atoms in molecules and in giant structures are held together by strong covalent, ionic, or metallic bonds. Molecules and separate atoms are attracted to one another by other, weaker forces called intermolecular forces. Inter means between. If the intermolecular forces are strong enough, then molecules are held closely enough together to be liquids or even solids.

Intermolecular forces

There are three types of intermolecular forces:

*   **van der Waals forces**
    act between *all* atoms and molecules.   weakest

*   **Dipole–dipole forces**
    act only between certain types of molecules.

*   **Hydrogen bonding**
    acts only between certain types of molecules.   strongest

Dipole–dipole forces

Dipole moments

Polarity is the property of a particular bond, see Topic 3.4, but molecules with polar bonds may have a dipole moment. This sums up the effect of the polarity of *all* the bonds in the molecule.

In molecules with more than one polar bond, the effects of each bond may cancel, leaving a molecule with no dipole moment. The effects may also add up and so reinforce each other. It depends on the shape of the molecule.

For example, carbon dioxide is a linear molecule and the dipoles cancel.

$$\delta^- O = C^{\delta+} = O^{\delta-}$$

Tetrachloromethane is tetrahedral and here too the dipoles cancel.

But in dichloromethane the dipoles do not cancel because of the shape of the molecule.

Dipole–dipole forces act between molecules that have permanent dipoles. For example, in the hydrogen chloride molecule, chlorine is more electronegative than hydrogen. So the electrons are pulled towards the chlorine atom rather than the hydrogen atom. The molecule therefore has a dipole and is written $H^{\delta+}–Cl^{\delta-}$.

Study tip

Do not confuse intermolecular forces with covalent bonds, which are at least 10 times stronger.

Two molecules which both have dipoles will attract one another, see Figure 1.

Whatever their starting positions, the molecules with dipoles will 'flip' to give an arrangement where the two molecules attract.

van der Waals forces

All atoms and molecules are made up of positive and negative charges even though they are neutral overall. These charges produce very weak electrostatic attractions between all atoms and molecules. These are called van der Waals forces.

How do van der Waals forces work?

Imagine a helium atom. It has two positive charges on its nucleus and two negatively charged electrons. The atom as a whole is neutral but at any moment in time the electrons could be anywhere, see Figure 2. This means the distribution of charge is changing *at every instant*.

Any of the arrangements in Figure 2 mean the atom has a dipole at that moment. An instant later, the dipole may be in a different direction. But, almost certainly the atom *will* have a dipole at any point in time, even though any particular dipole will be just for an instant – a temporary dipole. This dipole then affects the electron distribution in nearby atoms, so that they are attracted to the original helium atom for that instant. The original atom has induced dipoles in the nearby atoms, as shown Figure 3 in which the electron distribution is shown as a cloud.

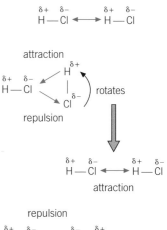

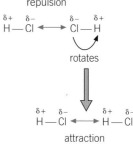

▲ **Figure 1** *Two polar molecules, such as hydrogen chloride, will always attract one another*

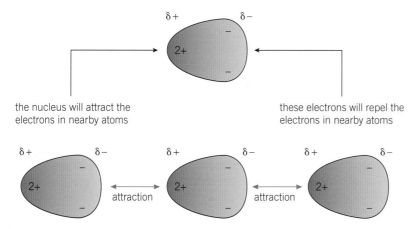

the nucleus will attract the electrons in nearby atoms

these electrons will repel the electrons in nearby atoms

▲ **Figure 3** *Instantaneous dipoles induce dipoles in nearby atoms*

▲ **Figure 2** *These are just a few of the possible arrangements of the two electrons in helium. Remember, electrons are never in a fixed position*

As the electron distribution of the original atom changes, it will induce new dipoles in the atoms around it, which will be attracted to the original one. These forces are sometimes called instantaneous dipole–induced dipole forces, but this is rather a mouthful. The more usual name is van der Waals forces after the Dutch scientist, Johannes van der Waals.

- van der Waals forces act between *all* atoms or molecules at all times.
- They are in addition to any other intermolecular forces.
- The dipole is caused by the changing position of the electron cloud, so the more electrons there are, the larger the instantaneous dipole will be.

Therefore the size of the van der Waals forces increases with the number of electrons present. This means that atoms or molecules with large atomic or molecular masses produce stronger van der Waals forces than atoms or molecules with small atomic or molecular masses.

This explains why:

- the boiling points of the noble gases increase as the atomic numbers of the noble gases increase
- the boiling points of hydrocarbons increase with increased chain length.

Hydrogen bonding

Hydrogen bonding is a special type of intermolecular force with some characteristics of dipole–dipole attraction and some of a covalent bond. It consists of a hydrogen atom 'sandwiched' between two very electronegative atoms. There are conditions that have to be present for hydrogen bonding to occur. You need a very electronegative atom with a lone pair of electrons covalently bonded to a hydrogen atom. Water molecules fulfil these conditions. Oxygen is much more electronegative than hydrogen so water is polar, see Figure 4.

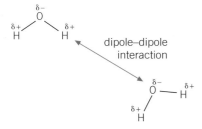

▲ **Figure 4** *Dipole attraction between water molecules*

You would expect to find weak dipole–dipole attractions (as shown between hydrogen chloride in Figure 1) but in this case the intermolecular bonding is much stronger for two reasons:

1   The oxygen atoms in water have lone pairs of electrons.

2   In water the hydrogen atoms are highly electron deficient. This is because the oxygen is very electronegative and attracts the shared electrons in the bond towards it. The hydrogen atoms in water are positively charged and very small. These exposed protons have a very strong electric field because of their small size.

The lone pair of electrons on the oxygen atom of another water molecule is strongly attracted to the electron deficient hydrogen atom.

This strong intermolecular force is called a hydrogen bond. Hydrogen bonds are considerably stronger than dipole–dipole attractions, though much weaker than a covalent bond. They are usually represented by dashes – – –, as in Figure 5.

▲ **Figure 5** *Hydrogen bond between water molecules*

When do hydrogen bonds form?

Water is not the only example of hydrogen bonding. In order to form a hydrogen bond there must be the following:

- a hydrogen atom that is bonded to a very electronegative atom. This will produce a strong partial positive charge on the hydrogen atom.
- a very electronegative atom with a lone pair of electrons. These will be attracted to the partially charged hydrogen atom in another molecule and form the bond.

The only atoms that are electronegative enough to form hydrogen bonds are oxygen, O, nitrogen, N, and fluorine, F. For example, ammonia molecules, NH_3, form hydrogen bonds with water molecules, see Figure 6.

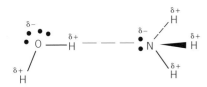

▲ **Figure 6** *Hydrogen bond between a water molecule and an ammonia molecule*

The nitrogen–hydrogen–oxygen system is linear. This is because the pair of electrons in the O—H covalent bond repels those in the hydrogen bond between nitrogen and hydrogen. This linearity is always the case with hydrogen bonds.

The boiling points of the hydrides

The effect of hydrogen bonding between molecules can be seen if you look at the boiling points of hydrides of elements of Group 4, 5, 6, and 7 plotted against the period number, see Figure 7.

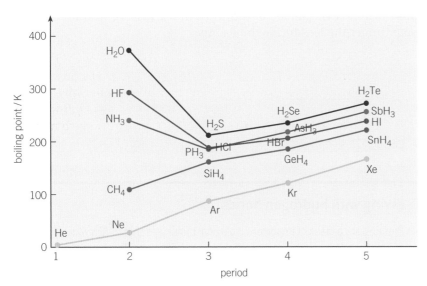

▲ **Figure 7** *Boiling points of the hydrides of Group 4, 5, 6, and 7 elements with the noble gases for comparison*

The noble gases show a gradual increase in boiling point because the only forces acting between the atoms are van der Waals forces and these increase with the number of electrons present.

The boiling points of water, H_2O, hydrogen fluoride, HF, and ammonia, NH_3, are all higher than those of the hydrides of the other elements in their group, whereas you would expect them to be lower if only van der Waals forces were operating. This is because hydrogen bonding is present between the molecules in each of these compounds and these stronger intermolecular forces of attraction make the molecules more difficult to separate. Oxygen, nitrogen, and fluorine are the three elements that are electronegative enough to make hydrogen bonding possible.

The importance of hydrogen bonding

Although hydrogen bonds are only about 10% of the strength of covalent bonds, their effect can be significant – especially when there are a lot of them. The very fact that they are weaker than covalent bonds, and can break or make under conditions where covalent bonds are unaffected, is very significant.

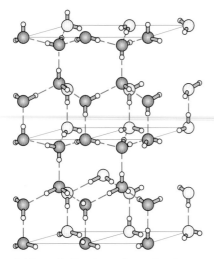

▲ **Figure 8** *The three-dimensional network of covalent bonds (grey) and hydrogen bonds (red) in ice. The blue lines are only construction lines.*

The structure and density of ice

In water in its liquid state, the hydrogen bonds break and reform easily as the molecules are moving about. When water freezes, the water molecules are no longer free to move about and the hydrogen bonds hold the molecules in fixed positions. The resulting three-dimensional structure, shown in Figure 8, resembles the structure of diamond, see Topic 3.7.

In order to fit into this structure, the molecules are slightly less closely packed than in liquid water. This means that ice is less dense than water and forms on top of ponds rather than at the bottom. This insulates the ponds and enables fish to survive through the winter. This must have helped life to continue, in the relative warmth of the water under the ice, during the Ice Ages.

Living with hydrogen bonds

Proteins are a class of important biological molecules that fulfil a wide variety of functions in living things, including enzyme catalysts. The exact shape of a protein molecule is vital to its function. Proteins are long chain molecules with lots of $C{=}O$ and N—H groups which can form hydrogen bonds. These hydrogen bonds hold the protein chains into fixed shapes. One common shape is the protein chain that forms a spiral (helix), as shown here.

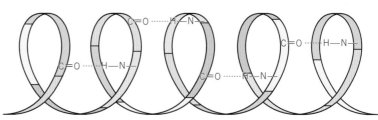

Another example is the beta-pleated sheet. Here protein chains line up side by side, held in position by hydrogen bonds to form a two-dimensional sheet. The protein that forms silk has this structure.

The relative weakness of hydrogen bonds means that the shapes of proteins can easily be altered. Heating proteins much above body temperature starts to break hydrogen bonds and causes the protein to lose its shape and thus its function. This is why enzymes lose their effect as catalysts when heated – the protein is denatured. You can see this when frying an egg. The clear liquid protein albumen is transformed into an opaque white solid.

Ironing

When you iron clothes, the iron provides heat to break hydrogen bonds in the crumpled material and pressure to force the molecules into new positions so that the material is flat. When you remove the iron, the hydrogen bonds reform and hold the molecules in these new positions, keeping the fabric flat.

DNA

Another vital biological molecule is DNA (deoxyribonucleic acid) (Figure 10). It is the molecule that stores and copies genetic information that makes offspring resemble their parents. This molecule exists as a double-stranded helix. The two strands of the spiral are held together by hydrogen bonds. When cells divide or replicate, the hydrogen bonds break (but the covalently bonded main chains stay unchanged). The two separate helixes then act as templates for a new helix to form on each, so you end up with a copy of the original helix.

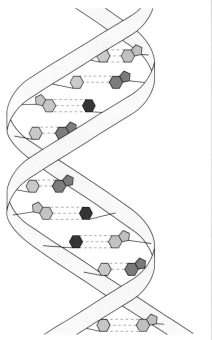

▲ **Figure 10** *The DNA double helix is held together by hydrogen bonds*

Summary questions

1 Place the following elements in order of the strength of the van der Waals forces between the atoms (weakest first): Ar, He, Kr, Ne. Explain your answer.

2 Identify which one of the following molecules *cannot* have dipole–dipole forces acting between them—H_2O, HCl, H_2

3 Explain why hexane is a liquid at room temperature whereas butane is a gas.

4 Explain why covalent molecules are gases, liquids, or solids with low-melting temperature.

5 Draw two hydrogen bromide molecules to show how they would be attracted together by dipole–dipole forces.

6 Identify in which of the following does hydrogen bonding *not* occur between molecules: H_2O, NH_3, HBr, HF

7 Explain why hydrogen bonds do not form between:

   a methane molecules, CH_4

   b tetrachloromethane molecules, CCl_4.

8 Draw a dot-and-cross diagram for a molecule of water.

   a State how many lone pairs it has.

   b State how many hydrogen atoms it has.

   c Explain why water molecules form on average two hydrogen bonds per molecule, whereas the ammonia molecule, NH_3, forms only one.

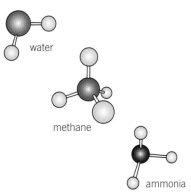

▲ **Figure 1** *The shapes of water, methane, and ammonia molecules*

> **Hint**
>
> It is acceptable to draw electron diagrams that show electrons in the outer shells only.

> **Hint**
>
> Notice that in neither $BeCl_2$ nor BF_3 does the central atom have a full outer main electron level.

Molecules are three-dimensional and they come in many different shapes (Figure 1).

Electron pair repulsion theory

You have seen that electrons in molecules exist in pairs in volumes of space called orbitals. You can predict the shape of a simple covalent molecule, for example, one consisting of a central atom surrounded by a number of other atoms, by using the ideas that:

- each pair of electrons around an atom will repel all other electron pairs
- the pairs of electrons will therefore take up positions as far apart as possible to minimise repulsion.

This is called the **electron pair repulsion theory**.

Electron pairs may be a shared pair or a lone pair.

The shape of a simple molecule depends on the number of pairs of electrons that surround the central atom. To work out the shape of any molecule you first need to draw a dot-and-cross diagram to find the number of pairs of electrons.

Two pairs of electrons

If there are two pairs of electrons around the atom, the molecule will be *linear*. The furthest away from each other the two pairs can get is *180°* apart. Beryllium chloride, which is a covalently bonded molecule in the gas phase, despite being a metal–non-metal compound, is an example of this.

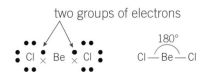

two groups of electrons

Three pairs of electrons

If there are three pairs of electrons around the central atom, they will be *120°* apart. The molecule is planar and is called *trigonal planar*. Boron trifluoride is an example of this.

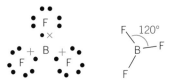

Four pairs of electrons

If there are four pairs of electrons, they are furthest apart when they are arranged so that they point to the four corners of a *tetrahedron*. This shape, with one atom positioned at the centre, is called *tetrahedral*, see Figure 2.

Methane, CH_4, is an example. The carbon atom is situated at the centre of the tetrahedron with the hydrogen atoms at the vertices. The angles here are *109.5°*. This is a three-dimensional, not planar, arrangement so the sum of the angles can be more than 360°.

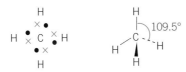

The ammonium ion is also tetrahedral. It has four groups of electrons surrounding the nitrogen atom. The fact that the ion has an overall charge does not affect the shape.

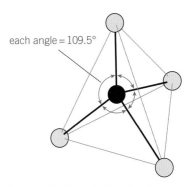

▲ **Figure 2** *A tetrahedron has four points and four faces*

Five pairs of electrons

If there are five pairs of electrons, the shape usually adopted is that of a *trigonal bipyramid*. Phosphorus pentachloride, PCl_5, is an example.

Six pairs of electrons

If there are six pairs of electrons, the shape adopted is *octahedral*, with bond angles of *90°*. The sulfur hexafluoride, SF_6, molecule is an example of this.

Molecules with lone pairs of electrons

Some molecules have unshared (lone) pairs of electrons. These are electrons that are not part of a covalent bond. The lone pairs affect the shape of the molecule. Always watch out for the lone pairs in your dot-and-cross diagram because otherwise you might overlook their effect. Ammonia and water are good examples of molecules where lone pairs affect the shape.

Ammonia, NH₃

Ammonia has four pairs of electrons and one of the groups is a lone pair.

With its four pairs of electrons around the nitrogen atom, the ammonia molecule has a shape based on a tetrahedron. However, there are only three 'arms' so the shape is that of a *triangular pyramid*.

Another way of looking at this is that the *electron pairs* form a tetrahedron but the bonds form a triangular pyramid. (There is an atom at each vertex but, unlike the tetrahedral arrangement, no atom in the centre.)

Bonding pair–lone pair repulsion

The angles of a regular tetrahedron, see Figure 2, are all 109.5° but lone pairs affect these angles. In ammonia, for example, the *bonding* pairs of electrons are attracted towards the nitrogen nucleus and also the hydrogen nucleus. However, the *lone* pair is attracted only by the nitrogen nucleus and is therefore pulled closer to it than the shared pairs. So repulsion between a lone pair of electrons and a bonding pair of electrons is greater than that between two bonding pairs. This effect squeezes the hydrogen atoms together, reducing all the H–N–H angles. The approximate rule of thumb is 2° per lone pair, so the bond angles in ammonia are approximately 107°:

Water, H₂O

Look at the dot-and-cross diagram for water.

There are four pairs of electrons around the oxygen atom so, as with ammonia, the shape is based on a tetrahedron. However, two of the 'arms' of the tetrahedron are lone pairs that are not part of a bond. This results in a *V-shaped* or angular molecule. As in ammonia the electron pairs form a tetrahedron but the bonds form a V-shape. With two lone pairs, the H–O–H angle is reduced to 104.5°.

Chlorine tetrafluoride ion, ClF_4^-

The dot-and-cross diagram for this ion is as shown:

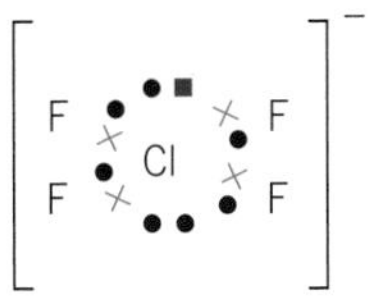

There are four bonding pairs of electrons and two lone pairs. One of the lone pairs contains an electron that has been donated to it, so the charge on the ion is negative (−1). This electron is shown as a square in the dot-and-cross diagram. This means that there are six pairs of electrons around the chlorine atom – four bonds and two lone pairs. The shape is therefore based on an octahedron in which two arms are not part of a bond.

As lone pairs repel the most, they adopt a position furthest apart. This leaves a flat square-shaped ion described as *square planar*. The lone pairs are above and below the plane, as shown here.

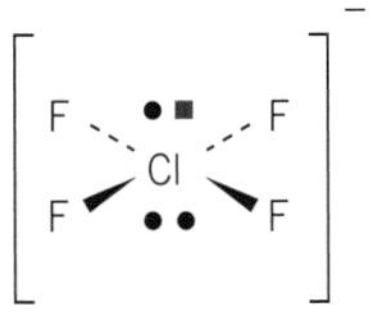

A summary of the repulsion between electron pairs

bonding pair–bonding pair   ↓

lone pair–bonding pair   repulsion increases

lone pair–lone pair   ↓

Summary questions

1   Draw a dot-and-cross diagram for NF_3 and predict its shape.

2   Explain why NF_3 has a different shape from BF_3.

3   Draw a dot-and-cross diagram for the molecule silane, SiH_4, and describe its shape.

4   State the H—Si—H angle in the silane molecule

5   Predict the shape of the H_2S molecule *without* drawing a dot-and-cross diagram.

Specification reference: 3.1.3

Learning objectives:

→ State the energy changes that occur when solids melt and liquids vaporise.

→ Explain the values of enthalpies of melting (fusion) and vaporisation are.

→ Explain the physical properties of ionic solids, metals, macromolecular solids, and molecular solids in terms of their detailed structures and bonding.

→ List the three types of strong bonds.

→ List the three types of intermolecular forces.

→ Describe how melting temperatures and structure are related.

→ Describe how electrical conductivity is related to bonding.

One of the key ideas of science is that matter, which is anything with mass, is made of tiny particles – it is particulate. These particles are in motion, which means they have kinetic energy. To understand the differences between the three states of matter – gas, liquid, and solid – you need to be able to explain the energy changes associated with changes between these physical states.

The three states of matter

Table 1 sets out the simple model used for the three states of matter.

▼ **Table 1** *The three states of matter*

	Solid	Liquid	Gas
arrangement of particles	regular	random	random
evidence	Crystal shapes have straight edges. Solids have definite shapes.	None direct but a liquid changes shape to fill the bottom of its container.	None direct but a gas will fill its container.
spacing	close	close	far apart
evidence	Solids are not easily compressed.	Liquids are not easily compressed.	Gases are easily compressed.
movement	vibrating about a point	rapid 'jostling'	rapid
evidence	Diffusion is very slow. Solids expand on heating.	Diffusion is slow. Liquids evaporate.	Diffusion is rapid. Gases exert pressure
models			

Energy changes on heating

Heating a solid

When you first heat a solid and supply energy to the particles, it makes them vibrate more about a fixed position. This slightly increases the average distance between the particles and so the solid expands.

Turning a solid to liquid (melting – also called fusion)

In order to turn a solid – with its ordered, closely packed, vibrating particles – into a liquid – where the particles are moving randomly but still closely together – you have to supply more energy. This energy is needed to weaken the forces that act between the particles, holding them together in the solid state. The energy needed is called the latent heat of melting, or more correctly the **enthalpy change of melting**. While a solid is melting, the temperature does not change because the heat energy provided is absorbed as the forces between particles are weakened.

Enthalpy is the heat energy change measured under constant pressure whilst *temperature* depends on the average kinetic energy of the particles and is therefore related to their speed – the greater the energy, the faster they go.

Heating a liquid
When you heat a liquid, you supply energy to the particles which makes them move more quickly – they have more kinetic energy. On average, the particles move a little further apart so liquids also expand on heating.

Turning a liquid to gas (boiling – also called vaporisation)
In order to turn a liquid into a gas, you need to supply enough energy to break all the intermolecular forces between the particles. A gas consists of particles that are far apart and moving independently. The energy needed is called the latent heat of vaporisation or more correctly the **enthalpy change of vaporisation**. As with melting, there is no temperature change during the process of boiling.

Heating a gas
As you heat a gas, the particles gain kinetic energy and move faster. They get much further apart and so gases expand a great deal on heating.

Crystals

Crystals are solids. The particles have a regular arrangement and are held together by forces of attraction. These could be strong bonds – covalent, ionic, or metallic – or weaker intermolecular forces – van der Waals, dipole–dipole, or hydrogen bonds. The strength of the forces of attraction between the particles in the crystal affects the physical properties of the crystals. For example, the stronger the force, the higher the melting temperature and the greater the enthalpy of fusion (the more difficult they are to melt). There are four basic crystal types – ionic, metallic, molecular, and macromolecular.

Ionic crystals
Ionic compounds have strong electrostatic attractions between oppositely charged ions. Sodium chloride, NaCl, is a typical ionic crystal, see Topic 3.1. Ionic compounds have high melting points. This is a result of the strong electrostatic attractions which extend throughout the structure. These require a lot of energy to break in order for the ions to move apart from each other. For example, the melting point of sodium chloride is 801 °C (1074 K).

Metallic crystals
Metals exist as a lattice of positive ions embedded in a delocalised sea of electrons, see Topic 3.3. Again the attraction of positive to negative extends throughout the crystal. The high melting temperature is a result of these strong metallic bonds.

Molecular crystals
Molecular crystals consist of molecules held in a regular array by intermolecular forces. Covalent bonds *within* the molecules hold the atoms together but they do not act *between* the molecules.

> **Hint**
> The enthalpy change of melting is sometimes called the enthalpy change of fusion.

> **Synoptic link**
> You will learn more about enthalpy in Topic 4.2, Enthalpy.

distance between a pair of covalently bonded iodine atoms = 0.267 nm

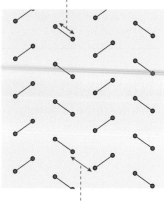

distance between a pair of iodine molecules (held by van der Waals forces) = 0.354 nm

▲ **Figure 1** *The arrangement of an iodine crystal*

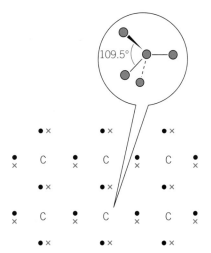

▲ **Figure 2** *A dot-and-cross diagram showing the bonding in diamond*

▲ **Figure 3** *A three-dimensional diagram of diamond*

Intermolecular forces are much weaker than covalent, ionic or metallic bonds, so molecular crystals have low melting temperatures and low enthalpies of melting.

Iodine (Figure 1) is an example of a molecular crystal. A strong covalent bond holds pairs of iodine atoms together to form I_2 molecules. Since iodine molecules have a large number of electrons, the van der Waals forces are strong enough to hold the molecules together as a solid. But van der Waals forces are much weaker than covalent bonds, giving iodine the following properties:

- crystals are soft and break easily
- low melting temperature (114 °C, 387 K) and sublimes readily to form gaseous iodine molecules.
- does not conduct electricity because there are no charged particles to carry charge.

Macromolecular crystals

Covalent compounds are not always made up of small molecules. In some substances the covalent bonds extend throughout the compound and have the typical property of a giant structure held together with strong bonds – a high melting temperature. There are many examples of macromolecular crystals, including diamond and graphite.

Diamond and graphite

Diamond and graphite are both made of the element carbon only. They are called polymorphs or allotropes of carbon. They are very different materials because their atoms are differently bonded and arranged. They are examples of macromolecular structures.

Diamond

Diamond consists of pure carbon with covalent bonding between every carbon atom. The bonds spread throughout the structure, which is why it is a giant structure.

A carbon atom has four electrons in its outer shell. In diamond, each carbon atom forms four single covalent bonds with other carbon atoms, as shown in Figure 2. These four electron pairs repel each other, following the rules of the electron pair repulsion theory. In three dimensions the bonds actually point to the corners of a tetrahedron (with bond angles of 109.5°).

Each carbon atom is in an identical position in the structure, surrounded by four other carbon atoms. Figure 3 shows this three-dimensional arrangement.

The atoms form a giant three-dimensional lattice of strong covalent bonds, which is why diamond has the following properties:

- very hard material (one of the hardest known)
- very high melting temperature, over 3700 K
- does not conduct electricity because there are no free charged particles to carry charge.

Graphite

Graphite also consists of pure carbon but the atoms are bonded and arranged differently from diamond. Graphite has two sorts of bonding – strong covalent and the weaker van der Waals forces.

In graphite, each carbon atom forms three single covalent bonds to other carbon atoms. As predicted by bonding electron pair repulsion theory, these form a flat trigonal arrangement, sometimes called trigonal planar, with a bond angle of 120° (Figure 4). This leaves each carbon atom with a 'spare' electron in a p-orbital that is not part of the three single covalent bonds.

This arrangement produces a two-dimensional layer of linked hexagons of carbon atoms, rather like a chicken-wire fence (Figure 5).

The p-orbitals with the 'spare' electron merge above and below the plane of the carbon atoms in each layer. These electrons can move anywhere within the layer. They are delocalised. This adds to the strength of the bonding and is rather like the **delocalised** sea of electrons in a metal, but in two dimensions only.

These delocalised electrons are what make graphite conduct electricity (very rare for a non-metal). They can travel freely through the material, though graphite will only conduct along the hexagonal planes, not at right angles to them.

There is no covalent bonding *between* the layers of carbon atoms. They are held together by the much weaker van der Waals forces, see Figure 5. This weak intermolecular force of attraction means that the layers can slide across one another making graphite soft and flaky. It is the lead in pencils. The flakiness allows the graphite layers to transfer from the pencil to the paper.

- Graphite is a soft material.
- It has a very high melting temperature and in fact it breaks down before it melts. This is because of the strong network of covalent bonds, which make it a giant structure.
- It conducts electricity along the planes of the hexagons.

Giant footballs
More recently a number of other forms of pure carbon have been discovered. Chemists found the first one whilst they were looking for molecules in outer space. The structures of these new forms of carbon include closed cages of carbon atoms and also tubes called nanotubes. The most famous is buckminsterfullerene, C_{60}, in which atoms are arranged in a football-like shape (Figure 6). Harry Kroto and colleagues received the Nobel Prize for the discovery. Now, scientists are investigating many uses for these new materials.

Bonding – summary
There are three types of *strong* bonding that hold atoms together – ionic, covalent, and metallic. All three involve the outer electrons of the atoms concerned.

- In covalent bonding, the electrons are shared between atoms.
- In ionic bonding, electrons are transferred from metal atoms to non-metal atoms.
- In metallic bonding, electrons are spread between metal atoms to form a lattice of ions held together by delocalised electrons.

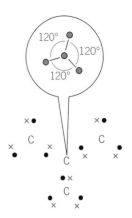

▲ **Figure 4** *A dot-and-cross diagram showing the three covalent bonds in graphite*

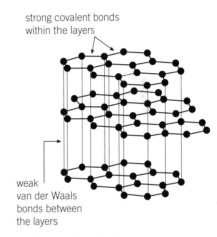

▲ **Figure 5** *Van der Waals forces between the layers of carbon atoms in graphite*

strong covalent bonds within the layers

weak van der Waals bonds between the layers

Hint

It is now believed that molecules such as oxygen can slide in between the layers of carbon and it is this that allows them to slide.

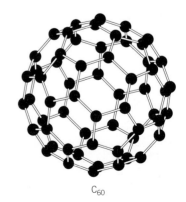

C_{60}

▲ **Figure 6** *Buckminsterfullerene – also called buckyballs*

If you know what the compound is, you can usually tell the type of bonding from the types of atoms that it contains:

- metal atoms only – metallic bonding
- metal and non-metal – ionic bonding
- non-metal atoms only – covalent bonding.

The three types of bonding give rise to different properties.

Electrical conductivity

The property that best tells us what sort of bonding you have is electrical conductivity. Metals and alloys (an alloy is a mixture of metals) conduct electricity well, in both the solid and liquid states due to their metallic bonding. The current is carried by the delocalised electrons that hold the metal ions together, see Figure 7.

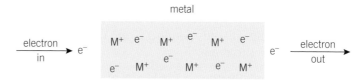

▲ **Figure 7** *The conduction of electricity by a metal*

Ionic compounds only conduct electricity in the liquid state (or when dissolved in water). They do *not* conduct when they are solid. The current is carried by the movement of ions towards the electrode of opposite charge. The ions are free to move when the ionic compound is liquid or dissolved in water. In the solid state they are fixed rigidly in position in the ionic lattice, Figure 8.

Generally, convalently bonded substances do not conduct electricity in either the solid or liquid state. This is because there are no charged particles to carry the current. Covalent compounds are often insoluble in water but some react to form ions, for example, ethanoic acid (present in vinegar). The solutions can then conduct electricity.

You can therefore decide what type of bonding a substance has by looking at how it conducts electricity. This is summarised in Table 2.

▼ **Table 2** *The pattern of electrical conductivity tells us about the type of bonding*

Type of bonding	Electrical conductivity		
	solid	liquid	aqueous solution
metallic	✓	✓	does not dissolve but may react
ionic	✗	✓	✓
covalent	✗	✗	✗ (but may react)

solid

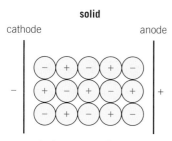

cathode   anode

the ions are not free to move in the solid state

liquid

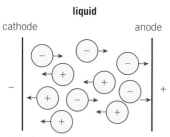

cathode   anode

the ions are free to move and the compound conducts electricity

▲ **Figure 8** *Ionic liquids conduct electricity, ionic solids do not*

Structure – summary

Structure describes the arrangement in which atoms, ions, or molecules are held together in space. There are four main types – simple molecular, macromolecular (giant covalent), giant ionic, and metallic.

> **Hint**
>
> Note there are some covalently bonded substances that *do* conduct electricity, for example, graphite.

- A **simple molecular** structure is composed of small molecules – small groups of atoms strongly held together by covalent bonding. The forces of attraction *between* molecules are much weaker (often over 50 times weaker than a covalent bond) and are called intermolecular forces. Examples of molecules include Cl_2, H_2O, H_2SO_4, and NH_3.

- A **macromolecular** structure is one in which large numbers of atoms are linked in a regular three-dimensional arrangement by covalent bonds. Examples include diamond and silicon dioxide (silica), the main constituent of sand.

- A **giant ionic** structure consists of a lattice of positive ions each surrounded in a regular arrangement by negative ions and vice versa.

- A **metallic** structure consists of a regular lattice of positively charged metal ions held together by a cloud of delocalised electrons.

Macromolecular, giant ionic, and metallic structures are often called giant structures because they have regular three-dimensional arrangements of atoms in contrast to simple molecular structures.

Melting and boiling points

The property that best tells us if a structure is giant or simple molecular is the melting (or boiling) point

- Simple molecular compounds have low melting (and boiling) points.
- Giant structures generally have high melting (and boiling) points.

If a compound has a low melting (and boiling) point, it has a simple molecular structure. All molecular compounds are covalently bonded. So all compounds with low melting (and boiling) points must have covalent bonding.

However, a compound with covalent bonding may have either a giant structure or a simple molecular structure and therefore may have either a high or low melting (and boiling) point.

Intermolecular forces

When you melt and boil simple molecular compounds, you are breaking the intermolecular forces *between* the molecules, not the covalent bonds *within* them. So the strength of the intermolecular forces determines the melting (and boiling) points.

There are three types of intermolecular force. In order of increasing strength, these are:

- van der Waals, which act between all atoms
- dipole–dipole forces, which act between molecules with permanent dipoles: $X^{\delta+}-Y^{\delta-}$
- hydrogen bonds, which act between the molecules formed when highly electronegative atoms (oxygen, nitrogen, and fluorine) and hydrogen atoms are covalently bonded.

> **Hint**
>
> Generally any substance with a high melting point also has a high boiling point. However, there are some substances, such as iodine, that sublime – they turn directly from solid to vapour.

Table 3 is a summary of the different properties of substances with covalent, ionic, and metallic bonding.

▼ **Table 3** *Summary of properties of substances with covalent, ionic, and metallic bonding*

	Structure	Bond	Melting point, T_m	Electrical conductivity		
				Solid	Liquid	Aqueous solution
	giant	ionic	high	no	yes	yes
	giant (macromolecular)	covalent	high	no (except graphite and graphene)	no	no
	simple molecular	covalent	low	no	no	no (but may react)
	giant	metallic	high	yes	yes	– does not dissolve but may react

Summary questions

1 Describe what is the difference between a macromolecular crystal and a molecular crystal in terms of the following.

   a bonding   **b** properties

2 Phosphorus consists of P_4 molecules and has a melting point of 317 K while sulfur, S_8, has a melting point of 386 K. Explain this difference in terms of bonding.

3 Explain why graphite can be used as a lubricant.

4 Explain how graphite conducts electricity. How does it conduct differently from metals?

5 Explain why both diamond and graphite have high melting points.

6 The table below gives some information about four substances.

   a Identify which substances have giant structures.

   b Identify which substance is a gas at room temperature.

   c Identify which substance is a metal.

   d Identify which substances are covalently bonded.

   e Identify which substance has ionic bonding.

   f Identify which substance is a macromolecule.

Substance	Melting point / K (°C)	Boiling point / K (°C)	Electrical conductivity	
			solid	liquid
A	1356 (1083)	2840 (2567)	good	good
B	91 (−182)	109 (−164)	poor	poor
C	1996 (1723)	2503 (2230)	poor	good
D	1266 (993)	1968 (1695)	poor	poor

Practice questions

1   Phosphorus exists in several different forms, two of which are white phosphorus and red phosphorus. White phosphorus consists of P_4 molecules, and melts at 44 °C.

Red phosphorus is macromolecular, and has a melting point above 550 °C.

Explain what is meant by the term *macromolecular*. By considering the structure and bonding present in these two forms of phosphorus, explain why their melting points are so different.

(5 marks)
AQA, 2006

2   (a)   Predict the shapes of the SF_6 molecule and the $AlCl_4^-$ ion. Draw diagrams of these species to show their three-dimensional shapes. Name the shapes and suggest values for the bond angles. Explain your reasoning.

(8 marks)

   (b)   Perfume is a mixture of fragrant compounds dissolved in a volatile solvent. When applied to the skin the solvent evaporates, causing the skin to cool for a short time. After a while, the fragrance may be detected some distance away. Explain these observations.

(4 marks)
AQA, 2003

3   Fritz Haber, a German chemist, first manufactured ammonia in 1909. Ammonia is very soluble in water.
   (a)   State the strongest type of intermolecular force between one molecule of ammonia and one molecule of water.

(1 mark)

   (b)   Draw a diagram to show how one molecule of ammonia is attracted to one molecule of water. Include all partial charges and all lone pairs of electrons in your diagram.

(3 marks)

   (c)   Phosphine, PH_3, has a structure similar to ammonia. In terms of intermolecular forces, suggest the main reason why phosphine is almost insoluble in water.

(1 mark)
AQA, 2013

4   The following equation shows the reaction of a phosphine molecule, PH_3, with an H^+ ion.
$$PH_3 + H^+ \rightarrow PH_4^+$$
   (a)   Draw the shape of the PH_3 molecule. Include any lone pairs of electrons that influence the shape.

(1 mark)

   (b)   State the type of bond that is formed between the PH_3 molecule and the H^+ ion. Explain how this bond is formed.

(2 marks)

   (c)   Predict the bond angle in the PH_4^+ ion.

(1 mark)

   (d)   Although phosphine molecules contain hydrogen atoms, there is no hydrogen bonding between phosphine molecules. Suggest an explanation for this.

(1 mark)
AQA, 2012

5   There are several types of crystal structure and bonding shown by elements and compounds.
   (a)   (i)   Name the type of bonding in the element sodium.

(1 mark)

      (ii)   Use your knowledge of structure and bonding to draw a diagram that shows how the particles are arranged in a crystal of sodium. You should identify the particles and show a minimum of six particles in a two-dimensional diagram.

(2 marks)
AQA, 2011

Learning objectives:

→ Define the terms endothermic and exothermic.

Specification reference: 3.1.4

Hint

The unit of energy is the joule, J. One joule represents quite a small amount of heat energy. For example, in order to boil water for a cup of tea you would need about 80 000 J which is 80 kJ.

Most chemical reactions give out or take in energy as they proceed. The amount of energy involved when a chemical reaction takes place is important for many reasons. For example:

- you can measure the energy values of fuels
- you can calculate the energy requirements for industrial processes
- you can work out the theoretical amount of energy to break bonds and the amount of energy released when bonds are made
- it helps to predict whether or not a reaction will take place.

The energy involved may be in different forms – light, electrical, or most usually heat.

Thermochemistry

Thermochemistry is the study of heat changes during chemical reactions.

- When a chemical reaction takes place, chemical bonds break and new ones are formed.
- Energy must be *put in* to break bonds and energy is *given out* when bonds are formed, so most chemical reactions involve an energy change.
- The overall change may result in energy being given out or taken in.
- At the end of the reaction, if energy has been given out, the reaction is **exothermic**.
- At the end of the reaction, if energy has been taken in, the reaction is **endothermic**.

Exothermic and endothermic reactions

Some reactions give out heat as they proceed. These are called *exothermic* reactions. Neutralising an acid with an alkali is an example of an exothermic reaction.

Some reactions take in heat from their surroundings to keep the reaction going. These are called *endothermic* reactions. The breakdown of limestone (calcium carbonate) to lime (calcium oxide) and carbon dioxide is an example of an endothermic reaction – it needs heat to proceed.

Another example of an endothermic reaction is heating copper sulfate. Blue copper sulfate crystals have the formula $CuSO_4.5H_2O$. The water molecules are bonded to the copper sulfate. In order to break these bonds and make white, anhydrous copper sulfate, heat energy must be supplied (Figure 1). This reaction takes in heat so it is endothermic.

▲ **Figure 1** *Heating copper sulfate*

$$CuSO_4.5H_2O \rightarrow CuSO_4 + 5H_2O$$

blue copper sulfate   white anhydrous copper sulfate   water

When you add water to anhydrous copper sulfate, the reaction gives out heat.

$$CuSO_4 \quad + \quad 5H_2O \quad \rightarrow \quad CuSO_4.5H_2O$$

white anhydrous copper sulfate   water   blue copper sulfate

In this direction the reaction is exothermic.

It is *always* the case that a reaction that is endothermic in one direction is exothermic in the reverse direction.

Quantities

The amount of heat given out or taken in during a chemical reaction depends on the quantity of reactants. This energy is usually measured in kilojoules per mole, $kJ\,mol^{-1}$. To avoid any confusion about quantities you need to give an equation. For example, in the combustion of methane, CH_4 one mole of methane reacts with two moles of oxygen:

$$CH_4(g) + 2O_2(g) \rightarrow CO_2(g) + 2H_2O(l)$$

890 kJ are given out when one mole of methane burns in two moles of oxygen.

Useful enthalpy changes

When fuels are burnt there is a large heat output. These are very exothermic reactions.

For example, coal is mostly carbon. Carbon gives out 393.5 kJ when one mole, 12 g, is burnt completely so that the most highly oxidised product is formed. This is carbon dioxide and not carbon monoxide. Carbon dioxide is the only product.

$$C(s) + O_2(g) \rightarrow CO_2(g)$$

As you saw above, natural gas, methane, gives out 890 kJ when one mole is burnt completely to carbon dioxide and water.

Physiotherapists often treat sports injuries with cold packs. these produce 'coldness' by an endothermic reaction such as:

$$NH_4NO_3(s) + (aq) \rightarrow NH_4NO_3(aq)$$

This absorbs $26\,kJ\,mol^{-1}$ of heat energy.

Hint √x̄

The expression mol^{-1} is a shorthand for 'per mole' and could also be written /mol. So kJ/mol has the same meaning as $kJ\,mol^{-1}$. Also note that the state symbols such as (g), meaning the gaseous state, are used. These are also important here.

The energy values of fuels

One important practical application of the study of thermochemistry is that it enables us to compare the efficiency of different fuels. Most of the fuels used today for transport (petrol for cars, diesel for cars and lorries, kerosene for aviation fuel, etc.) are derived from crude oil. This is a resource that will eventually run out so chemists are actively studying alternatives. Possible replacements include ethanol and methanol, both of which can be made from plant material, and hydrogen, which can be made by the electrolysis of water.

Theoretical chemists refer to the energy given out when a fuel burns completely as its heat (or enthalpy) of combustion. They measure this energy in kilojoules per mole ($kJ\,mol^{-1}$) because this compares the same number of *molecules* of each fuel. For use as fuels, the energy given out per *gram* of fuel burned, or the *energy density* of a fuel, is more important.

Some approximate values are given in the Table 1.

▼ **Table 1** *Enthalpy of combustion for various fuels*

Fuel	Enthalpy of combustion / $kJ\,mol^{-1}$	Mass of 1 mole / g	Energy density / $kJ\,g^{-1}$
petrol (pure octane)	−5500	114	48.2
ethanol	−1370	46	29.8
methanol	−730	32	22.8
hydrogen	−242	2	121.0

Notice that petrol stores significantly more energy per gram than either ethanol or methanol. This is a factor that will be significant for vehicles fuelled by either of these alcohols.

At first sight, hydrogen's energy density seems amazing. However, there is a catch. The other three fuels are liquids whereas hydrogen is a gas. Although hydrogen stores lots of energy per gram, a gram of gaseous hydrogen takes up a lot of space because of the low density of gases. How to store hydrogen efficiently is a challenge for designers.

1 Write a balanced symbol equation for the combustion of methanol, CH_3OH.
2 How do the product(s) of combustion vary between hydrogen and the other fuels?
3 What environmental significance does this have?

Summary questions

1 Natural gas, methane, CH_4, gives out 890 kJ when one mole is burnt completely.

$$CH_4(g) + 2O_2(g) \rightarrow CO_2(g) + 2H_2O(l)$$

Calculate how much heat would be given out when 8 g of methane is burnt completely.

2 The following reaction does not take place under normal conditions.

$$CO_2(g) + 2H_2O(l) \rightarrow CH_4(g) + 2O_2(g)$$

If it did, would you expect it to be exothermic or endothermic?

3 Explain your answer to question **2**.

4 Approximately how much methane would have to be burnt to provide enough heat to boil a cup of tea? Choose from a, b, or c.

   a 16 g   b 1.6 g   c 160 g

4.2 Enthalpy

The amount of heat given out or taken in by a reaction varies with the conditions – temperature, pressure, concentration of solutions, and so on. This means that you must state the conditions under which measurements are made. For example, you normally measure heat changes at constant atmospheric pressure.

Enthalpy change, ΔH

When you measure a heat change at constant pressure, it is called an **enthalpy change**.

Enthalpy has the symbol H so enthalpy changes are given the symbol ΔH. The Greek letter Δ (delta) is used to indicate a *change* in any quantity.

There are standard conditions for measuring enthalpy changes:

- pressure of 100 kPa (approximately normal atmospheric pressure)
- temperature of 298 K (around normal room temperature, 25 °C).

(The standard state of an element is the state in which it exists at 298 K and 100 kPa.)

When an enthalpy change is measured under standard conditions, it is written as $\Delta H^{\ominus}_{298}$ although usually the 298 is left out. $\Delta H^{\ominus}$ is pronounced delta H standard.

It may seem strange to talk about measuring heat changes at a constant temperature because heat changes normally *cause* temperature changes. The way to think about this is to imagine the reactants at 298 K, see Figure 1. Mix the reactants and heat is produced (this is an exothermic reaction). This heat is given out to the surroundings.

A reaction is not thought of as being over until the *products* have *cooled back to 298 K*. The heat given out to the surroundings while the reaction mixture cools is the enthalpy change for the reaction, $\Delta H^{\ominus}$.

- In an exothermic reaction the products end up with less heat energy than the starting materials because they have lost heat energy when they heated up their surroundings. This means that ΔH is negative. It is therefore given a negative sign.

Some endothermic reactions that take place in aqueous solution absorb heat from the water and cool it down, for example, dissolving ammonium nitrate in water. Again you don't think of the reaction as being over until the *products* have *warmed up to the temperature at which they started*.

In this case the solution has to take in heat from the surroundings to do this. Unless you remember this, it can seem strange that a reaction that is absorbing heat, initially gets cold.

- In an endothermic reaction the products end up with more energy than the starting materials, so ΔH is positive. It is therefore given a positive sign.

Pressure affects the amount of heat energy given out by reactions that involve gases. If a gas is given out, some energy is required to push

Hint 🧪

Chemists often use flasks open to the atmosphere to measure heat changes. The reaction is then carried out at atmospheric pressure. This varies slightly from day to day. Because these slight daily variations are small, this is only a small source of systematic error.

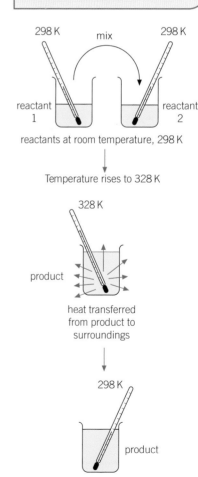

▲ **Figure 1** *A reaction giving out heat at 298 K*

Hint

Don't be confused by the different terms. Heat is a form of energy so a heat change can also be described as an energy change. An enthalpy change is still an energy change but it is measured under stated conditions of temperature and pressure.

away the atmosphere. The greater the atmospheric pressure, the more energy is used for this. This means that less energy remains to be given out as heat by the reaction. This is why it is important to have a standard of pressure for measuring energy changes.

The physical states of the reactants and products

The physical states (gas, liquid, or solid) of the reactants and products also affect the enthalpy change of a reaction. For example, heat must be put in to change liquid to gas and is given out when a gas is changed to a liquid. This means that you must always include state symbols in your equations.

For example, hydrogen burns in oxygen to form water but there are two possibilities:

1   forming liquid water

$$H_2(g) + \frac{1}{2}O_2(g) \rightarrow H_2O(l) \qquad \Delta H\ -285.8\ \text{kJ mol}^{-1}$$

2   forming steam

$$H_2(g) + \frac{1}{2}O_2(g) \rightarrow H_2O(g) \qquad \Delta H\ -241.8\ \text{kJ mol}^{-1}$$

The difference in ΔH represents the amount of heat needed to turn one mole of water into steam.

Hint

One way of making sure that both reactants are at the same temperature is simply to leave them in the same room for some time.

Enthalpy level diagrams

Enthalpy level diagrams, sometimes called energy level diagrams, are used to represent enthalpy changes. They show the relative enthalpy levels of the reactants (starting materials) and the products. The vertical axis represents enthalpy, and the horizontal axis, the extent of the reaction. You are usually only interested in the beginning of the reaction, 100% reactants, and the end of the reaction, 0% reactants (and 100% products), so the horizontal axis is usually left without units.

Figure 2 shows a general enthalpy diagram for an exothermic reaction (the products have less enthalpy than the reactants) and Figure 3 shows an endothermic reaction (the products have more enthalpy than the reactants).

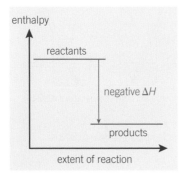

▲ **Figure 2** *Enthalpy diagram for an exothermic reaction*

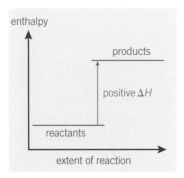

▲ **Figure 3** *Enthalpy diagram for an endothermic reaction*

Summary questions

1   Consider this reaction:

$$CH_4(g) + 2O_2(g) \rightarrow CO_2(g) + 2H_2O(l) \qquad \Delta H_{298}^{\ominus} = -890\ \text{kJ mol}^{-1}$$

  a   State what the symbol Δ means.

  b   State what the symbol H means.

  c   State what the 298 indicates.

  d   State what the minus sign indicates.

  e   Explain whether the reaction is exothermic or endothermic.

  f   Draw an enthalpy diagram to show the reaction.

The general name for the enthalpy change for any reaction is the standard molar enthalpy change of reaction $\Delta H^{\ominus}$. It is measured in kilojoules per mole, $kJ\,mol^{-1}$ (molar means 'per mole'). You write a balanced symbol equation for the reaction and then find the heat change for the quantities in moles given by this equation.

For example, ΔH for $2NaOH + H_2SO_4 \rightarrow Na_2SO_4 + 2H_2O$ is the enthalpy change when two moles of NaOH react with one mole of H_2SO_4.

Standard enthalpies

Some commonly used enthalpy changes are given names, for example, the enthalpy change of formation $\Delta_f H^{\ominus}$ and the enthalpy change of combustion $\Delta_c H^{\ominus}$. Both of these quantities are useful when calculating enthalpy changes for reactions. In addition, $\Delta_c H^{\ominus}$s are relatively easy to measure for compounds that burn readily in oxygen. Their formal definitions are as follows:

The **standard molar enthalpy of formation**, $\Delta_f H^{\ominus}$, is the enthalpy change when one mole of substance is formed from its constituent elements under standard conditions, all reactants and products being in their standard states.

The **standard molar enthalpy of combustion**, $\Delta_c H^{\ominus}$, is the enthalpy change when one mole of substance is completely burnt in oxygen under standard conditions, all reactants and products being in their standard states.

Heat and temperature

Temperature is related to the *average* kinetic energy of the particles in a system. As the particles move faster, their average kinetic energy increases and the temperature goes up. But it doesn't matter how many particles there are, temperature is independent of the *number* present. Temperature is measured with a thermometer.

Heat is a measure of the *total* energy of all the particles present in a given amount of substance. It *does* depend on how much of the substance is present. The energy of every particle is included. So a bath of lukewarm water has much more heat than a red hot nail because there are so many more particles in it. Heat always flows from high to low temperature, so heat will flow from the nail into the bath water, even though the water has much more heat than the nail.

Measuring the enthalpy change of a reaction

The enthalpy change of a reaction is the heat given out or taken in as the reaction proceeds. There is no instrument that measures heat directly. To measure the enthalpy *change* you arrange for the heat to be transferred into a particular mass of a substance, often water. Then you need to know three things:

1 mass of the substance that is being heated up or cooled down
2 temperature change
3 specific heat capacity of the substance.

Learning objectives:

→ Describe how enthalpy change is measured in a reaction.

→ Describe how you measure enthalpy changes more accurately.

→ Describe how you measure enthalpy changes in solution.

Specification reference: 3.1.4

Hint

The words heat and temperature are often used to mean the same thing in daily conversation but in science they are quite distinct and you must be clear about the difference.

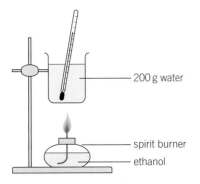

▲ **Figure 1** *A simple calorimeter*

Labels: 200 g water, spirit burner, ethanol

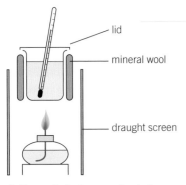

▲ **Figure 2** *An improved calorimeter*

Labels: lid, mineral wool, draught screen

The **specific heat capacity** c is the amount of heat needed to raise the temperature of 1 g of substance by 1 K. Its units are joules per gram per kelvin, or J g^{-1} K^{-1}. For example, the specific heat capacity of water is 4.18 J g^{-1} K^{-1}. This means that it takes 4.18 joules to raise the temperature of 1 gram of water by 1 kelvin. This is often rounded up to 4.2 J g^{-1} K^{-1}.

Then:

$$\text{enthalpy change } q \;=\; \frac{\text{mass of}}{\text{substance } m} \;\times\; \frac{\text{specific heat}}{\text{capacity } c} \;\times\; \frac{\text{temperature}}{\text{change } \Delta T}$$

The simple calorimeter or $q = mc\Delta T$

You can use the apparatus in Figure 1 to find the approximate enthalpy change when a fuel burns.

You burn the fuel to heat a known mass of water and then measure the temperature rise of the water. You assume that all the heat from the fuel goes into the water.

The apparatus used is called a **calorimeter** (from the Latin *calor* meaning heat).

> ## Worked example: Working out the enthalpy change
>
> The calorimeter in Figure 1 was used to measure the enthalpy change of combustion of methanol.
>
> $$CH_3OH(l) + 1\tfrac{1}{2}O_2(g) \rightarrow CO_2(g) + 2H_2O(l)$$
>
> 0.32 g (0.01 mol) of methanol was burnt and the temperature of the 200.0 g of water rose by 4.0 K.
>
> Heat change $= q = m \times c \times \Delta T$
>
> $$= 200.0 \times 4.2 \times 4.0 = 3360 \, J$$
>
> 0.01 mol gives 3360 J
>
> So 1 mol would give 336 000 J or 336 kJ
>
> $$\Delta_c H = -340 \, kJ \, mol^{-1} \text{ (negative because heat is given out)}$$

The simple calorimeter can be used to compare the $\Delta_c H$ values of a series of similar compounds because the errors will be similar for every experiment. However, you can improve the results by cutting down the heat loss as shown in Figure 2.

The flame calorimeter

The flame calorimeter, shown in Figure 3, is an improved version of the simple calorimeters used for measuring enthalpy changes of combustion. It incorporates the following features that are designed to reduce heat loss even further:

- the spiral chimney is made of copper
- the flame is enclosed
- the fuel burns in pure oxygen, rather than air.

Measuring enthalpy changes of reactions in solution

It is relatively easy to measure heat changes for reactions that take place in solution. The heat is generated in the solutions themselves and only has to be kept in the calorimeter. Expanded polystyrene beakers are often used for the calorimeters. These are good insulators (this reduces heat loss through their sides) and they have a low heat capacity so they absorb very little heat. The specific heat capacity of dilute solutions is usually taken to be the same as that of water, $4.2\,J\,g^{-1}\,K^{-1}$ (or more precisely $4.18\,J\,g^{-1}\,K^{-1}$).

Neutralisation reactions

Neutralisation reactions in solution are exothermic – they give out heat. When an acid is neutralised by an alkali the equation is:

$$acid + alkali \rightarrow salt + water$$

To find an enthalpy change for a reaction, you use the quantities in moles given by the balanced equation. For example, to find the molar enthalpy change of reaction for the neutralisation of hydrochloric acid by sodium hydroxide, the heat given out by the quantities in the equation needs to be found:

$HCl(aq)$	+	$NaOH(aq)$	$\rightarrow$	$NaCl(aq)$	+	$H_2O(l)$
hydrochloric acid		sodium hydroxide		sodium chloride		water
1 mol		1 mol		1 mol		1 mol

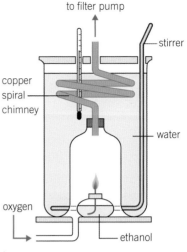

▲ **Figure 3** *A flame calorimeter*

labels: to filter pump, stirrer, copper spiral chimney, water, oxygen, ethanol

Worked example: Enthalpy change for a reaction

$50\,cm^3$ of $1.0\,mol\,dm^{-3}$ hydrochloric acid and $50\,cm^3$ of $1.0\,mol\,dm^{-3}$ sodium hydroxide solution were mixed in an expanded polystyrene beaker. The temperature rose by $6.6\,K$.

The total volume of the mixture is $100\,cm^3$. This has a mass of approximately $100\,g$ because the density of water and of dilute aqueous solutions is approximately $1\,g\,cm^{-3}$.

$$\begin{array}{c} enthalpy \\ change\ q \end{array} = \begin{array}{c} mass\ of \\ water\ m \end{array} \times \begin{array}{c} specific\ heat\ capacity \\ of\ solution\ c \end{array} \times \begin{array}{c} temperature \\ change\ \Delta T \end{array}$$

$$q = m \times c \times \Delta T$$
$$= 100 \times 4.2 \times 6.6 = 2772\,J$$

$$\begin{array}{c} number\ of\ moles \\ of\ acid\ (and\ also \\ of\ alkali)\ n \end{array} = \frac{concentration\ c\ (mol\ dm^{-3}) \times volume\ V\ (cm^3)}{1000}$$

$$= 1.0 \times \frac{50}{1000} = 0.05\,mol$$

so 1 mol would give $\frac{2772}{0.05}\,J = 55\,440\,J = 55.44\,kJ$

$$\Delta H = -55.44\ kJ\ mol^{-1}$$
$$\Delta H = -55\,kJ\,mol^{-1}\ (to\ 2\ s.f.)$$

The sign of ΔH is negative because heat is given out.

Hint

Remember to use the *total* volume of the mixture, $100\,cm^3$. A common mistake is to use $50\,cm^3$.

Displacement reactions

A metal that is more reactive than another will displace the less reactive one from a compound. If the compound will dissolve in water, this reaction can be investigated using a polystyrene beaker as before.

For example, zinc will displace copper from a solution of copper sulfate. The reaction is exothermic.

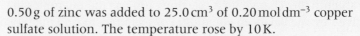

$$Zn(s) + CuSO_4(aq) \rightarrow ZnSO_4(aq) + Cu(s)$$

1 mol   1 mol   1 mol   1 mol

From the equation one mole of zinc reacts with one mole of copper sulfate.

> **Study tip** $\sqrt{x}$
>
> You should be able to rearrange $q = m \times c \times \Delta T$ to find any of the quantities in terms of the others:
>
> $\Delta T = \dfrac{q}{mc}$
>
> $m = \dfrac{q}{c} \Delta T$
>
> $c = \dfrac{q}{m} \Delta T$

Worked example: Enthalpy change in a displacement reaction

0.50 g of zinc was added to 25.0 cm³ of 0.20 mol dm⁻³ copper sulfate solution. The temperature rose by 10 K.

$$q = m \times c \times \Delta T$$
$$= 25 \times 4.2 \times 10 = 1050 \text{ J}$$

A_r zinc = 65.4, so 0.50 g of zinc is $\dfrac{0.50}{65.4}$ moles = 0.0076 moles

number of moles of copper sulfate in solution = $\dfrac{c \times V}{1000}$

where c is concentration in mol dm⁻³ and V is volume in cm³

$$= 0.20 \times \frac{25.0}{1000} = 0.005 \text{ mol}$$

This means that the zinc was in excess; 0.005 mol of each reactant has taken part in the reaction, leaving some unreacted zinc behind.

Therefore, 1 mole of zinc would produce $\dfrac{1050}{0.005}$ J = 210 000 J.

So, ΔH for this reaction is −210 kJ mol⁻¹ (to 2 s.f.).

The sign of ΔH is negative because heat is given out.

Allowing for heat loss

Although expanded polystyrene cups are good insulators, some heat will still be lost from the sides and top leading to low values for enthalpy changes measured by this method. This can be allowed for by plotting a cooling curve. As an example, the measurement of the heat of neutralisation of hydrochloric acid and sodium hydroxide is repeated using a cooling curve.

Before the experiment, all the apparatus and both solutions are left to stand in the laboratory for some time. This ensures that they all reach the same temperature, that of the laboratory itself.

Then proceed as follows:

1 Place 50 cm³ of 1.0 mol dm⁻³ hydrochloric acid in one polystyrene cup and 50 cm³ of 1.0 mol dm⁻³ sodium hydroxide solution in another.

2 Using a thermometer that reads to 0.1 °C, take the temperature of each solution every 30 seconds for four minutes to confirm

▲ **Figure 6** *Polystyrene beakers make good calorimeters because they are good insulators and have low heat capacities*

that both solutions remain at the same temperature, that of the laboratory. A line of 'best fit' is drawn through these points. It is likely there will be very small variations around the line of best fit, indicating random errors.

3 Now pour one solution into the other and stir, continuing to record the temperature every 30 seconds for a further six minutes.

The results are shown on the graph in Figure 7. The experiment can also be done using an electronic temperature sensor and data logging software to plot the graph directly.

On mixing, the temperature rises rapidly as the reaction gives out heat, and then drops slowly and regularly as heat is lost from the polystyrene cup. To find the best estimate of the temperature immediately after mixing, you draw the best straight line through the graph points after mixing and extrapolate back to the time of mixing. This gives a temperature rise of 6.9 °C.

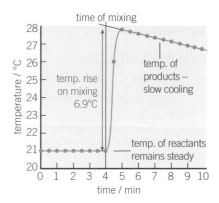

▲ **Figure 7** *Graph to show temperature as a neutralisation reaction proceeds*

The calculation is as before.

$$q = m \times c \times \Delta T = 100 \times 4.2 \times 6.9 = 2898 \, \text{J}$$

The number of moles of acid (and alkali) was 0.05 mol (as before).

So 1 mol would give $\dfrac{2898}{0.05} \, \text{J} = 57\,960 \, \text{J} = 57.96 \, \text{kJ}$

$$\Delta_{\text{neut}} H = -58 \, \text{kJ mol}^{-1} \text{ (to 2 s.f.)}$$

The sign of ΔH is negative because heat is given out.

Summary questions

1 ✔ 0.74 g (0.010 mol) of propanoic acid was burnt in the simple calorimeter like that described above for the combustion of methanol. The temperature rose by 8.0 K. Calculate the value this gives for the enthalpy change of combustion of propanoic acid.

2 50.0 cm³ of 2.00 mol dm⁻³ sodium hydroxide and 50.0 cm³ of 2.00 mol dm⁻³ hydrochloric acid were mixed in an expanded polystyrene beaker. The temperature rose by 11.0 K.

a ✔ Calculate ΔH for the reaction.

b Describe how this value will compare with the accepted value for this reaction.

c Explain your answer to **b**.

3 Consider the expression $q = mc\Delta T$

a State what the term q represents.

b State what the term m represents.

c State what the term c represents.

d State what the term ΔT represents.

Learning objectives:

→ Describe how to find enthalpy changes that cannot be measured directly.

Specification reference: 3.1.4

The enthalpy changes for some reactions cannot be measured directly. To find these you use an indirect approach. Chemists use enthalpy changes that they can measure to work out enthalpy changes that they cannot. In particular, it is often easy to measure enthalpies of combustion. To do this, chemists use Hess's law, first stated by Germain Hess, a Swiss-born Russian chemist, born in 1802.

Hess's law

Hess's law states that the enthalpy change for a chemical reaction is the same, whatever route is taken from reactants to products.

This is a consequence of a more general scientific law, the Law of Conservation of Energy, which states that energy can never be created or destroyed. So, provided the starting and finishing points of a process are the same, the energy change must be the same. If not, energy would have been created or destroyed.

Using Hess's law

To see what Hess's law means, look at the following example where ethyne, C_2H_2, is converted to ethane, C_2H_6, by two different routes. How can we find the enthalpy of reaction?

Route 1: The reaction takes place directly – ethyne reacts with two moles of hydrogen to give ethane.

$$C_2H_2(g) + 2H_2(g) \rightarrow C_2H_6(g) \qquad \Delta H_1 = ?$$
$$\text{ethyne} \qquad\qquad\qquad \text{ethane}$$

Route 2: The reaction takes place in two stages.

a Ethyne, C_2H_2, reacts with one mole of hydrogen to give ethene, C_2H_4.

$$C_2H_2(g) + H_2(g) \rightarrow C_2H_4(g) \qquad \Delta H_2 = -176\,kJ\,mol^{-1}$$
$$\text{ethyne} \qquad\qquad \text{ethene}$$

b Ethene, C_2H_4, then reacts with a second mole of hydrogen to give ethane, C_2H_6

$$C_2H_4(g) + H_2(g) \rightarrow C_2H_6(g) \qquad \Delta H_3 = -137\,kJ\,mol^{-1}$$
$$\text{ethene} \qquad\qquad \text{ethane}$$

Hess's law tells us that the total energy change is the same whichever route you take – direct or via ethene (or, in fact, by any other route). You can show this on a diagram called a **thermochemical cycle**.

$$H-C\equiv C-H(g) + 2H_2(g) \xrightarrow[1.]{\Delta H_1\ (?)} H-\underset{\underset{H}{|}}{\overset{\overset{H}{|}}{C}}-\underset{\underset{H}{|}}{\overset{\overset{H}{|}}{C}}-H(g)$$

ΔH_2 (−176 kJ mol⁻¹) 2.

ΔH_3 (−137 kJ mol⁻¹) 3.

$$\underset{\underset{H}{}}{\overset{\overset{H}{}}{C}}=\underset{\underset{H}{}}{\overset{\overset{H}{}}{C}}\ (g)$$

+ H₂ (g)

Hess's law means that: $\Delta H_1 = \Delta H_2 + \Delta H_3$

The actual figures are: $\Delta H_2 = -176\,\text{kJ}\,\text{mol}^{-1}$

$\Delta H_3 = -137\,\text{kJ}\,\text{mol}^{-1}$

So $\Delta H_1 = (-176) + (-137) = -313$ kJ mol⁻¹

This method of calculating ΔH_1 is fine if you know the enthalpy changes for the other two reactions. There are certain enthalpy changes that can be looked up for a large range of compounds. These include the enthalpy change of formation, $\Delta_f H^\ominus$, and enthalpy change of combustion, $\Delta_c H^\ominus$. In practice, many $\Delta_f H^\ominus$s are calculated from $\Delta_c H^\ominus$s via Hess's law cycles.

Using the enthalpy changes of formation $\Delta_f H^\ominus$

The enthalpy of formation, $\Delta_f H^\ominus$, is the enthalpy change when one mole of compound is formed from its constituent elements under standard conditions, all reactants and products being in their standard states.

Another theoretical way to convert ethyne to ethane could be via the elements carbon and hydrogen.

- Ethyne is first converted to its elements, carbon and hydrogen. This is the reverse of formation and the enthalpy change is the *negative* of the enthalpy of formation. This is a general rule. The reverse of a reaction has the negative of its ΔH value. It is in fact a consequence of Hess's law.

- Then the carbon and hydrogen react to form ethane. This is the enthalpy of formation for ethane.

Hess's law tells us that $\Delta H_1 = \Delta H_4 + \Delta H_5$

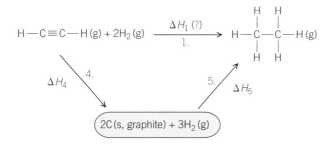

Hint

Graphite is the most stable form of carbon (another form is diamond). It has a special state symbol (s, graphite).

Hint

For an element, $\Delta_f H^\ominus$ is zero by definition.

ΔH_5 is the enthalpy of formation, $\Delta_f H^\ominus$, of ethane whilst reaction 4 is the reverse of the formation of ethyne.

The values you need are: $\Delta_f H^\ominus(C_2H_2) = +228\,kJ\,mol^{-1}$

and $\Delta_f H^\ominus(C_2H_6) = -85\,kJ\,mol^{-1}$

So $\Delta H_4 = -228\,kJ\,mol^{-1}$

(remember to change the sign)

$\Delta H_5 = -85\ kJ\ mol^{-1}$

Therefore $\Delta H_1 = -228 + -85 = -313\,kJ\,mol^{-1}$

This was the result you got from the previous method, as you should expect from Hess's law.

Notice that in reaction 4 there are two moles of hydrogen 'spare' as only one of the three moles of hydrogen is involved. These two moles of hydrogen remain in their standard states and so no enthalpy change is invoved.

$C_2H_2(g) \rightarrow 2C(s, graphite) + H_2(g)$ is the reaction you are considering, but you have:

$$C_2H_2(g) + 2H_2(g) \rightarrow 2C(s, graphite) + 3H_2(g)$$

However, this makes no difference. The 'extra' hydrogen is *not* involved in the reaction and it does not affect ΔH.

Summary questions

1 Use the values of $\Delta_f H^\ominus$ in the table to calculate $\Delta H^\ominus$ for each of the reactions below using a thermochemical cycle.

a $CH_3COCH_3(l) + H_2(g) \rightarrow CH_3CH(OH)CH_3(l)$

b $C_2H_4(g) + Cl_2(g) \rightarrow C_2H_4Cl_2(l)$

c $C_2H_4(g) + HCl(g) \rightarrow C_2H_5Cl(l)$

d $Zn(s) + CuO(s) \rightarrow ZnO(s) + Cu(s)$

e $Pb(NO_3)_2(s) \rightarrow PbO(s) + 2NO_2(g) + \frac{1}{2}O_2(g)$

Compound	$\Delta_f H^\ominus$/ kJ mol^{-1}
$CH_3COCH_3(l)$	−248
$CH_3CH(OH)CH_3(l)$	−318
$C_2H_4(g)$	+52
$C_2H_4Cl_2(l)$	−165
$C_2H_5Cl(l)$	−137
$HCl(g)$	−92
$CuO(s)$	−157
$ZnO(s)$	−348
$Pb(NO_3)_2(s)$	−452
$PbO(s)$	−217
$NO_2(g)$	+33

The enthalpy change of combustion, $\Delta_c H^{\ominus}$, is the enthalpy change when one mole of substance is completely burnt in oxygen under standard conditions.

Thermochemical cycles using enthalpy changes of combustion

Look again at the thermochemical cycle used to find $\Delta H^{\ominus}$ for the reaction between ethyne and hydrogen to form ethane.

$$C_2H_2(g) + 2H_2(g) \rightarrow C_2H_6(g)$$

This time use enthalpy changes of combustion. In this case you can go via the combustion products of the three substances – carbon dioxide and water.

All three substances – ethyne, hydrogen, and ethane – burn readily. This means their enthalpy changes of combustion can be easily measured. The thermochemical cycle is:

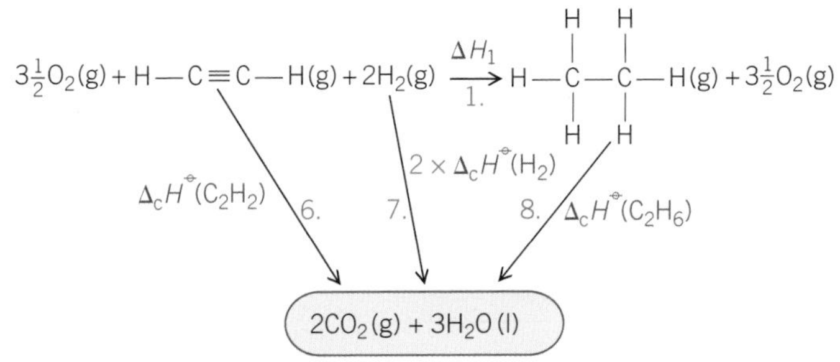

Putting in the values:

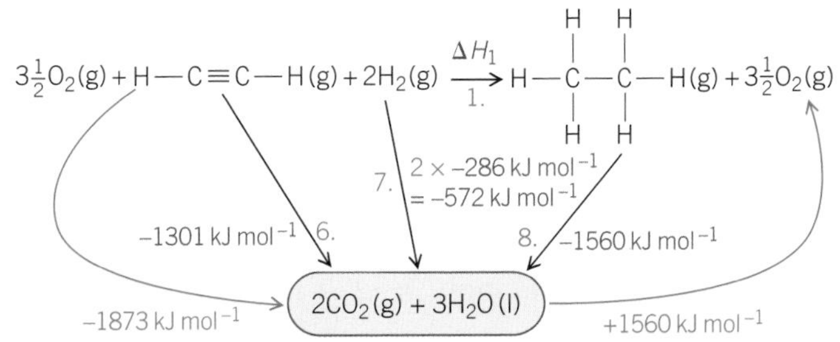

To get the enthalpy change for reaction 1 you must go round the cycle in the direction of the red arrows. This means reversing reaction 8 so you must change its sign.

So $\Delta H_1 = -1873 + 1560\,\text{kJ}\,\text{mol}^{-1}$

$\Delta H_1 = -313\,\text{kJ}\,\text{mol}^{-1}$ once again, the same answer as before

Notice that in reaction 1 there are $3\frac{1}{2}$ moles of oxygen on either side of the equation. They take no part in the reaction and do not affect the value of ΔH.

Learning objectives:
→ Describe how the enthalpy change of combustion can be used to find the enthalpy change of a reaction.

Specification reference: 3.1.4

Study tip √x̄

Remember to multiply by the number of moles of reagents involved in each step.

Hint

- Both reactions 6 and 7 have to occur to get from the starting materials to the combustion products. Do not forget the hydrogen.

- In this case there are two moles of hydrogen so you need *twice* the value of $\Delta_c H^{\ominus}$ which refers to one mole of hydrogen.

  $\Delta_c H^{\ominus}(C_2H_2) = -1301\,\text{kJ}\,\text{mol}^{-1}$
  $\Delta_c H^{\ominus}(H_2) = -286\,\text{kJ}\,\text{mol}^{-1}$
  $\Delta_c H^{\ominus}(C_2H_6) = -1560\,\text{kJ}\,\text{mol}^{-1}$

Finding $\Delta_f H^\ominus$ from $\Delta_c H^\ominus$

Enthalpy changes of formation of compounds are often difficult or impossible to measure directly. This is because the reactants often do not react directly to form the compound that you are interested in.

For example, the following equation represents the formation of ethanol from its elements.

$$2C(s, \text{graphite}) + 3H_2(g) + \tfrac{1}{2}O_2(g) \rightarrow C_2H_5OH(l)$$

This does not take place. However, all the species concerned will readily burn in oxygen so their enthalpy changes of combustion can be measured. The thermochemical cycle you need is:

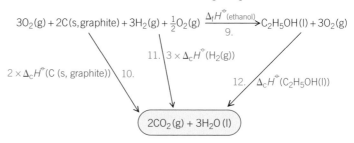

Hint

The values we need are:

$\Delta_c H^\ominus(C(s, \text{graphite})) = -393.5 \text{ kJ mol}^{-1}$

$\Delta_c H^\ominus(H_2(g)) = -285.8 \text{ kJ mol}^{-1}$

$\Delta_c H^\ominus(C_2H_5OH(l)) = -1367.3 \text{ kJ mol}^{-1}$

Putting in the values:

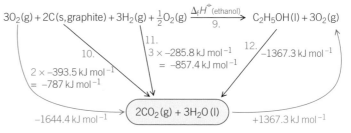

Note that in reaction 9 there are three moles of oxygen on either side of the equation that take no part in the reaction. This means that they do not affect the value of ΔH.

Note also that $\Delta_c H^\ominus(C(s, \text{graphite}))$ is the same as $\Delta_f H^\ominus(CO_2(g))$ and $\Delta_c H^\ominus(H_2(g))$ is the same as $\Delta_f H^\ominus(H_2O(l))$.

To get the enthalpy change for reaction 9, you must go round the cycle in the direction of the red arrows. This means reversing reaction 12 so you must change its sign.

So, $\Delta H_9 = -1664.4 + 1367.3 \text{ kJ mol}^{-1} = -277.1 \text{ kJ mol}^{-1}$

So, $\Delta_f H^\ominus(C_2H_5OH(l)) = -277.1 \text{ kJ mol}^{-1}$

Summary questions

1 🔢 Calculate $\Delta H^\ominus$ for the reaction by thermochemical cycles:

H─C─C(l) + H₂(g) ⟶ H─C─C─O─H

a via $\Delta_f H^\ominus$ values   **b** via $\Delta_c H^\ominus$ values

Compound	$\Delta H_f^\ominus$ / kJ mol^{-1}	$\Delta H_c^\ominus$ / kJ mol^{-1}
CH$_3$CHO	−192	−1167
H$_2$	–	−286
CH$_3$CH$_2$OH	−277	−1367

You can use **enthalpy diagrams** rather than thermochemical cycles to represent the enthalpy changes in chemical reactions. These show the energy (enthalpy) levels of the reactants and products of a chemical reaction on a vertical scale, so you can compare their energies. If a substance is of lower energy than another, you say it is energetically more stable.

The enthalpy of elements

So far you have considered enthalpy *changes*, not absolute values. When drawing enthalpy diagrams you need a zero to work from. You can then give absolute numbers to the enthalpies of different substances.

> **The enthalpies of all elements in their standard states (i.e., the states in which they exist at 298 K and 100 kPa) are taken as zero. (298 K and 100 kPa are approximately normal room conditions.)**

This convention means that the standard state of hydrogen, for example, is H_2 and not H, because hydrogen exists as H_2 at room temperature and pressure.

Pure carbon can exist in a number of forms at room temperature including graphite, diamond, and buckminsterfullerene (buckyballs). These are called **allotropes**. Graphite is the most stable of these and is taken as the standard state of carbon. It is given the special state symbol (s, graphite), so C(s, graphite) represents graphite.

Thermochemical cycles and enthalpy diagrams

Here are two examples of reactions, with their enthalpy changes presented both as thermochemical cycles and as enthalpy diagrams.

Example 1

What is $\Delta H^\ominus$ for the change from methoxymethane to ethanol? (The compounds are a pair of isomers – they have the same formula but different structures, see Figure 1.)

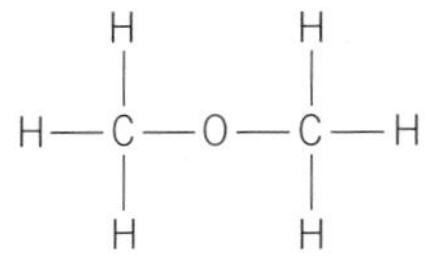

The standard molar enthalpy changes of formation of the two compounds are:

$$CH_3OCH_3 \quad \Delta_f H^\ominus = -184 \, kJ \, mol^{-1}$$

$$C_2H_5OH \quad \Delta_f H^\ominus = -277 \, kJ \, mol^{-1}$$

Using a thermochemical cycle

The following steps are shown in red on the thermochemical cycle.

1 Write an equation for the reaction.
2 Write down the elements in the two compounds with the correct quantities of each.
3 Put in the $\Delta_f H^\ominus$ values with arrows showing the direction – *from* elements *to* compounds.
4 Put in the arrows to go from starting materials to products via the elements (the red arrows).

Learning objectives:
→ Describe what an enthalpy diagram is.
→ State what is used as the zero for enthalpy changes.

Specification reference: 3.1.4

methoxymethane

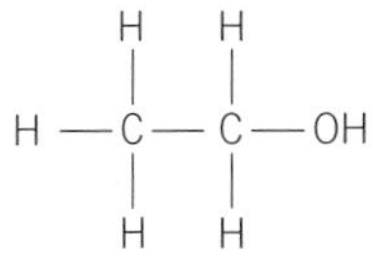

ethanol

▲ **Figure 1** *Isomers of C_2H_6O*

5 Reverse the sign of $\Delta_f H^\ominus$ if the red arrow is in the opposite direction to the black arrow.

6 Go round the cycle in the direction of the red arrows and add up the $\Delta H^\ominus$ values as you go.

Hess's law shows that this is the same as $\Delta H^\ominus$ for the direct reaction.

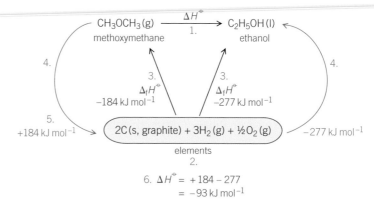

▲ **Figure 2** *Thermochemical cycle for the formation of ethanol from methoxymethane*

Using an enthalpy diagram

The following steps are shown in red on the enthalpy diagram.

1 Draw a line at level 0 to represent the elements.

2 Look up the values of $\Delta_f H^\ominus$ for each compound and enter these on the enthalpy diagrams, taking account of the signs – negative values are below 0, positive values are above.

3 Find the difference in levels between the two compounds. This represents the difference in their enthalpies.

4 $\Delta H^\ominus$ is the difference in levels *taking account of the direction of change*. Up is positive and down is negative. From methoxymethane to ethanol is *down* so the sign is negative. From ethanol to methoxymethane the sign of $\Delta H^\ominus$ would be positive.

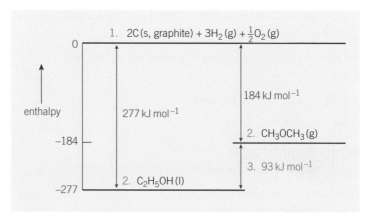

▲ **Figure 3** *The enthalpy diagram for the formation of ethanol from methoxymethane*

Notice how the enthalpy level diagram makes it much clearer than the thermochemical cycle that ethanol has less energy than methoxymethane. This means that it is the more energetically stable compound. The values of $\Delta H^\ominus$ for the reaction are the same whichever method you use.

Example 2

To find $\Delta H^{\ominus}$ for the reaction $NH_3(g) + HCl(g) \rightarrow NH_4Cl(s)$

The standard molar enthalpy changes of formation of the compounds are:

$NH_3 \qquad \Delta_f H^{\ominus} = -46\,kJ\,mol^{-1}$

$HCl \qquad \Delta_f H^{\ominus} = -92\,kJ\,mol^{-1}$

$NH_4Cl \quad \Delta_f H^{\ominus} = -314\,kJ\,mol^{-1}$

Using a thermochemical cycle

The thermochemical cycle for the formation of ammonium chloride is shown in Figure 4.

1 Write an equation for the reaction.
2 Write down the elements that make up the two compounds with the correct quantities of each.
3 Put in the $\Delta_f H^{\ominus}$ values with arrows showing the direction, that is, from elements to compounds.
4 Put in the arrows going from the starting materials to products via the elements (the red arrows).
5 Reverse the sign of $\Delta_f H^{\ominus}$ if the red arrow is in the opposite direction to the black arrow(s).
6 Go round the cycle in the direction of the red arrows and add up the values of $\Delta H^{\ominus}$ as you go.

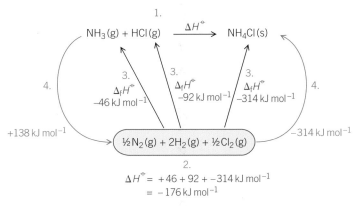

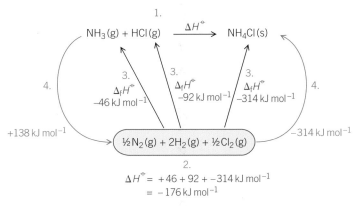
(Figure content)

▲ **Figure 4** *Thermochemical cycle for the formation of ammonium chloride*

> ## Hint $\sqrt{x}$
>
> You can use a short cut to save drawing an enthalpy diagram or a thermochemical cycle. The enthalpy change of a reaction is the sum of the enthalpies of formation of all the products minus the sum of the enthalpies of formation of all the reactants. In this example
> $\Delta H^{\ominus} = -314 - (-46 + (-92))$
> $\quad\quad = -314 - (-138)$
> $\quad\quad = -176\,kJ\,mol^{-1}.$
>
> If you use this short cut, you must be *very* careful of the signs.

➕ Using an enthalpy diagram

The following steps are shown in red on the enthalpy diagram.

1 Draw a line at level 0 to represent the elements.
2 Draw in NH_4Cl 314 kJ mol^{-1} below this.
3 Draw a line representing ammonia 46 kJ mol^{-1} below the level of the elements. (There is still $\frac{1}{2}H_2$ and $\frac{1}{2}Cl_2$ left unused.)
4 Draw a line 92 kJ mol^{-1} below ammonia. This represents hydrogen chloride.
5 Find the difference in levels between the $(NH_3 + HCl)$ line and the NH_4Cl one. This represents $\Delta H^{\ominus}$ for the reaction. As the change from $(NH_3 + HCl)$ to NH_4Cl is down, $\Delta H^{\ominus}$ must be negative.

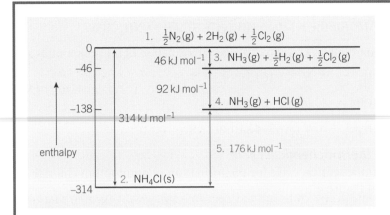

Notice how the enthalpy level diagram makes it much clearer than the thermochemical cycle that ammonium chloride is more energetically stable than the gaseous mixture of ammonia and hydrogen chloride. This is part of the reason why ammonia and hydrogen chloride react readily to form ammonium chloride. The values of $\Delta H^\ominus$ for the reaction are the same whichever method you use.

What would be the enthalpy change when solid ammonium chloride decomposes into the gases ammonia and hydrogen chloride?

+176 kJ mol^{-1}

Summary questions

1 $\sqrt{x}$ Use the values of $\Delta_f H^\ominus$ in the table to calculate $\Delta H^\ominus$ for each of the reactions below using enthalpy diagrams.

a $CH_3COCH_3(l) + H_2(g) \rightarrow CH_3CH(OH)CH_3(l)$

b $C_2H_4(g) + Cl_2(g) \rightarrow C_2H_4Cl_2(l)$

c $C_2H_4(g) + HCl(g) \rightarrow C_2H_5Cl(l)$

d $Zn(s) + CuO(s) \rightarrow ZnO(s) + Cu(s)$

e $Pb(NO_3)_2(s) \rightarrow PbO(s) + 2NO_2(g) + \frac{1}{2}O_2(g)$

Compound	$\Delta_f H^\ominus$/ kJ mol^{-1}
$CH_3COCH_3(l)$	−248
$CH_3CH(OH)CH_3(l)$	−318
$C_2H_4(g)$	+52
$C_2H_4Cl_2(l)$	−165
$C_2H_5Cl(l)$	−137
$HCl(g)$	−92
$CuO(s)$	−157
$ZnO(s)$	−348
$Pb(NO_3)_2(s)$	−452
$PbO(s)$	−217
$NO_2(g)$	+33

4.7 Bond enthalpies

$\Delta_c H^\ominus$ is the enthalpy change of combustion. If you plot $\Delta_c H^\ominus$ against the number of carbon atoms in the molecule, for straight chain alkanes, you get a straight line graph, see Figure 1. For methane (with one carbon), $\Delta_c H^\ominus$ is the enthalpy change for:

$$CH_4(g) + 2O_2(g) \rightarrow CO_2(g) + 2H_2O(l)$$

The straight line means that $\Delta_c H^\ominus$ changes by the same amount for each extra carbon atom in the chain.

Each alkane differs from the previous one by one CH_2 group, that is, there is one extra C—C bond in the molecule and two extra C—H bonds. This suggests that you can assign a definite amount of energy to a particular bond. This is called the bond enthalpy.

Bond enthalpies

You have to put in energy to break a covalent bond – this is an endothermic change. **Bond dissociation enthalpy** is defined as the enthalpy change required to break a covalent bond with all species in the gaseous state. The same amount of energy is given out when the bond is formed – this is an exothermic change. However, the same bond, for example C—H, may have slightly different bond enthalpies in different molecules, but you usually use the average value. This value is called the **mean bond enthalpy** (often called the bond energy). The fact that you get out the same amount of energy when you make a bond, as you put in to break it is an example of Hess's law.

As mean bond enthalpies are averages, calculations using them for specific compounds will only give approximate answers. However, they are useful, and quick and easy to use. Mean bond enthalpies have been calculated from Hess's law cycles. They can be looked up in data books and databases.

The H—H bond energy is the energy required to separate the two atoms in a hydrogen molecule in the gas phase into separate gaseous atoms.

$$H_2(g) \rightarrow 2H(g) \quad \Delta H^\ominus = +436\,\text{kJ}\,\text{mol}^{-1}$$

The C—H mean bond energy in methane is one quarter of the energy for the following process, in which four bonds are broken.

$$CH_4(g) \rightarrow C(g) + 4H(g) \quad \Delta H^\ominus = +1664\,\text{kJ}\,\text{mol}^{-1}$$

So the mean (or average) C—H bond energy in methane $= \dfrac{1664}{4}$

$$= +416\,\text{kJ}\,\text{mol}^{-1}$$

Using mean bond enthalpies to calculate enthalpy changes of reaction

You can use mean bond enthalpies to work out the enthalpy change of reactions, for example:

$$\underset{\text{ethane}}{C_2H_6(g)} + \underset{\text{chlorine}}{Cl_2(g)} \rightarrow \underset{\text{chloroethane}}{C_2H_5Cl(g)} + \underset{\text{hydrogen chloride}}{HCl(g)}$$

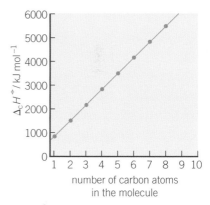

▲ **Figure 1** $\Delta_c H^\ominus$ plotted against the number of carbon atoms in the alkane

▼ **Table 1** *Mean bond enthalpies*

Bond	Bond enthalpy / kJ mol^{-1}
C—H	413
C—C	347
Cl—Cl	243
C—Cl	346
Cl—H	432
Br—Br	193
Br—H	366
C—Br	285

The mean bond enthalpies you will need for this example are given in Table 1.

The steps are as follows:

1   First draw out the molecules and show all the bonds. (Formulae drawn showing all the bonds are called displayed formulae.)

2   Now imagine that all the bonds in the *reactants* break leaving separate atoms. Look up the bond enthalpy for each bond and add them all up. This will give you the total energy that must be *put in* to break the bonds and form separate atoms.

You need to *break* these bonds:

$$\begin{array}{lll}
6 \times \text{C—H} & 6 \times 413\,\text{kJ mol}^{-1} & = 2478\,\text{kJ mol}^{-1} \\
1 \times \text{C—C} & 1 \times 347\,\text{kJ mol}^{-1} & = 347\,\text{kJ mol}^{-1} \\
1 \times \text{Cl—Cl} & 1 \times 243\,\text{kJ mol}^{-1} & = 243\,\text{kJ mol}^{-1} \\
 & & \mathbf{= 3068\,kJ\,mol^{-1}}
\end{array}$$

So 3068 kJ mol^{-1} must be *put in* to convert ethane and chlorine to separate hydrogen, chlorine, and carbon atoms.

3   Next imagine the separate atoms join together to give the *products*. Add up the bond enthalpies of the bonds that must form. This will give you the total enthalpy *given out* by the bonds forming.

You need to *make* these bonds:

$$\begin{array}{lll}
5 \times \text{C—H} & 5 \times 413\,\text{kJ mol}^{-1} & = 2065\,\text{kJ mol}^{-1} \\
1 \times \text{C—C} & 1 \times 347\,\text{kJ mol}^{-1} & = 347\,\text{kJ mol}^{-1} \\
1 \times \text{C—Cl} & 1 \times 346\,\text{kJ mol}^{-1} & = 346\,\text{kJ mol}^{-1} \\
1 \times \text{Cl—H} & 1 \times 432\,\text{kJ mol}^{-1} & = 432\,\text{kJ mol}^{-1} \\
 & & \mathbf{= 3190\,kJ\,mol^{-1}}
\end{array}$$

So 3190 kJ mol^{-1} is *given out* when you convert the separate hydrogen, chlorine, and carbon atoms to chloroethane and hydrogen chloride.

The difference between the energy put in to break the bonds and the energy given out to form bonds is the approximate enthalpy change of the reaction.

The difference is 3190 − 3068 = 122 kJ mol^{-1}.

4   Finally work out the sign of the enthalpy change. If more energy was put in than was given out, the enthalpy change is positive (the reaction is endothermic). If more energy was given out than was put in the enthalpy change is negative (the reaction is exothermic).

In this case, more enthalpy is given out than put in, so the reaction is exothermic and $\Delta H = -122\,\text{kJ mol}^{-1}$

Note that in practice it would be impossible for the reaction to happen like this. However, Hess's law tells us that you will get the same answer whatever route you take, real or theoretical.

A shortcut

You can often shorten mean bond enthalpy calculations:

$$H-\overset{\overset{H}{|}}{\underset{\underset{H}{|}}{C}}-\overset{\overset{H}{|}}{\underset{\underset{H}{|}}{C}}-H(g) + Cl-Cl(g) \longrightarrow H-\overset{\overset{H}{|}}{\underset{\underset{H}{|}}{C}}-\overset{\overset{H}{|}}{\underset{\underset{H}{|}}{C}}-Cl(g) + H-Cl(g)$$

Only the bonds drawn in red make or break during the reaction so you only need to break: $1 \times$ C—H $= 413\,kJ\,mol^{-1}$

$$1 \times \text{Cl—Cl} = 243\,kJ\,mol^{-1}$$

Total energy put in $= \mathbf{656\,kJ\,mol^{-1}}$

You only need to make: $1 \times$ C—Cl $= 346\,kJ\,mol^{-1}$

$$1 \times \text{H—Cl} = 432\,kJ\,mol^{-1}$$

Total energy given out $= \mathbf{778\,kJ\,mol^{-1}}$

The difference is $\quad\quad 778 - 656 = 122\,kJ\,mol^{-1}$

More energy is given out than taken in so

$$\Delta H = -122\,kJ\,mol^{-1} \text{ (as before)}$$

Comparing the result with that from a thermochemical cycle

This is only an approximate value. This is because the bond enthalpies are averages whereas in a compound any bond has a specific value for its enthalpy. You can find an accurate value for $\Delta H^{\ominus}$ by using a thermochemical cycle as shown here:

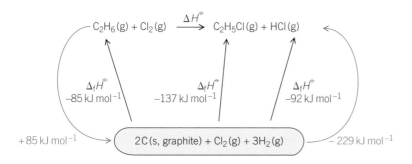

Remember $Cl_2(g)$ is an element so its $\Delta_f H^{\ominus}$ is zero.

$$\Delta H^{\ominus} = 85 - 229\,kJ\,mol^{-1}$$

$\Delta H^{\ominus} = -144\,kJ\,mol^{-1}$ (compared with $-122\,kJ\,mol^{-1}$ calculated from bond enthalpies)

This difference is typical of what might be expected using mean bond enthalpies. The answer obtained from the thermochemical cycle is the 'correct' one because all the $\Delta_f H^{\ominus}$ values have been obtained from the actual compounds involved.

Mean bond enthalpy calculations also allow us to calculate an approximate value for $\Delta_f H$ for a compound that has never been made.

Synoptic link

Bond enthalpies give a measure of the strength of bonds, and can help to predict which bond in a molecule is most likely to break. However, this is not the only factor, the polarity of the bond is also important – see Topic 3.5, Forces acting between molecules, and Topic 13.2, Nucleophilic substitution in halogenoalkanes.

Summary questions

These questions are about the reaction:

$$CH_3CH_3 + Br_2 \rightarrow CH_3CH_2Br + HBr$$

1 Draw out the displayed structural formulae of all the products and reactants so that all the bonds are shown.

2 a Identify the bonds that have to be broken to convert the reactants into separate atoms.

   b How much energy does this take?

3 a Identify the bonds that have to be made to convert separate atoms into the products.

   b How much energy does this take?

4 Describe what the difference is between the energy put in to break bonds and the energy given out when the new bonds are formed.

5 a State what is $\Delta H^{\ominus}$ for the reaction (this requires a sign).

   b Identify if the reaction in part **a** is endothermic or exothermic.

Practice questions

1   A student used Hess's Law to determine a value for the enthalpy change that occurs when anhydrous copper(II) sulfate is hydrated.
This enthalpy change was labelled ΔH_{exp} by the student in a scheme of reactions.

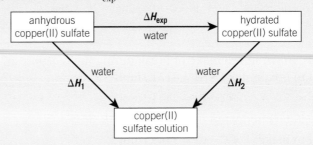

(a)   State Hess's Law.   *(1 mark)*

(b)   Write a mathematical expression to show how ΔH_{exp}, ΔH_1, and ΔH_2 are related to each other by Hess's Law.   *(1 mark)*

(c)   Use the mathematical expression that you have written in part **(b)**, and the data book values for the two enthalpy changes ΔH_1 and ΔH_2 shown, to calculate a value for ΔH_{exp}

$$\Delta H_1 = -156 \text{ kJ mol}^{-1}$$

$$\Delta H_2 = +12 \text{ kJ mol}^{-1}$$

(d)   The student added 0.0210 mol of pure anhydrous copper(II) sulfate to 25.0 cm³ of deionised water in an open polystyrene cup. An exothermic reaction occurred and the temperature of the water increased by 14.0 °C.

(i)   Use these data to calculate the enthalpy change, in kJ mol⁻¹, for this reaction of copper(II) sulfate. This is the student value for ΔH_1

In this experiment, you should assume that all of the heat released is used to raise the temperature of the 25.0 g of water. The specific heat capacity of water is 4.18 J K⁻¹ g⁻¹.   *(3 marks)*

(ii) Suggest **one** reason why the student value for ΔH_1 calculated in part **(d) (i)** is less accurate than the data book value given in part **(c)**.   *(1 mark)*

(e)   Suggest **one** reason why the value for ΔH_{exp} **cannot** be measured directly.

AQA, 2013

2   Hydrazine, N_2H_4, decomposes in an exothermic reaction. Hydrazine also reacts exothermically with hydrogen peroxide when used as a rocket fuel.

(a)   Write an equation for the decomposition of hydrazine into ammonia and nitrogen only.   *(1 mark)*

(b)   State the meaning of the term mean bond enthalpy.   *(2 marks)*

(c)   Some mean bond enthalpies are given in the table.

Mean bond enthalpy/ kJ mol⁻¹	N—H	N—N	N≡N	O—H	O—O
	388	163	944	463	146

Use these data to calculate the enthalpy change for the gas-phase reaction between hydrazine and hydrogen peroxide.

$$\underset{H}{\overset{H}{N}}=\underset{H}{\overset{H}{N}} \quad + \; 2 \; H—O—O—H \longrightarrow N\equiv N \; + \; 4 \; H—O—H$$

AQA, 2013

3 Hess's Law is used to calculate the enthalpy change in reactions for which it is difficult to determine a value experimentally.

(a) State the meaning of the term enthalpy change.

(1 mark)

(b) State Hess's Law.

(1 mark)

(c) Consider the following table of data and the scheme of reactions.

Reaction	Enthalpy change/kJ mol^{-1}
$HCl(g) \rightarrow H^+(aq) + Cl^-(aq)$	-75
$H(g) + Cl(g) \rightarrow HCl(g)$	-432
$H(g) + Cl(g) \rightarrow H^+(g) + Cl^-(g)$	$+963$

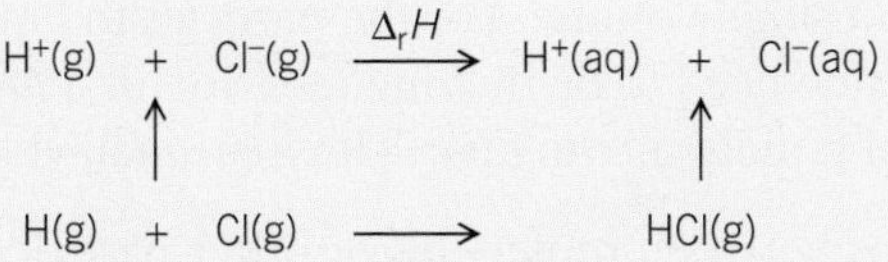

$$H^+(g) \quad + \quad Cl^-(g) \quad \xrightarrow{\Delta_r H} \quad H^+(aq) \quad + \quad Cl^-(aq)$$
$$\uparrow \qquad\qquad\qquad\qquad\qquad\qquad \uparrow$$
$$H(g) \quad + \quad Cl(g) \quad \longrightarrow \qquad HCl(g)$$

Use the data in the table, the scheme of reactions, and Hess's Law to calculate a value for $\Delta_r H$

(3 marks)

AQA, 2010

4 (a) Define the term *standard enthalpy of combustion*, $\Delta_c H^{\ominus}$.

(3 marks)

(b) Use the mean bond enthalpy data from the table and the equation given below to calculate a value for the standard enthalpy of combustion of propene. All substances are in the gaseous state.

Bond	C=C	C—C	C—H	O=O	O=C	O—H
Mean bond enthalpy/ kJ mol^{-1}	612	348	412	496	743	463

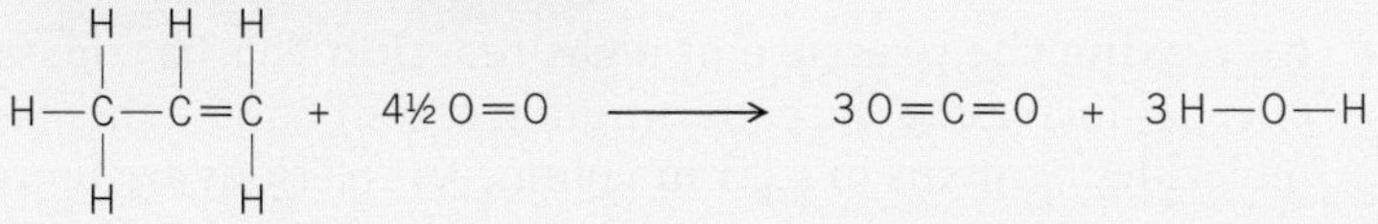

$$H-\overset{\overset{\displaystyle H}{|}}{\underset{\underset{\displaystyle H}{|}}{C}}-\overset{\overset{\displaystyle H}{|}}{C}=\overset{\overset{\displaystyle H}{|}}{\underset{\underset{\displaystyle H}{|}}{C} } \quad + \quad 4\tfrac{1}{2}\,O=O \quad \longrightarrow \quad 3\,O=C=O \quad + \quad 3\,H-O-H$$

(3 marks)

(c) State why the standard enthalpy of formation, $\Delta_c H^{\ominus}$, of oxygen is zero.

(1 mark)

(d) Use the data from the table below to calculate a more accurate value for the standard enthalpy of combustion of propene.

Compound	$C_3H_6(g)$	$CO_2(g)$	$H_2O(g)$
Standard enthalpy of formation, $\Delta_f H^{\ominus}$ / kJ mol^{-1}	$+20$	-394	-242

(3 marks)

(e) Explain why your answer to part (b) is a less accurate value than your answer to part (d).

(2 marks)

AQA, 2006

Learning objectives:

→ Describe what must happen before a reaction will take place.

→ Explain why all collisions do not result in a reaction.

Specification reference: 3.1.5

> **Hint**
>
> A rough rule for many chemical reactions is that if the temperature goes up by 10 K (10 °C), the rate of reaction approximately doubles.

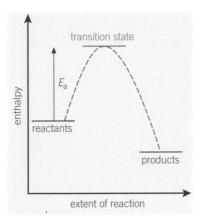

▲ **Figure 1** *An exothermic reaction with a large activation energy, E_a*

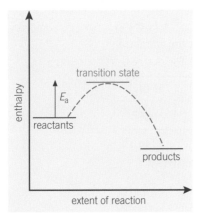

▲ **Figure 2** *An exothermic reaction with a small activation energy, E_a*

Kinetics is the study of the factors that affect rates of chemical reactions – how quickly they take place. There is a large variation in reaction rates. 'Popping' a test tube full of hydrogen is over in a fraction of a second, whilst the complete rusting away of an iron nail could take several years. Reactions can be sped up or slowed down by changing the conditions.

Collision theory

For a reaction to take place between two particles, they must collide with enough energy to break bonds. The collision must also take place between the parts of the molecule that are going to react together, so orientation also has a part to play. To get a lot of collisions you need a lot of particles in a small volume. For the particles to have enough energy to break bonds they need to be moving fast. So, for a fast reaction rate you need plenty of rapidly moving particles in a small volume.

Most collisions between molecules or other particles do not lead to reaction. They either do not have enough energy, or they are in the wrong orientation.

Factors that affect the rate of chemical reactions

The following factors will increase the rate of a reaction.

- **Increasing the temperature** This increases the speed of the molecules, which in turn increases both their energy and the number of collisions.

- **Increasing the concentration of a solution** If there are more particles present in a given volume then collisions are more likely and the reaction rate would be faster. However, as a reaction proceeds, the reactants are used up and their concentration falls. So, in most reactions the rate of reaction drops as the reaction goes on.

- **Increasing the pressure of a gas reaction** This has the same effect as increasing the concentration of a solution – there are more molecules or atoms in a given volume so collisions are more likely.

- **Increasing the surface area of solid reactants** The greater the *total* surface area of a solid, the more of its particles are available to collide with molecules in a gas or a liquid. This means that breaking a solid lump into smaller pieces increases the rate of its reaction because there are more sites for reaction.

- **Using a catalyst** A catalyst is a substance that can change the rate of a chemical reaction without being chemically changed itself.

Activation energy

Only a very small proportion of collisions actually result in a reaction.

For a collision to result in a reaction, the molecules must have a certain minimum energy, enough to start breaking bonds. The minimum energy needed to start a reaction is called the **activation energy** and has the abbreviation E_a.

You can include the idea of activation energy on an enthalpy diagram that shows the course of a reaction.

Exothermic reactions

Figure 1 shows the reaction profile for an exothermic reaction with a large activation energy. This reaction will take place extremely slowly at room temperature because very few collisions will have sufficient energy to bring about a reaction.

Figure 2 shows the reaction profile for an exothermic reaction with a small activation energy. This reaction will take place rapidly at room temperature because many collisions will have enough energy to bring about a reaction.

The situation is a little like a ball on a hill, see Figure 3. A small amount of energy is needed in Figure 3a, to set the ball rolling, whilst a large amount of energy is needed in Figure 3b.

The species that exists at the top of the curve of an enthalpy diagram is called a **transition state** or **activated complex**. Some bonds are in the process of being made and some bonds are in the process of being broken. Like the ball at the very top of the hill, it has extra energy and is unstable.

Endothermic reactions

Endothermic reactions are those in which the products have more energy than the reactants. An endothermic reaction, with activation energy E_a, is shown in Figure 4. The transition state has been labelled.

Notice that the activation energy is measured from the reactants to the top of the curve.

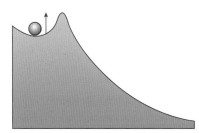

a with low activation energy

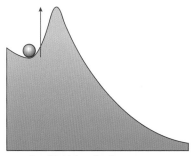

b with high activation energy

▲ **Figure 3** *Ball on a mountainside models*

Hint

Species is a term used by chemists to refer to an atom, molecule, or ion.

Summary questions

1 List five factors that affect the speed of a chemical reaction.

Use the reaction profile in the figure below to answer questions **2** and **3**:

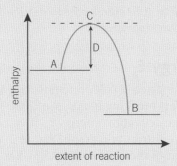

extent of reaction

2 **a** What is A?

b What is B?

c What is C?

d What is D?

3 **a** Identify whether the enthalpy profile represents an endothermic or an exothermic reaction.

b Explain your answer to part **a**.

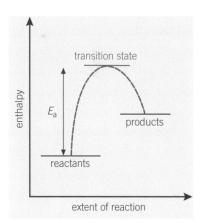

▲ **Figure 4** *An endothermic reaction with activation energy E_a*

Learning objectives:

→ Define activation energy.

→ Explain how temperature affects the number of molecules with energy equal to or more than the activation energy.

→ Explain why a small increase in temperature has a large effect on the rate of a reaction.

Specification reference: 3.1.5

The particles in any gas (or solution) are all moving at different speeds – a few are moving slowly, a few very fast but most are somewhere in the middle. The energy of a particle depends on its speed so the particles also have a range of energies. If you plot a graph of energy against the fraction of particles that have that energy, you end up with the curve shown in Figure 1. This particular shape is called the **Maxwell–Boltzmann distribution** – it tells us about the distribution of energy amongst the particles.

- No particles have zero energy.
- Most particles have intermediate energies – around the peak of the curve.
- A few have very high energies (the right-hand side of the curve). In fact there is no upper limit.
- Note also that the average energy is not the same as the most probable energy.

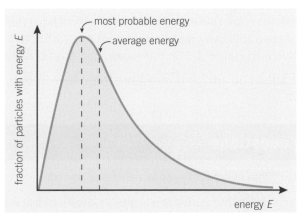

▲ **Figure 1**   *The distribution of the energies of particles. The area under the graph represents the total number of particles*

Activation energy E_a

For a reaction to take place, a collision between particles must have enough energy to start breaking bonds, see Topic 5.1. This amount of energy is called the activation energy E_a. If you mark E_a on the Maxwell–Boltzmann distribution graph, Figure 2, then the area under the graph to the right of the activation energy line represents the number of particles with enough energy to react.

The need for the activation energy to be present before a reaction takes place explains why not all reactions that are exothermic occur spontaneously at room temperature.

For example, fuels are mostly safe at room temperature, as in a petrol station. But a small spark may provide enough energy to start the combustion reaction. The heat given out by the initial reaction is enough to supply the activation energy for further reactions. Similarly the chemicals in a match head are quite stable until the activation energy is provided by friction.

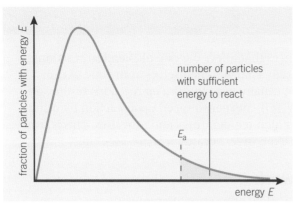

▲ **Figure 2** *Only particles with energy greater than E_a can react*

Even the high temperature of a single spark can set off a reaction. This is why if you smell gas, you must not even turn on a light. The electrical connection provided by the switch could produce enough energy to begin an explosion.

The effect of temperature on reaction rate

The shape of the Maxwell–Boltzmann graph changes with temperature, as shown in Figure 3.

At higher temperatures the peak of the curve is lower and moves to the right. The number of particles with very high energy increases. The total area under the curve is *the same* for each temperature because it represents the total number of particles.

The shaded areas to the right of the E_a line represent the number of molecules that have greater energy than E_a at each temperature.

The graphs show that at higher temperatures more of the molecules have energy greater than E_a so a higher percentage of collisions will result in reaction. This is why reaction rates increase with temperature. In fact, a small increase in temperature produces a large increase in the number of particles with energy greater than E_a.

Also, the total *number* of collisions in a given time increases a little as the particles move faster. However, this is not as important to the rate of reaction as the increase in the number of *effective* collisions (those with energy greater than E_a).

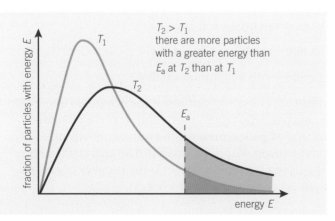

▲ **Figure 3** *The Maxwell–Boltzmann distribution of the energies of the same number of particles at two temperatures*

Summary questions

1 Use Figure 4 to answer the following questions:

  a What is the axis labelled A?

  b What is the axis labelled B?

  c What does area C represent?

  d If the temperature is increased, what happens to the peak of the curve?

  e If the temperature is increased, what happens to E_a?

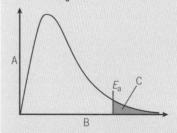

▲ **Figure 4** *The Maxwell–Boltzmann distribution of energies of particles at a particular temperature, with the activation energy, E_a marked*

Learning objectives:

→ State the definition of a catalyst.

→ Describe how a catalyst affects activation energy.

→ Describe how a catalyst affects enthalpy change.

Specification reference: 3.1.5

Catalysts are substances that affect the rate of chemical reactions without being chemically changed themselves at the end of the reaction. Catalysts are usually used to *speed up* reactions so they are important in industry. It is cheaper to speed up a reaction by using a catalyst than by using high temperatures and pressures. This is true, even if the catalyst is expensive, because it is not used up.

How catalysts work

Catalysts work because they provide a different pathway for the reaction, one with a lower activation energy. Therefore they reduce the activation energy of the reaction (the minimum amount of energy that is needed to start the reaction). You can see this on the enthalpy diagrams in Figure 1.

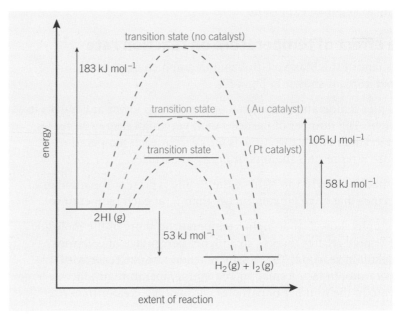

▲ **Figure 1** *The decomposition of hydrogen iodide with different catalysts*

For example, for the decomposition of hydrogen iodide:

$$2HI(g) \rightarrow H_2(g) + I_2(g)$$

$E_a = 183$ kJ mol^{-1} (without a catalyst)

$E_a = 105$ kJ mol^{-1} (with a gold catalyst)

$E_a = 58$ kJ mol^{-1} (with a platinum catalyst)

You can see what happens when you lower the activation energy if you look at the Maxwell–Boltzmann distribution curve in Figure 2. The area that is shaded pink represents the number of effective collisions that can happen without a catalyst. The area shaded blue, plus the area that is shaded pink, represents the number of effective collisions that can takes place with a catalyst.

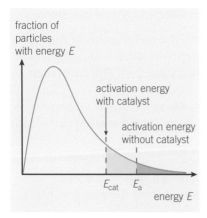

▲ **Figure 2** *With a catalyst the extra particles in the blue area react*

Catalysts do not affect the enthalpy change of the reactions, nor do they affect the position of equilibrium in a reversible reaction, see Topic 6.1.

▼ **Table 1** *Examples of catalysts*

Reaction	Catalyst	Use
$N_2(g) + 3H_2(g) \rightarrow 2NH_3(g)$ Haber process	iron	making fertilisers
$4NH_3 + 5O_2 \rightarrow 4NO + 6H_2O$ Ostwald process for making nitric acid	platinum and rhodium	making fertilisers and explosives
$H_2C{=}CH_2 + H_2 \rightarrow CH_3CH_3$ hardening of fats with hydrogen	nickel	making margarine
cracking hydrocarbon chains from crude oil	aluminium oxide and silicon dioxide zeolite	making petrol
catalytic converter reactions in car exhausts	platinum and rhodium	removing polluting gases
$H_2C{=}CH_2 + H_2O \rightarrow CH_3CH_2OH$ hydration of ethene to produce ethanol	H^+ absorbed on solid silica phosphoric acid, H_3PO_4	making ethanol – a fuel additive, solvent, and chemical feedstock
$CH_3CO_2H(l) + CH_3OH(l) \rightarrow CH_3CO_2CH_3(aq) + H_2O(l)$ esterification	H^+	making solvents

Different catalysts work in different ways – most were discovered by trial and error.

Catalytic converters

All new petrol-engine cars are now equipped with catalytic converters in their exhaust systems. These reduce the levels of a number of polluting gases.

The catalytic converter is a honeycomb, made of a ceramic material coated with platinum and rhodium metals – the catalysts. The honeycomb shape provides an enormous surface area, on which the reactions take place, so a little of these expensive metals goes a long way.

As they pass over the catalyst, the polluting gases react with each other to form less harmful products by the following reactions:

carbon monoxide + nitrogen oxides → nitrogen + carbon dioxide

hydrocarbons + nitrogen oxides → nitrogen + carbon dioxide + water

Synoptic link

You will learn more about catalytic converters in Topic 12.4, Combustion of alkanes.

The reactions take place on the surface of the catalyst in two steps:

1   The gases first form weak bonds with the metal atoms of the catalyst – this process is called **adsorption**. This holds the gases in just the right position for them to react together. The gases then react on the surface.

2   The products then break away from the metal atoms – this process is called **desorption**. This frees up room on the catalyst surface for more gases to take their place and react.

The strength of the weak bonds holding the gases onto the metal surface is critical. They must be strong enough to hold the gases for long enough to react, but weak enough to release the products easily.

Zeolites

Zeolites are *minerals* that have a very open pore structure that ions or molecules can fit into. Zeolites confine molecules in small spaces, which causes changes in their structure and reactivity. More than 150 zeolite types have been synthesized and 48 naturally occurring zeolites are known. Synthetic zeolites are widely used as catalysts in the petrochemical industry.

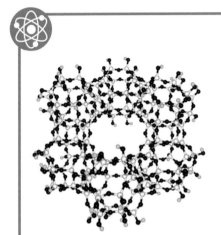

▲ **Figure 3** *Part of the structure of a synthetic zeolite*

Hardening fats

Unsaturated fats, used in margarines for example, are made more solid or hardened when hydrogen is added across some of the double bonds. This is done by bubbling hydrogen into the liquid fat which has a nickel catalyst mixed with it. The nickel is filtered off after the reaction. This allows the manufacturer to tailor the spreadability of the margarine.

▲ **Figure 4** *Margarine*

Catalysts and the ozone layer

Until recently, a group of apparently unreactive compounds called chlorofluorocarbons (CFCs) were used for a number of applications such as solvents, aerosol propellants, and in expanded polystyrene foams. They escaped high into the atmosphere where they remain because they are so relatively unreactive. This is partly due to the strength of the carbon–halogen bonds.

CFCs do eventually decompose to produce separate chlorine atoms. These act as catalysts in reactions that bring about the destruction of ozone, O_3. Ozone is important because it forms a layer in the atmosphere of the Earth that acts as a shield. The layer prevents too much ultraviolet radiation from reaching the Earth's surface.

The overall reaction is shown below:

$$O_3(g) + O(g) \xrightarrow{\text{chlorine atom catalyst}} 2O_2(g)$$

Nitrogen monoxide acts as a catalyst in a similar way to chlorine atoms.

International agreements, such as the 1987 Montreal Protocol, have resulted in CFCs being phased out. Unfortunately there is still a reservoir of them remaining from before these agreements. Chemists have developed, and continue to work on, suitable substitutes for CFCs that do not result in damage to the upper atmosphere. These include hydrochlorofluorocarbons and hydrofluorocarbons. Former United Nations Secretary General, Kofi Annan, has referred to the Montreal Protocol as "perhaps the single most successful international agreement to date".

Summary questions

1 The following questions refer to Figure 5.

   a What are labels A, B, C, R, and P?

   b What do the distances from D to R and from C to R represent?

   c Is the reaction exothermic or endothermic?

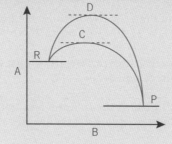

▲ **Figure 5** *A profile for a reaction with and without a catalyst*

1   The gas-phase reaction between hydrogen and chlorine is very slow at room temperature.

$$H_2(g) + Cl_2(g) \rightarrow 2HCl(g)$$

(a) Define the term *activation energy*.

(2 marks)

(b) Give **one** reason why the reaction between hydrogen and chlorine is very slow at room temperature.

(1 mark)

(c) Explain why an increase in pressure, at constant temperature, increases the rate of reaction between hydrogen and chlorine.

(2 marks)

(d) Explain why a small increase in temperature can lead to a large increase in the rate of reaction between hydrogen and chlorine.

(2 marks)

(e) Give the meaning of the term *catalyst*.

(1 mark)

(f) Suggest **one** reason why a solid catalyst for a gas-phase reaction is often in the form of a powder.

(1 mark)

AQA, 2006

2   The diagram below represents a Maxwell–Boltzmann distribution curve for the particles in a sample of a gas at a given temperature. The questions below refer to this sample of particles.

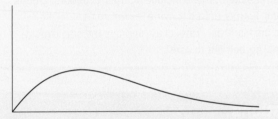

(a) Label the axes on a copy of the diagram.

(2 marks)

(b) On the diagram draw a curve to show the distribution for this sample at a **lower** temperature.

(2 marks)

(c) In order for two particles to react they must collide. Explain why most collisions do not result in a reaction.

(1 mark)

(d) State one way in which the collision frequency between particles in a gas can be increased without changing the temperature.

(1 mark)

(e) Suggest why a small increase in temperature can lead to a large increase in the reaction rate between colliding particles.

(2 marks)

(f) Explain in general terms how a catalyst works.

(2 marks)

AQA, 2004

3 The diagram shows the Maxwell–Boltzmann distribution of molecular energies in a gas at two different temperatures.

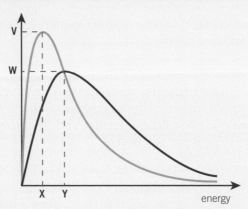

(a) One of the axes is labelled. Complete the diagram by labelling the other axis.

(1 mark)

(b) State the effect, if any, of a solid catalyst on the shape of either of these distributions.

(1 mark)

(c) State the letter, **V**, **W**, **X**, or **Y**, that represents the most probable energy of the molecules at the lower temperature.

(1 mark)

(d) Explain what must happen for a reaction to occur between molecules of two different gases.

(2 marks)

(e) Explain why a small increase in temperature has a large effect on the initial rate of a reaction.

(1 mark)

AQA, 2012

4 The diagram shows the Maxwell–Boltzmann distribution for a sample of gas at a fixed temperature.

E_a is the activation energy for the decomposition of this gas.

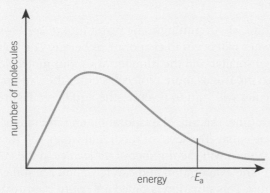

E_{mp} is the most probable value for the energy of the molecules.

(a) On the appropriate axis of this diagram, mark the value of E_{mp} for **this** distribution. On this diagram, sketch a new distribution for the same sample of gas at a **lower** temperature.

(3 marks)

(b) With reference to the Maxwell–Boltzmann distribution, explain why a decrease in temperature decreases the rate of decomposition of this gas.

(2 marks)

AQA, 2013

Learning objectives:
→ State the definition of a reversible reaction.
→ State what is meant by chemical equilibrium.
→ Explain why all reactions do not go to completion.
→ Explain what happens when equilibrium has been reached.

Specification reference: 3.1.5

Chemists usually think of a reaction as starting with the reactants and ending with the products.

$$reactants \rightarrow products$$

However, some reactions are reversible. For example, when you heat blue hydrated copper sulfate it becomes white anhydrous copper sulfate as the water of crystallisation is driven off. The white copper sulfate returns to blue if you add water.

$$CuSO_4.5H_2O \rightleftharpoons CuSO_4 + 5H_2O$$

blue hydrated white anhydrous
copper sulfate copper sulfate

However, something different would happen if you were to do this reaction in a closed container. As soon as the products are formed they react together and form the reactants again, so that instead of reactants *or* products you get a mixture of both. Eventually you get a mixture in which the proportions of all three components remain constant. This mixture is called an **equilibrium mixture**.

Setting up an equilibrium

You can understand how an equilibrium mixture is set up by thinking about what happens in a physical process, like the evaporation of water. This is easier to picture than a chemical change.

First imagine a puddle of water out in the open. Some of the water molecules at the surface will move fast enough to escape from the liquid and evaporate. Evaporation will continue until all the water is gone.

But think about putting some water into a *closed* container. At first the water will begin to evaporate as before. The volume of the liquid will get smaller and the number of vapour molecules in the gas phase will go up. But as more molecules enter the vapour, some gas-phase molecules will start to re-enter the liquid, see Figure 1.

After a time, the rate of evaporation and the rate of condensation will become equal. The level of the liquid water will then stay exactly the same and so will the number of molecules in the vapour and in the liquid. The evaporation and condensation are still going on but *at the same rate*. This situation is called a **dynamic equilibrium** and is one of the key ideas of this topic.

In fact, you could have started by filling the empty container with the same mass of water vapour as you originally had liquid water. The vapour would begin to condense and, in time, would reach exactly the same equilibrium position.

a

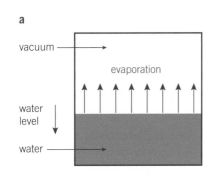

b

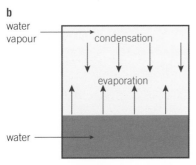

▲ **Figure 1 a** *Water will evaporate into an empty container. Eventually the rates of evaporation and condensation will be the same*
b *Equilibrium is set up*

The conditions for equilibrium

Although the system used here is very simple, you can pick out four conditions that apply to *all* equilibria:

- Equilibrium can only be reached in a **closed system** (one where the reactants and products can't escape). The system does not have to be sealed. For example, a beaker may be a closed system for a reaction that takes place in a solvent, as long as the reactants, products, and solvent do not evaporate.

- Equilibrium can be approached from *either direction* (in figure 1, from liquid or from vapour) and the final equilibrium position will be the same (as long as conditions, such as temperature and pressure, stay the same).

- Equilibrium is a dynamic process. It is reached when the *rates* of two opposing processes, which are going on all the time (in figure 1, evaporation and condensation), *are the same*.

- You know that equilibrium has been reached when the macroscopic properties of the system do not change with time. These are properties like density, concentration, colour, and pressure – properties that do not depend on the total quantity of matter.

A reversible reaction that can reach equilibrium is denoted by the symbol ⇌, for example:

liquid water ⇌ water vapour

$$H_2O(l) \rightleftharpoons H_2O(g)$$

Chemical equilibria

The same principles that you have found for a physical change also apply to chemical equilibria such as:

$$A + B \rightleftharpoons C + D$$
reactants products

- Imagine starting with A and B only. At the start of the reaction the forward rate is fast, because A and B are plentiful. There is no reverse reaction because there is no C and D.

- Then as the concentrations of C and D build up, the reverse reaction speeds up. At the same time the concentrations of A and B decrease so the forward reaction slows down.

- A point is reached where exactly the same number of particles are changing from A + B to C + D as are changing from C + D to A + B. Equilibrium has been reached.

One important point to remember is that an equilibrium mixture can have *any* proportions of reactants and products. It is not necessarily half reactants and half products, though it could be. The proportions may be changed depending on the conditions of the reaction, such as temperature, pressure, and concentration. But at any given constant conditions the proportions of reactants and products do not change.

Summary questions

1 For each of the following statements about all equilibria, say whether it is true or false.

a Once equilibrium is reached the concentrations of the reactants and the products do not change.

b At equilibrium the forward and the backward reactions come to a halt.

c Equilibrium is only reached in a closed system.

d An equilibrium mixture always contains half reactants and half products.

2 What can be said about the rates of the forward and the backward reactions when equilibrium is reached?

Some industrial processes, like the production of ammonia or sulfuric acid, have reversible reactions as a key step. In closed systems these reactions would produce equilibrium mixtures containing both products and reactants. In principle, you would like to increase the proportion of products. For this reason it is important to understand how to control equilibrium reactions.

The equilibrium mixture

It is possible to change the proportion of reactants to products in an equilibrium mixture. In this way you are able to obtain a greater yield of the products. This is called changing the *position* of equilibrium.

- If the proportion of products in the equilibrium mixture is increased, the equilibrium is moved to the right, or in the forward direction.

- If the proportion of reactants in the equilibrium mixture is increased, the equilibrium is moved to the left, or in the backward direction.

You can often move the equilibrium position to the left or right by varying conditions like temperature, the concentration of species involved, or the pressure (in the case of reactions involving gases).

Le Châtelier's principle

Le Châtelier's principle is useful because it gives us a rule. It tells us whether the equilibrium moves to the right or to the left when the conditions of an equilibrium mixture are changed.

It states:

If a system at equilibrium is disturbed, the equilibrium moves in the direction that tends to reduce the disturbance.

So in other words, if any factor is changed which affects the equilibrium mixture, the position of equilibrium will shift so as to oppose the change.

Le Châtelier's principle does not tell us *how far* the equilibrium moves so you cannot predict the *quantities* involved.

Changing concentrations

If you *increase* the concentration of one of the reactants, Le Châtelier's principle says that the equilibrium will shift in the direction that tends to *reduce* the concentration of this reactant. Look at the reaction:

$$A(aq) + B(aq) \rightleftharpoons C(aq) + D(aq)$$

Suppose you add some extra A. This would increase the concentration of A. The only way that this system can reduce the concentration of A, is by some of A reacting with B (so forming more C and D). So, adding more A uses up more B, produces more C and D, and moves the equilibrium to the right. You end up with a greater proportion of

▲ **Figure 1** *Henri-Louis Le Châtelier was a French chemist who first put forward his 'Loi de stabilité déquilibre chimique' in 1884*

products in the reaction mixture than before you added A. The same thing would happen if you added more B.

You could also remove C as it was formed. The equilibrium would move to the right to produce more C (and D) using up A and B. The same thing would happen if you removed D as soon as it was formed.

Changing the overall pressure

Pressure changes only affect reactions involving gases. Changing the overall pressure will only change the position of equilibrium of a gaseous reaction if there are a different number of molecules on either side of the equation.

An example of a such a reaction is:

$$N_2O_4(g) \rightleftharpoons 2NO_2(g)$$

dinitrogen tetraoxide	nitrogen dioxide
1 mole	2 moles
colourless	brown

Increasing the pressure of a gas means that there are more molecules of it in a given volume – it is equivalent to increasing the concentration of a solution.

If you increase the pressure on this system, Le Châtelier's principle tells us that the position of equilibrium will move to decrease the pressure. This means that it will move to the left because fewer molecules exert less pressure. In the same way if you decrease the pressure, the equilibrium will move to the right – molecules of N_2O_4 will decompose to form molecules of NO_2, thereby increasing the pressure.

Dinitrogen tetraoxide is a colourless gas and nitrogen dioxide is brown. You can investigate this in the laboratory, by setting up the equilibrium mixture in a syringe. If you decrease the pressure, by pulling out the syringe barrel, you can watch as the equilibrium moves to the right because the colour of the mixture gets browner, see Figure 2.

Note that if there is the same number of moles of gases on both sides of the equation, then pressure has no effect on the equilibrium position. For example:

$$H_2(g) + I_2(g) \rightleftharpoons 2HI(g)$$

2 moles	2 moles

The equilibrium position will not change in this reaction when the pressure is changed so the proportions of the three gases will stay the same.

Changing temperature

Reversible reactions that are exothermic (give out heat) in one direction are endothermic (take in heat) in the other direction, see Topic 4.4. The size of the enthalpy is the same in both directions but the sign changes.

Example 1

Suppose you increase the temperature of an equilibrium mixture that is exothermic in the forward direction. An example is:

$$2SO_2(g) + O_2(g) \rightleftharpoons 2SO_3(g) \qquad \Delta H^\ominus = -197 \text{ kJ mol}^{-1}$$

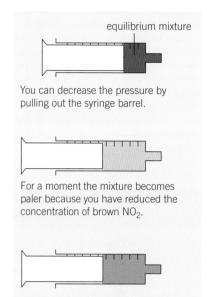

equilibrium mixture

You can decrease the pressure by pulling out the syringe barrel.

For a moment the mixture becomes paler because you have reduced the concentration of brown NO_2.

After a few moments the mixture becomes darker brown as the equilibrium moves to the right and more brown NO_2 is formed.

▲ **Figure 2** $N_2O_4(g) \rightleftharpoons 2NO_2(g)$ *The equilibrium moves to the right as you decrease the pressure*

Hint

Increasing the pressure or decreasing the volume of a mixture of gases increases the concentration of all the reactants and products by the same amount, not just one of them.

Hint

The *rate* at which equilibrium is reached *will* be speeded up by increasing the pressure, as there will be more collisions in a given time.

Study tips

- The term move forwards and move to the right mean the same thing in this context.

- The term move backwards and move to the left mean the same thing in this context.

The negative sign of $\Delta H^{\ominus}$ means that heat is given out when sulfur dioxide and oxygen react to form sulfur trioxide in the forward direction. This means that heat is absorbed as the reaction goes in the reverse direction, that is, to the left.

Le Châtelier's principle tells us that if you increase the temperature, the equilibrium moves in the direction that cools the system down. To do this it will move in the direction which absorbs heat (is endothermic), that is, to the left. The equilibrium mixture will then contain a greater proportion of sulfur dioxide and oxygen than before. In the same way, if we cool the mixture the equilibrium will move to the right and increase the proportion of sulfur trioxide.

Example 2

The effect of temperature on the dinitrogen tetraoxide/nitrogen dioxide equilibrium can also be investigated using the same apparatus you used to investigate the effect of pressure on this reaction. The reaction is endothermic as it proceeds from dinitrogen tetraoxide to nitrogen dioxide (the forward direction).

$$N_2O_4(g) \rightleftharpoons 2NO_2(g) \qquad \Delta H^{\ominus} = +58 \text{ kJ mol}^{-1}$$

The gas mixture is contained in a syringe as before. The syringe is then immersed in warm water along with another syringe containing the same volume of air for comparison. The plunger of the syringe containing air will rise as the air expands. The plunger of the syringe containing the N_2O_4 / NO_2 mixture will also rise but by a greater amount. This indicates that more molecules of gas have been formed in this syringe. This is because the equilibrium has moved to the right; each molecule of N_2O_4 that disappears produces two molecules of NO_2. This is consistent with Le Châtelier's principle. When the mixture is warmed up, the equilibrium moves in the endothermic direction, that is, it absorbs heat which tends to cool the mixture down.

You should be able to predict the colour change that you would see during this experiment and also what would happen if the experiment were repeated in ice water.

Catalysts

Catalysts have no effect on the position of equilibrium so they do not alter composition of the equilibrium mixture. They work by producing an alternative route for the reaction, which has a lower activation energy of the reaction, see Topic 5.3. This affects the forward and back reactions equally.

Although catalysts have no effect on the position of equilibrium, that is, the yield of the reaction, they do allow equilibrium to be reached more quickly and are therefore important in industry.

Summary questions

1 In which of the following reactions will the position of equilibrium be affected by changing the pressure?

   Explain your answers.

   a $2SO_2(g) + O_2(g) \rightleftharpoons 2SO_3(g)$

   b $CH_3CO_2H(aq) \rightleftharpoons CH_3CO_2^-(aq) + H^+(aq)$

   c $H_2(g) + CO_2(g) \rightleftharpoons H_2O(g) + CO(g)$

2 Consider the following equilibrium reaction.

   $N_2(g) + 3H_2(g) \rightleftharpoons 2NH_3(g)$
   $\Delta H^{\ominus} = -92 \text{ kJ mol}^{-1}$

   a What would be the effect on the equilibrium position of heating the reaction? Choose from 'move to the right', 'move to the left' and 'no change'.

   b What would be the effect on the equilibrium position of adding an iron catalyst? Choose from move to the right', 'move to the left' and 'no change'.

   c What effect would an iron catalyst have on the reaction?

   d To get the maximum yield of ammonia in this reaction would a high or low pressure be best? Explain your answer.

A number of industrial processes involve reversible reactions. In these cases, the yield of the reaction is important and Le Châtelier's principle can be used to help find the best conditions for increasing it. However, yield is not the only consideration. Sometimes a low temperature would give the best yield but this would slow the reaction down. The costs of building and running a plant that operates at high temperatures and pressures must also be taken into account. In most cases a compromised set of conditions is used. This topic looks at the industrial production of three important chemicals.

Learning objectives:

→ Explain why compromises are made when deciding how to get the best yield in industry.

Specification reference: 3.1.5

Ammonia, NH_3

Ammonia is an important chemical in industry. World production is over 140 million tonnes each year. Around 80% is used to make fertilisers like ammonium nitrate, ammonium sulfate, and urea. The rest is used to make synthetic fibres (including nylon), dyes, explosives, and plastics like polyurethane.

Making ammonia

Nitrogen and hydrogen react together by a reversible reaction which, at equilibrium, forms a mixture of nitrogen, hydrogen, and ammonia:

$$N_2(g) + 3H_2(g) \rightleftharpoons 2NH_3(g) \qquad \Delta H^\ominus = -92 \text{ kJ mol}^{-1}$$

The percentage of ammonia obtained *at equilibrium* depends on temperature and pressure as shown in Figure 1. The graph shows that low temperature and high pressure would give close to 100% conversion whereas low pressure and high temperature would give almost no ammonia.

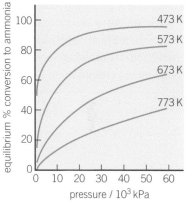

▲ **Figure 1** *Equilibrium % conversion of nitrogen and hydrogen to ammonia under different conditions*

1 Explain how Le Châtelier's principle predicts that the highest conversion to ammonia is obtained at **a** low temperature **b** high pressure.

The Haber process

Almost all ammonia is made by the Haber process, in which the reaction above is the key step. The process was developed by the German chemist Fritz Haber and the chemical engineer Carl Bosch in the early years of the 20th century. It allowed Germany to make explosives and fertilisers. This prolonged the First World War because, at that time, the source of nitrogen for these products was nitrates from South America. These could be blockaded by the navies of Britain and its allies.

The raw materials

The raw materials for the Haber process are air (which provides the nitrogen), water, and natural gas (methane, CH_4). These provide the hydrogen by the following reaction:

$$CH_4(g) + H_2O(g) \rightarrow CO(g) + 3H_2(g)$$

Hint

What is the atom economy (see Topic 2.6) of the reaction that forms ammonia? Why do you not need to do a calculation?

Hint

This reaction can be classified as an addition reaction. Other reactions can be classified as substitution (where one atom or group of atoms are replaced by another) or elimination (where an atom or group of atoms are removed from the starting material). What can you say about the atom economies of these types of reaction?

The nitrogen and hydrogen are fed into a converter in the ratio of 1 : 3 and passed over an iron catalyst.

Most plants run at a pressure of around 20 000 kPa (around 200 atmospheres) and a temperature of about 670 K. This is a lower pressure and a higher temperature than would give the maximum conversion.

2   Use the graph in Figure 1 to find the equilibrium percentage conversion to ammonia at 20 000 kPa and 673 K.

3   Suggest why these compromise conditions are used.

Nitrogen and hydrogen flow continuously over the catalyst so the gases do not spend long enough in contact with the catalyst to reach equilibrium; there is about 15% conversion to ammonia. The ammonia is cooled so that it becomes liquid and is piped off. Any nitrogen and hydrogen that is not converted into ammonia is fed back into the reactor.

The catalyst is iron in pea-sized lumps (to increase the surface area). It lasts about five years before it becomes poisoned by impurities in the gas stream and has to be replaced.

Uses of ammonia

Eighty per cent of ammonia is used to make fertilisers including ammonium sulfate, ammonium nitrate, and urea. In the first two cases, ammonia (an alkali) is reacted with an acid to make a salt. In the case of ammonium nitrate, the acid used is nitric acid, which is itself made from ammonia.

The next largest use of ammonia is in making nylon. Other uses include the manufacture of explosives, drugs, and dyes.

Ethanol, C_2H_5OH

Ethanol is the alcohol in alcoholic drinks and, as such, has been produced by mankind for thousands of years by fermentation from sugars such as glucose using the enzymes in yeast as a catalyst:

$$C_6H_{12}O_6(aq) \rightarrow 2C_2H_5OH(aq) + 2CO_2(g)$$
$$\text{glucose} \qquad \text{ethanol}$$

Ethanol also has many industrial uses, for example, for making cosmetics, drugs, detergents, inks, and as a motor fuel. UK production is around 330 000 tonnes per year. At present, the main source of ethanol for industrial use is ethene from crude oil. This is obtained by fractional distillation and then cracking.

- Ethanol is made by the hydration (adding water) to ethene.

- The reaction is reversible.

- It is speeded up by a catalyst of phosphoric acid absorbed on silica.

- The equation is:

$$H_2C{=}CH_2(g) + H_2O(g) \rightleftharpoons CH_3CH_2OH(g) \qquad \Delta H^{\ominus} = -46 \text{ kJ mol}^{-1}$$
$$\text{ethene} \qquad\qquad\qquad \text{ethanol}$$

- The reactants and products are all gaseous at the temperature used.

Applying Le Châtelier's principle to this equilibrium predicts that the maximum yield will be produced with:

- a high pressure, which will force the equilibrium to move to the right, to the side with fewer molecules

- a low temperature, which will force the equilibrium to move to the right to give out heat

- excess steam, which will force the equilibrium to the right to reduce the steam concentration.

However, there are practical problems:

- High pressure tends to cause the ethene to polymerise (to poly(ethene)).

- High pressure increases the costs of building the plant and the energy costs of running it.

- Low temperature will reduce the reaction rate and therefore how quickly equilibrium is reached, although this is partially compensated for by the use of a catalyst.

- Too much steam dilutes the catalyst.

In practice, conditions of about 570 K and 6500 kPa pressure are used. These give a conversion to ethanol of only about 5% but the unreacted ethene is separated from the reaction mixture and recycled over the catalyst again and again until about 95% conversion is obtained.

Methanol, CH_3OH

Methanol is used principally as a chemical feedstock, that is, as a starting material for making other chemicals. In particular it is used in the manufacture of methanal (formaldehyde) which in turn is used to make plastics such as Bakelite. Methanol is also used in the manufacture of other plastics such as terylene and perspex. Methanol may also be used (alone or added to petrol) as a motor fuel. It was manufactured for this use in Germany during the Second World War when crude oil supplies were limited by bombing. Indycars in the USA run on pure methanol, which has an advantage over petrol because methanol fires can be put out with water. Each year, 33 million tonnes of methanol are made worldwide, mostly from the reversible reaction of hydrogen and carbon monoxide using a copper catalyst:

$$CO(g) + 2H_2(g) \rightleftharpoons CH_3OH(g) \qquad \Delta H^{\ominus} = -91 \text{ kJ mol}^{-1}$$

The starting gas mixture is called synthesis gas and is made by reacting methane or propane with steam.

Le Châtelier's principle tells us that the methanol synthesis reaction will give the highest yield at low temperature and high pressure (as is the case for the ethanol synthesis reaction). But again, compromise conditions are used. In practice a temperature of around 500 K and a pressure of 10 000 kPa produces around 5–10% yield.

Summary questions

1 The platinum catalyst for the oxidation of ammonia to nitric acid is used in the form of a fine gauze. Suggest why it is used in this form.

2 Explain why ethanol produced by fermentation is a renewable resource while ethanol produced from ethene is not.

3 Is methanol made from synthesis gas a renewable resource? Explain your answer.

Learning objectives:

→ Define the expression reversible reaction.

→ Define the term chemical equilibrium.

→ State the definition of an equilibrium constant and describe describe how it is determined.

Specification reference: 3.1.6

Synoptic link

There is more about titrations in Topic 2.5, Balanced equations and relates calculations.

▲ **Figure 1** *Titrating the ethanoic acid to investigate the equilibrium position*

Hint

The concentration of a solution is the number of moles of solute dissolved in 1 dm³ of solution. A square bracket around a formula is shorthand for 'concentration of that substance in mol dm⁻³'.

Study tip √x̄

Make sure that you are able to calculate K_c from given data.

As you have seen, reactions are reversible and do not go to completion, but instead end up as an equilibrium mixture of reactants and products. A reversible reaction that can reach equilibrium is indicated by the symbol ⇌. In this topic you see how you can tackle equilibrium reactions mathematically. You will deal only with homogeneous systems – those where all the reactants and products are in the same phase, for example, all liquids.

The equilibrium constant K_c

Many reactions are reversible and will reach equilibrium with time. The reaction between ethanol, C_2H_5OH, and ethanoic acid, CH_3CO_2H, to produce ethyl ethanoate, $CH_3CO_2C_2H_5$, (an ester) and water is typical.

If ethanol and ethanoic acid are mixed in a flask (stoppered to prevent evaporation) and left for several days with a strong acid catalyst, an equilibrium mixture is obtained in which *all four* substances are present. You can write:

$$C_2H_5OH\ (l)\ +\ CH_3CO_2H\ (l)\ \rightleftharpoons\ CH_3CO_2C_2H_5(l)\ +\ H_2O(l)$$

ethanol   ethanoic acid   ethyl ethanoate   water

The mixture may be analysed by titrating the ethanoic acid with standard alkali (allowing for the amount of acid catalyst added). It is possible to do this without significantly disturbing the equilibrium mixture because the reversible reaction is much slower than the titration reaction.

The titration allows us to work out the number of moles of ethanoic acid in the equilibrium mixture. From this you can calculate the number of moles of the other components (and from this their concentrations if the total volume of the mixture is known).

If several experiments are done with different quantities of starting materials, it is always found that the ratio:

$$\frac{[CH_3CO_2C_2H_5(l)]_{eqm}\ [H_2O(l)]_{eqm}}{[CH_3CO_2H(l)]_{eqm}\ [C_2H_5OH(l)]_{eqm}}$$

has a constant value, provided the experiments are done at the same temperature. The subscript 'eqm' means that the concentrations have been measured when equilibrium has been reached.

For *any* reaction that reaches an equilibrium we can write the equation in the form:

$$a\mathrm{A} + b\mathrm{B} + c\mathrm{C} \rightleftharpoons x\mathrm{X} + y\mathrm{Y} + z\mathrm{Z}$$

Then the expression $\dfrac{[X]_{eqm}^{\ x}[Y]_{eqm}^{\ y}[Z]_{eqm}^{\ z}}{[A]_{eqm}^{\ a}[B]_{eqm}^{\ b}[C]_{eqm}^{\ c}}$ is constant, provided the

temperature is constant. We call this constant, K_c. This expression can be applied to *any* reversible reaction. K_c is called the **equilibrium constant** and is different for different reactions. It changes with temperature. The units of K_c vary, and you must work them out for each reaction by cancelling out the units of each term, for example:

$$2A + B \rightleftharpoons 2C \qquad K_c = \frac{[C]^2}{[A]^2[B]}$$

Units are:
$$\frac{(mol\ dm^{-3})^2}{(mol\ dm^{-3})^2(mol\ dm^{-3})} = \frac{1}{mol\ dm^{-3}} = mol^{-1}dm^3$$

The value of K_c is found by experiment for any particular reaction at a given temperature.

To find the value of K_c for the reaction between ethanol and ethanoic acid

0.10 mol of ethanol is mixed with 0.10 mol of ethanoic acid and allowed to reach equilibrium. The total volume of the system is made up to 20.0 cm³ (0.020 dm³) with water. By titration, it is found that 0.033 mol ethanoic acid is present once equilibrium is reached.

From this you can work out the number of moles of the other components present at equilibrium:

At start

$$C_2H_5OH\ (l)\ +\ CH_3CO_2H\ (l) \rightleftharpoons CH_3CO_2C_2H_5(l)\ +\ H_2O(l)$$
$$\quad 0.10\ mol \qquad\quad 0.10\ mol \qquad\qquad 0\ mol \qquad\qquad 0\ mol$$

You know that there are 0.033 mol of CH_3CO_2H at equilibrium. This means that:

- there must also be 0.033 mol of C_2H_5OH at equilibrium. (The equation tells you that they react 1:1 and you know we started with the same number of moles of each.)
- $(0.10 - 0.033) = 0.067$ mol of CH_3CO_2H has been used up. The equation tells you that when 1 mol of CH_3CO_2H is used up, 1 mol each of $CH_3CO_2C_2H_5$ and H_2O are produced. So, there must be 0.067 mol of each of these.

At equilibrium

$$C_2H_5OH\ (l)\ +\ CH_3CO_2H\ (l) \rightleftharpoons CH_3CO_2C_2H_5(l)\ +\ H_2O(l)$$
$$\quad 0.033\ mol \qquad\quad 0.033\ mol \qquad\qquad 0.067\ mol \qquad\qquad 0.067\ mol$$

You need the *concentrations* of the components at equilibrium. As the volume of the system is 0.020 dm³ these are:

$$C_2H_5OH\ (l)\ +\ CH_3CO_2H\ (l) \rightleftharpoons CH_3CO_2C_2H_5(l)\ +\ H_2O(l)$$
$$\frac{0.033}{0.020}\ mol\ dm^{-3} \quad \frac{0.033}{0.020}\ mol\ dm^{-3} \quad \frac{0.067}{0.020}\ mol\ dm^{-3} \quad \frac{0.067}{0.020}\ mol\ dm^{-3}$$

Enter the concentrations into the equilibrium equation:

$$K_c = \frac{[CH_3CO_2C_2H_5(l)][H_2O(l)]}{[CH_3CO_2H(l)][C_2H_5OH(l)]}$$

$$K_c = \frac{[0.067/0.020\ mol\ dm^{-3}][0.067/0.020\ mol\ dm^{-3}]}{[0.033/0.020\ mol\ dm^{-3}][0.033/0.020\ mol\ dm^{-3}]} = 4.1$$

The units all cancel out, and the volumes (0.020 dm³) cancel out, so in this case you didn't need to know the volume of the system, so $K_c = 4.1$. In this case, K_c has no units.

See Section 5, Mathematical skills, if you are not sure about cancelling units.

Maths link

See Section 5, Mathematical skills, if you are not sure about cancelling units.

Study tip

It is acceptable to omit the 'eqm' subscripts unless they are specifically asked for.

Summary questions

1 Write down the expressions for the equilibrium constant for the following:

   a $A + B \rightleftharpoons C$

   b $2A + B \rightleftharpoons C$

   c $2A + 2B \rightleftharpoons 2C$

2 Work out the units for K_c for question 1a to c.

3 For the reaction between ethanol and ethanoic acid, at a different temperature to the example above, the equilibrium mixture was found to contain 0.117 mol of ethanoic acid, 0.017 mol of ethanol, 0.083 mol ethyl ethanoate and 0.083 mol of water.

   a Calculate K_c

   b Why do you not need to know the volume of the system to calculate K_c in this example?

   c Is the equilibrium further to the right or to the left compared with the worked example above?

Learning objectives:

→ Describe how K_c is used to work out the composition of an equilibrium mixture.

Specification reference: 3.1.6

A reaction that has reached equilibrium at a given temperature will be a mixture of reactants and products. You can use the equilibrium expression to calculate the composition of this mixture.

Worked example: Calculating the composition of a reaction mixture

The reaction of ethanol and ethanoic acid is:

$$C_2H_5OH \, (l) \; + \; CH_3CO_2H \, (l) \; \rightleftharpoons \; CH_3CO_2C_2H_5(l) \; + \; H_2O(l)$$

ethanol ⁣ ethanoic acid ⁣ ethyl ethanoate ⁣ water

You know that at equilibrium:

$$K_c = \frac{[CH_3CO_2C_2H_5(l)][H_2O(l)]}{[CH_3CO_2H(l)][C_2H_5OH(l)]}$$

Suppose that $K_c = 4.0$ at the temperature of our experiment and you want to know how much ethyl ethanoate you could produce by mixing one mol of ethanol and one mol of ethanoic acid. Set out the information as shown below:

Equation: $C_2H_5OH \, (l) + CH_3CO_2H \, (l) \rightleftharpoons CH_3CO_2C_2H_5(l) + H_2O(l)$

	ethanol	ethanoic acid	ethyl ethanoate	water
At start:	1 mol	1 mol	1 mol	1 mol
At equilibrium:	$(1-x)$ mol	$(1-x)$ mol	x mol	x mol

You do not know how many moles of ethyl ethanoate will be produced, so you call this x. The equation tells us that x mol of water will also be produced and, in doing so, x mol of both ethanol and ethanoic acid will be used up. So the amount of each of these remaining at equilibrium is $(1 - x)$ mol.

These figures are in moles, but you need concentrations in mol dm^{-3} to substitute in the equilibrium law expression. Suppose the volume of the system at equilibrium was V dm^{-3}. Then:

$$[C_2H_5OH(l)]_{eqm} = \frac{(1-x)}{V} \, mol \, dm^{-3}$$

$$[CH_3CO_2H(l)]_{eqm} = \frac{(1-x)}{V} \, mol \, dm^{-3}$$

$$[CH_3CO_2C_2H_5(l)]_{eqm} = \frac{x}{V} \, mol \, dm^{-3}$$

$$[H_2O(l)]_{eqm} = \frac{x}{V} \, mol \, dm^{-3}$$

These figures may now be put into the expression for K_c:

$$K_c = \frac{x/\cancel{V} \times x/\cancel{V}}{(1-x)/\cancel{V} \times (1-x)/\cancel{V}}$$

The V's cancel, so *in this case* you do not need to know the actual volume of the system.

$$4.0 = \frac{x \times x}{(1 - x) \times (1 - x)}$$

$$4.0 = \frac{x^2}{(1 - x)^2}$$

Taking the square root of both sides, you get:

$$2 = \frac{x}{(1 - x)}$$

$$2(1 - x) = x$$

$$2 - 2x = x$$

$$2 = 3x$$

$$x = \frac{2}{3}$$

So $\frac{2}{3}$ mol of ethyl ethanoate and $\frac{2}{3}$ mol of water is produced if the reaction reaches equilibrium, and the composition of the equilibrium mixture would be: ethanol $\frac{1}{3}$ mol, ethanoic acid $\frac{1}{3}$ mol, ethyl ethanoate $\frac{2}{3}$ mol, water $\frac{2}{3}$ mol.

Study tip

The volume of the reaction mixture will cancel for all systems with equal numbers of moles of products and reactants, so V is sometimes omitted. But, it is always better to include V and cancel it out later, so you will not forget it for systems where the Vs do not cancel out.

You can also use K_c to find the amount of a reactant needed to give a required amount of product.

Worked example: Calculating the amount of a reactant needed

For the following reaction in ethanol solution, $K_c = 30.0\ \text{mol}^{-1}\ \text{dm}^3$:

$$CH_3COCH_3 + HCH \rightleftharpoons CH_3C(CN)(OH)CH_3$$

propanone hydrogen cyanide 2-hydroxy-2-methylpropanenitrile

$$K_c = \frac{[CH_3C(CN)(OH)CH_3]}{[CH_3COCH_3][HCN]} = 30.0\ \text{mol}^{-1}\ \text{dm}^3$$

Suppose you are carrying out this reaction in 2.00 dm³ of ethanol. How much hydrogen cyanide is required to produce 1.00 mol of product if you start with 4.00 mol of propanone? Set out as before with the quantities at the start and at equilibrium.

At equilibrium, you want 1 mol of product. Let x be the number of moles of HCN required.

Equation:	CH_3COCH_3 +	HCN	$\rightleftharpoons CH_3C(CN)(OH)CH_3$
At start:	4.00 mol	x mol	0 mol
At equilibrium:	(4.00 – 1.00) mol	(x – 1.00) mol	1.00 mol
	3.00 mol	(x – 1.00) mol	1.00 mol

Hint

Notice that this equilibrium constant *does* have units, which can be found by cancelling the units of the concentrations in the expression for K_c.

These are the numbers of moles, but we need the *concentrations* to put in the equilibrium law expression. The volume of the solution is $2.00\,dm^3$ and the units for concentration are $mol\,dm^{-3}$ so you next divide each quantity by $2.00\,dm^3$.

So, at equilibrium

$$[CH_3COCH_3]_{eqm} = \frac{3.00}{2.00}\ mol\,dm^{-3}$$

$$[HCN]_{eqm} = \frac{(x-1.00)}{2.00}\ mol\,dm^{-3}$$

$$[CH_3C(CN)(OH)CH_3]_{eqm} = \frac{1.00}{2.00}\ mol\,dm^{-3}$$

Putting the figures into the equilibrium expression:

$$30.0^3\,mol^{-1}\,dm^3 = \frac{1.00/2.00\ \cancel{mol\,dm^{-3}}}{3.00/2.00\ \cancel{mol\,dm^{-3}} \times (x-1.00)/2.00\ mol\,dm^{-3}}$$

Cancelling through and rearranging we have:

$$30\left(\frac{3/2(x-1)}{2}\right) = \frac{1}{2}$$

$$45(x-1) = 1$$

$$45x = 46$$

$$x = \frac{46}{45} = 1.02$$

So, to obtain 1 mol of product you must start with 1.02 mol hydrogen cyanide, if the volume of the system is $2.00\,dm^3$.

In this example the volume of the system *does* make a difference, because this reaction does not have the same number of moles of products and reactants.

Summary questions

 1 Try reworking the problem above with the same conditions but:

   a with a volume of $1.00\,dm^3$ of ethanol

   b starting with 2.0 mol of propanone

   c to produce 2.0 mol of product.

As you have seen Le Châtelier's principle states that when a system at equilibrium is disturbed, the equilibrium position moves in the direction that will reduce the disturbance. You can use Le Châtelier's principle to predict the qualitative effect of changing temperature and concentration on the position of equilibrium.

In this topic you look at what underlies this by examining the effect of changing conditions on the equilibrium constant K_c.

The effect of changing temperature on the equilibrium constant

Changing the temperature changes the value of the equilibrium constant, K_c. Whether K_c increases or decreases depends on whether the reaction is exothermic or endothermic. What happens is summarised in Table 1.

▼ **Table 1** *The effect of changing temperature on equilibria*

Type of reaction	Temperature change	Effect on K_c	Effect on products	Effect on reactants	Direction of change of equilibrium
endothermic	decrease	decrease	decrease	increase	moves left
endothermic	increase	increase	increase	decrease	moves right
exothermic	increase	decrease	decrease	increase	moves left
exothermic	decrease	increase	increase	decrease	moves right

If the equilibrium constant K_c increases in value, the equilibrium moves to the right, that is, the forward direction (more product). If it decreases in value, the equilibrium moves to the left, that is, the backward direction (less product).

This is because the expression for K_c is always of the form $\dfrac{[products]}{[reactants]}$.

The general rule is that:

* For an exothermic reaction (ΔH is negative) increasing the temperature decreases the equilibrium constant.
* For an endothermic reaction (ΔH is positive) increasing the temperature increases the equilibrium constant.

So for an exothermic reaction, increasing the temperature will move the equilibrium to the left – for an endothermic reaction, increasing the temperature will move the equilibrium to the right.

The effect of changing concentration on the position of equilibrium

First remember that the equilibrium constant does not change unless the temperature changes.

Learning objectives:

→ Explain how Le Châtelier's principle can predict how changes in conditions affect the position of equilibrium.

→ Describe how the equilibrium constant is affected by changing the conditions of a reaction.

Specification reference: 3.1.6

Study tip

When the value for ΔH is given for a reversible reaction, it is taken to refer to the forward reaction, that is, left to right.

Study tip

Make sure that you know how to apply Le Châtelier's principle for all changes in conditions.

Look at the following example:

$$C_2H_5OH\ (l)\ +\ CH_3CO_2H\ (l)\ \rightleftharpoons\ CH_3CO_2C_2H_5\ (l)\ +\ H_2O\ (l)$$

ethanol   ethanoic acid   ethyl ethanoate   water

Le Châtelier's principle tells you that the equilibrium will react to any disturbance by moving in such a way as to reduce the disturbance.

Imagine you add more ethanol, thereby increasing its concentration. The only way this concentration can be reduced is by some of the ethanol reacting with ethanoic acid producing more ethyl ethanoate and water. Eventually a new equilibrium will be set up with relatively more of the products. The equilibrium has moved to the right (or in the forward direction).

Let us see how this works mathematically.

You know that:

$$K_c = \frac{[CH_3CO_2C_2H_5\,(l)][H_2O\,(l)]}{[CH_3CO_2H\,(l)][C_2H_5OH\,(l)]}$$

Remember that K_c remains constant, provided that temperature remains constant. Adding ethanol makes the bottom line of the **equilibrium law expression** larger. To restore the situation, some of the ethanol reacts with ethanoic acid reducing both the concentrations in the bottom line of the fraction. This produces more ethyl ethanoate and water, thus increasing the value in the top line of the fraction. The combined effect is to restore the fraction to the original value of K_c.

K_c and the position of equilibrium

The size of the equilibrium constant K_c can tell us about the composition of the equilibrium mixture. The equilibrium expression is always of the general form $\frac{\text{products}}{\text{reactants}}$. So:

- If K_c is much greater than 1, products predominate over reactants and the equilibrium position is over to the right.
- If K_c is much less than 1, reactants predominate and the equilibrium position is over to the left.

Reactions where the equilibrium constant is greater than 10^{10} are usually regarded as going to completion, whilst those with an equilibrium constant of less than 10^{-10} are regarded as not taking place at all.

Catalysts and the value of K_c

Catalysts have no effect whatsoever on the value of K_c and therefore the position of equilibrium. This is because they affect the rates of both forward and back reactions equally. They do this by reducing the activation energy for the reactions. They do however affect the *rate* at which equilibrium is attained – this is important in industrial processes.

Gaseous equilibria

Reversible reactions may take place in the gas phase as well as in solution. These include many reactions of industrial importance such as the manufacture of ammonia by the Haber process and a key stage

Synoptic link

Look back at Topic 5.1, Collision theory, to revise activation energy for reactions.

Synoptic link

The equilibrium constant can also be calculated for gases using partial pressures. This equilibrium constant has the symbol K_p. The equilibrium constant K_p is covered in the second year of A Level chemistry course.

of the Contact process for making sulfuric acid. Gaseous equilibria also obey the equilibrium law, but usually their concentrations are expressed in a different way using partial pressures rather than concentrations.

Summary questions

1   $A + B \rightleftharpoons C + D$ represents an exothermic reaction and

   $K_c = \dfrac{[C][D]}{[A][B]}$ In the above expression, what would happen to K_c:

   a   if the temperature were decreased

   b   if more A were added to the mixture

   c   if a catalyst were added?

2   The reaction of ethanol with ethanoic acid produces ethyl ethanoate and water.

   $$C_2H_5OH(l) + CH_3COOH(l) \rightleftharpoons CH_3COOC_2H_5(l) + H_2O(l)$$

   A student suggested that the yield of ethyl ethanoate, $CH_3COOC_2H_5$, could be increased by removing the water as it was formed.

   Explain, using the idea of K_c, why this suggestion is sensible.

3   These questions are about reversible reactions. Give the correct word from **increases/decreases/does not change** to fill in the blank for each statement.

   a   In an endothermic reaction K_c _____ when the temperature is increased.

   b   In an endothermic reaction K_c _____ when the concentration of the reactants is decreased.

   c   In an exothermic reaction K_c _____ when the temperature is decreased.

   d   In an exothermic reaction K_c _____ when the concentration of the reactants is increased.

   e   If a suitable catalyst is added to the reaction K_c _____.

1   Methanol can be synthesised from carbon monoxide by the reversible reaction shown below.

$$CO(g) + 2H_2(g) \rightleftharpoons CH_3OH(g) \qquad \Delta H = -91 \text{ kJ mol}^{-1}$$

The process operates at a pressure of 5 MPa and a temperature of 700 K in the presence of a copper-containing catalyst. This reaction can reach dynamic equilibrium.

(a) By reference to rates and concentrations, explain the meaning of the term *dynamic equilibrium*.

(2 marks)

(b) Explain why a high yield of methanol is favoured by high pressure.

(2 marks)

(c) Suggest **two** reasons why the operation of this process at a pressure much higher than 5 MPa would be very expensive.

(2 marks)

(d) State the effect of an increase in temperature on the equilibrium yield of methanol and explain your answer.

(3 marks)

(e) If a catalyst were not used in this process, the operating temperature would have to be greater than 700 K. Suggest why an increased temperature would be required.

(1 mark)

AQA, 2003

2   At high temperatures, nitrogen is oxidised by oxygen to form nitrogen monoxide in a reversible reaction as shown in the equation below.

$$N_2(g) + O_2(g) \rightleftharpoons 2NO(g) \qquad \Delta H^{\ominus} = +180 \text{ kJ mol}^{-1}$$

(a) In terms of electrons, give the meaning of the term *oxidation*.

(1 mark)

(b) State and explain the effect of an increase in pressure, and the effect of an increase in temperature, on the yield of nitrogen monoxide in the above equilibrium.

(6 marks)

AQA, 2006

3   Hydrogen is produced on an industrial scale from methane as shown by the equation below.

$$CH_4(g) + H_2O(g) \rightleftharpoons CO(g) + 3H_2(g) \qquad \Delta H^{\ominus} = +205 \text{ kJ mol}^{-1}$$

(a) State Le Châtelier's principle.

(1 mark)

(b) The following changes are made to this reaction at equilibrium. In each case, predict what would happen to the yield of hydrogen from a given amount of methane. Use Le Châtelier's principle to explain your answer.
(i)   The overall pressure is increased.
(ii) The concentration of steam in the reaction mixture is increased.

(6 marks)

(c) At equilibrium, a high yield of hydrogen is favoured by high temperature. In a typical industrial process, the operating temperature is usually less than 1200 K. Suggest two reasons why temperatures higher than this are not used.

(2 marks)

AQA, 2004

4   The equation for the formation of ammonia is shown below.

$$N_2(g) + 3H_2(g) \rightleftharpoons 2NH_3(g)$$

Experiment **A** was carried out starting with 1 mol of nitrogen and 3 mol of hydrogen at a constant temperature and a pressure of 20 MPa.

Curve **A** shows how the number of moles of ammonia present changed with time.

Curves **B**, **C**, and **D** refer to similar experiments, starting with 1 mol of nitrogen and 3 mol of hydrogen. In each experiment different conditions were used.

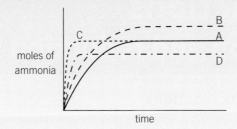

(a) On a copy of curve **A**, mark the point that represents the time at which equilibrium is first reached. Label this point **X**.

(1 mark)

(b) State Le Châtelier's principle.

(1 mark)

(c) Use Le Châtelier's principle to identify which one of the curves **B**, **C**, or **D** represents an experiment carried out at the same temperature as experiment **A** but at a higher pressure. Explain why this curve is different from curve **A**.

(4 marks)

(d) Identify which one of the curves **B**, **C**, or **D** represents an experiment in which the conditions are the same as in experiment **A** except that a catalyst is added to the reaction mixture. Explain your choice of curve.

(3 marks)

AQA, 2005

5　The reaction of methane with steam produces hydrogen for use in many industrial processes. Under certain conditions the following reaction occurs.

$$CH_4(g) + 2H_2O(g) \rightleftharpoons CO_2(g) + 4H_2(g) \qquad \Delta H = +165 \text{ kJ mol}^{-1}$$

√x̄ (a) Initially, 1.0 mol of methane and 2.0 mol of steam were placed in a flask and heated with a catalyst until equilibrium was established. The equilibrium mixture contained 0.25 mol of carbon dioxide.

(i) Calculate the amounts, in moles, of methane, steam and hydrogen in the equilibrium mixture.

(3 marks)

(ii) The volume of the flask was 5.0 dm³. Calculate the concentration, in mol dm⁻³, of methane in the equilibrium mixture.

(1 mark)

(b) The table below shows the equilibrium concentration of each gas in a different equilibrium mixture in the same flask and at temperature T.

gas	$CH_4(g)$	$H_2O(g)$	$CO_2(g)$	$H_2(g)$
concentration / mol dm⁻³	0.10	0.48	0.15	0.25

(i) Write an expression for the equilibrium constant, K_c, for this reaction.

(1 mark)

√x̄ (ii) Calculate a value for K_c at temperature T and give its units.

(3 marks)

(c) The mixture in part **(b)** was placed in a flask of volume greater than 5.0 dm³ and allowed to reach equilibrium at temperature T.
State and explain the effect on the amount of hydrogen.

(3 marks)

(d) Explain why the amount of hydrogen decreases when the mixture in part **(b)** reaches equilibrium at a lower temperature.

(2 marks)

AQA, 2010

Learning objectives:

→ Define a redox reaction in terms of oxygen or hydrogen transfer.

→ Define a redox reaction in terms of electron transfer.

→ Define a half equation.

Specification reference: 3.1.7

Redox reactions

The word redox is short for reduction–oxidation. Historically, **oxidation** was used for reactions in which oxygen was added.

In this reaction copper has been oxidised to copper oxide. Oxygen is called an **oxidising agent**.

$$Cu(s) + \frac{1}{2}O_2(g) \rightarrow CuO(s)$$

Reduction described a reaction in which oxygen was removed.

In this reaction copper oxide has been reduced and hydrogen is the **reducing agent**.

$$CuO(s) + H_2(g) \rightarrow Cu(s) + H_2O(l)$$

As hydrogen was often used to remove oxygen, the addition of hydrogen was called reduction.

In this reaction chlorine has been reduced because hydrogen has been added to it.

$$Cl_2(g) + H_2(g) \rightarrow 2HCl(g)$$

The reverse, where hydrogen was removed, was called oxidation.

Gaining and losing electrons – redox reactions

By describing what happens to the *electrons* in the above reactions, you get a much more general picture. When something is oxidised it loses electrons, and when something is reduced it gains electrons. Since **redox reactions** always involve the movement of electrons they are also called electron transfer reactions. You can see the transfer of electrons by separating a redox reaction into two half equations that show the gain and loss of electrons.

Worked example 1: Half equations

Look again at the reaction between copper and oxygen to form copper oxide:

$$Cu + \frac{1}{2}O_2 \rightarrow CuO$$

Copper oxide is an ionic compound so you can write the balanced symbol equation using ($Cu^{2+} + O^{2-}$) (instead of CuO) to show the ions present in copper oxide:

$$Cu + \frac{1}{2}O_2 \rightarrow (Cu^{2+} + O^{2-})$$

Next look at the copper. It has lost two electrons so it has been oxidised.

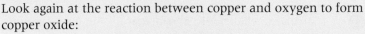

$$Cu - 2e^- \rightarrow Cu^{2+} \qquad \text{or} \qquad Cu \rightarrow Cu^{2+} + 2e^-$$

This is a **half equation**. It is usual to write half equations with plus electrons rather than minus electrons, that is:

$$Cu \rightarrow Cu^{2+} + 2e^- \text{ rather than}$$

$$Cu - 2e^- \rightarrow Cu^{2+}$$

Next look at the oxygen. It has gained two electrons so it has been reduced:

$$\frac{1}{2}O_2(g) + 2e^- \rightarrow O^{2-}$$

If you add the two half equations together, you end up with the original equation. Notice that the numbers of electrons cancel out.

$$Cu \rightarrow Cu^{2+} + \cancel{2e^-}$$

$$\frac{1}{2}O_2(g) + \cancel{2e^-} \rightarrow O^{2-}$$

$$Cu(s) + \frac{1}{2}O_2(g) \rightarrow (Cu^{2+} + O^{2-})(s)$$

Worked example 2: Half equations

When copper oxide reacts with magnesium, copper and magnesium oxide are produced:

$$CuO(s) + Mg(s) \rightarrow MgO(s) + Cu(s)$$

Write the equation with copper oxide as $(Cu^{2+} + O^{2-})$ and magnesium oxide as $(Mg^{2+} + O^{2-})$ to show the ions present.

$$(Cu^{2+} + O^{2-}) + Mg \rightarrow Cu + (Mg^{2+} + O^{2-})$$

Look at the copper. It has gained two electrons so it has been reduced.

$$Cu^{2+} + 2e^- \rightarrow Cu$$

Look at the magnesium. It has lost electrons so it has been oxidised.

$$Mg \rightarrow Mg^{2+} + 2e^-$$

Notice that the O^{2-} ion takes no part in the reaction. It is called a **spectator ion**.

If you add these half equations you get:

$$Cu^{2+} + Mg \rightarrow Cu + Mg^{2+}$$

This is the ionic equation for the redox reaction.

The definition of oxidation and reduction now used is:

Oxidation Is Loss of electrons.
Reduction Is Gain of electrons.

By this definition, magnesium is oxidised by *anything* that removes electrons from it (not just oxygen) leaving a positive ion. For example, chlorine oxidises magnesium:

$$Mg(s) + Cl_2(g) \rightarrow (Mg^{2+} + 2Cl^-)(s)$$

Hint

The phrase OIL RIG makes the definition of oxidation and reduction easy to remember.
Oxidation
Is
Loss (of electrons)
Reduction
Is
Gain (of electrons)

Look at the magnesium. It has lost electrons and has therefore been oxidised.

$$Mg \rightarrow Mg^{2+} + 2e^-$$

Look at the chlorine. It has gained electrons and has therefore been reduced.

$$Cl_2 + 2e^- \rightarrow 2Cl^-$$

And adding the two half equations together, the electrons cancel out:

$$Mg(s) + Cl_2(g) \rightarrow (Mg^{2+} + 2Cl^-)(s)$$

You may find that adding arrows to the equation, which show the transfer of electrons, helps keep track of them, as shown in Figure 1.

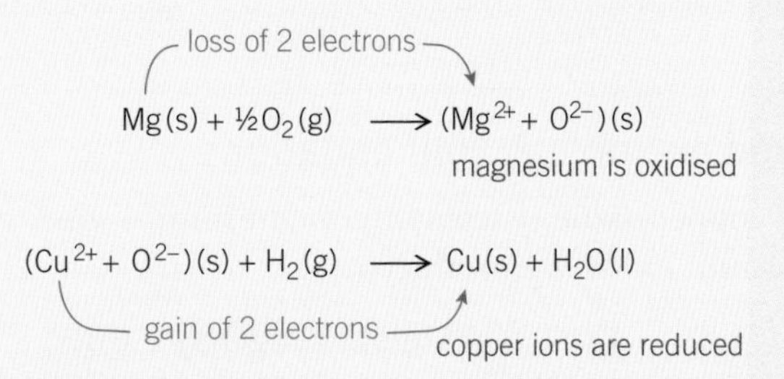

▲ **Figure 1** *Writing the electrons that are transferred helps to keep track of them*

In a chemical reaction, if one species is oxidised (loses electrons), another *must* be reduced (gains them).

Oxidising and reducing agents

It follows from the above that:

- reducing agents give away electrons – they are electron donors
- oxidising agents accept electrons.

Summary questions

1 The following questions are about the reaction:

$$Ca(s) + Br_2(g) \rightarrow (Ca^{2+} + 2Br^-)(s)$$

   a Which element has gained electrons?

   b Which element has lost electrons?

   c Which element has been oxidised?

   d Which element has been reduced?

   e Write the half equations for these redox reactions.

   f What is the oxidising agent?

   g What is the reducing agent?

Oxidation states

Oxidation states are used to see what has been oxidised and what has been reduced in a redox reaction. Oxidation states are also called oxidation numbers.

The idea of oxidation states

Each element in a compound is given an oxidation state. In an ionic compound the oxidation state simply tells us how many electrons it has lost or gained, compared with the element in its uncombined state. In a molecule, the oxidation state tells us about the distribution of electrons between elements of different electronegativity. The more electronegative element is given the negative oxidation state.

- Every element in its uncombined state has an oxidation state of zero.

- A positive number shows that the element has lost electrons and has therefore been oxidised. For example, Mg^{2+} has an oxidation state of +2.

- A negative number shows that the element has gained electrons and has therefore been reduced. For example Cl^- has an oxidation state of −1.

- The more positive the number, the more the element has been oxidised. The more negative the number, the more it has been reduced.

- The numbers always have a + or − sign unless they are zero.

Rules for finding oxidation states

The following rules will allow you to work out oxidation states:

1   Uncombined elements have oxidation state of 0.
2   Some elements always have the same oxidation state in all their compounds. Others usually have the same oxidation state. Table 1 gives the oxidation states of these elements.

▼ Table 1 *The usual oxidation states of some elements*

Element	Oxidation state in compound	Example
hydrogen, H	+1 (except in metal hydrides, e.g., NaH, where it is −1)	HCl
Group 1	always +1	NaCl
Group 2	always +2	$CaCl_2$
aluminium, Al	always +3	$AlCl_3$
oxygen, O	−2 (except in peroxides where it is −1, and the compound OF_2, where it is +2)	Na_2O
fluorine, F	Always −1	NaF
chlorine, Cl	−1 (except in compounds with F and O, where it has positive values)	NaCl

3   The sum of all the oxidation states in a compound = 0, since all compounds are electrically neutral.

Learning objectives:
→ Define an oxidation state.
→ Describe how oxidation states are worked out.
Specification reference: 3.1.7

4   The sum of the oxidation states of a complex ion, such as NH_4^+ or SO_4^{2-}, equals the charge on the ion.

5   In a compound the most electronegative element always has a negative oxidation state.

Working out oxidation states of elements in compounds

Start with the correct formula. Look for the elements whose oxidation states you know from the rules. Then deduce the oxidation states of any other element. Some examples are shown below.

Phosphorus pentachloride, PCl_5

Chlorine has an oxidation state of -1, so the phosphorus must be $+5$, to make the sum of the oxidation states zero.

Ammonia, NH_3

Hydrogen has an oxidation state of $+1$, so the nitrogen must be -3, to make the sum of the oxidation states zero. Also, nitrogen is more electronegative than hydrogen, so hydrogen must have a positive oxidation state.

Nitric acid, HNO_3

Each oxygen has an oxidation state of -2, making -6 in total.

Hydrogen has an oxidation state of $+1$.

So the nitrogen must be $+5$, to make the sum of the oxidation states zero.

Notice that nitrogen may have different oxidation states in different compounds. Here that nitrogen has a positive oxidation state because it is combined with a more electronegative element, oxygen.

Hydrogen sulfide, H_2S

Hydrogen has an oxidation state of $+1$, so the sulfur must be -2, to make the sum of the oxidation states zero.

Sulfate ion, SO_4^{2-}

Each oxygen has an oxidation state of -2, making -8 in total.

So the sulfur must be $+6$, to make the sum of the oxidation states equal to the charge on the ion.

Notice that sulfur may have different oxidation states in different compounds.

Black copper oxide, CuO

Oxygen has an oxidation state of -2, so the copper must be $+2$, to make the sum of the oxidation states zero.

Red copper oxide, Cu_2O

Oxygen has an oxidation state of -2, so each copper must be $+1$, to make the sum of the oxidation states zero.

Oxidation states are used in Roman numerals to distinguish between similar compounds in which the metal has a different oxidation state. So, black copper oxide is copper(II) oxide and red copper oxide is copper(I) oxide. These compounds are shown in Figure 1.

Study tip

You will need to know the rules for finding oxidation states.

▲ **Figure 1** *The two oxides of copper – copper(II) oxide (left) and copper(I) oxide (right)*

Superoxide – an unusual oxidation state

The idea of oxidation states (sometimes called oxidation numbers) is essentially 'book keeping' of electrons and can lead to some unusual outcomes. In the superoxide ion, each oxygen atom has an oxidation number of $-\frac{1}{2}$.

When potassium is heated in oxygen, a certain amount of an orange-yellow compound is formed with the formula KO_2 as well as other oxides K_2O_2 and K_2O. KO_2 is called potassium superoxide. Its yellow colour is unusual for a compound of an alkali metal and is due to the O_2^- ion.

Following the rules for working out oxidation numbers, K is +I so each oxygen atom must be $-\frac{1}{2}$. In the superoxide ion, there is a single covalent bond between the oxygen atoms and the negative charge resulting from the electron gained from the potassium ion is shared between the two oxygens so they do not have full outer shells.

$$\left[\overset{\times\,\times}{\underset{\times\,\times}{\times\,O}} \cdot \underset{\bullet\bullet}{\overset{\bullet\bullet}{O}} \cdot \right]^-$$

Potassium superoxide has been used in spacecraft to remove carbon dioxide (exhaled by the astronauts) and replace it with oxygen:

$$4KO_2 + 2CO_2 \rightarrow 2K_2CO_3 + 3O_2$$

1 Calculate the mass of CO_2 that can be absorbed and the mass of O_2 released by 284 g of KO_2.
2 Both sodium superoxide and rubidium superoxide undergo similar reactions. Suggest and explain which would be best for use in a spacecraft.

1 CO_2: 88 g, O_2: 96 g

Summary questions

1 Work out the oxidation states of each element in the following compounds:

   a $PbCl_2$

   b CCl_4

   c $NaNO_3$

2 In the reaction: $CuO + Mg \rightarrow Cu + MgO$, what are the oxidation states of oxygen before and after the reaction?

3 In the reaction: $2Cu + O_2 \rightarrow 2CuO$, what are the oxidation states of oxygen before and after the reaction?

4 In the reaction: $FeCl_2 + \frac{1}{2}Cl_2 \rightarrow FeCl_3$, what are the oxidation states of iron before and after the reaction?

5 Give the oxidation state of the following:

   a P in PO_4^{3-}

   b N in NO_3^-

   c N in NH_4^+

Using oxidation states in redox equations

You saw in Topic 7.1 that you can work out which element has been oxidised and which has been reduced in a redox reaction by considering electron transfer.

Remember that Oxidation is loss of electrons (OIL) and reduction is gain of electrons (RIG).

You can also use oxidation states to help you to understand redox reactions.

When an element is reduced, it gains electrons and its oxidation state goes down. In the reaction below, iron is reduced because its oxidation state has gone down from +3 to +2, whilst iodide is oxidised:

$$\overset{+3}{Fe^{3+}} + \overset{-1}{I^-} \rightarrow \overset{+2}{Fe^{2+}} + \overset{0}{\tfrac{1}{2}I_2}$$

Even in complicated reactions, you can see which element has been oxidised and which has been reduced when you put in the oxidation states:

$$\overset{+5 \; -2}{2IO_3^-} + \overset{+1 \; +4 \; -2}{5HSO_3^-} \rightarrow \overset{0}{I_2} + \overset{+6 \; -2}{5SO_4^{2-}} + \overset{+1}{3H^+} + \overset{+1 \; -2}{H_2O}$$

Iodine in IO_3^- is reduced (+5 to 0) and sulfur in HSO_3^- is oxidised (+4 to +6). The oxidation states of all the other atoms have not changed.

Balancing redox reactions

You can use the idea of oxidation states to help balance equations for redox reactions.

For an equation to be balanced:

- the numbers of atoms of each element on each side of the equation must be the same
- the total charge on each side of the equation must be the same.

Example 1: the thermite reaction

This is a strongly exothermic reaction in which aluminium reacts with iron(III) oxide to produce molten iron. It was used to weld railway lines.

The unbalanced equation is:

$$Fe_2O_3(s) + Al(s) \rightarrow Fe(l) + Al_2O_3(s)$$

Write the oxidation states above each element:

$$\overset{+3 \; -2}{Fe_2O_3(s)} + \overset{0}{Al(s)} \rightarrow \overset{0}{Fe(l)} + \overset{+3 \; -2}{Al_2O_3(s)}$$

If you look at the equation you can see that that only the iron and aluminium have changed their oxidation state. The oxygen is unchanged.

Each iron atom has been reduced by gaining three electrons so you can write the half equation:

$$Fe^{3+} + 3e^- \rightarrow Fe$$

Each aluminium atom has been oxidised by losing three electrons:

$$Al \rightarrow Al^{3+} + 3e^-$$

In the reaction, the number of electrons gained must equal the number of electrons lost. This means that there must be the same number of aluminium atoms as iron atoms. (The oxygen is a spectator ion.) You started with two iron atoms, so you must also have two aluminium atoms. The balanced equation is therefore:

$$Fe_2O_3(s) + 2Al(s) \rightarrow 2Fe(l) + Al_2O_3(s)$$

Example 2: aqueous solutions

Sometimes in aqueous solutions, species take part in redox reactions but are neither oxidised nor reduced. You must balance them separately. These include water molecules, H^+ ions (in acid solution), and OH^- ions (in alkaline solution). Oxidation states only help us to balance the species that are oxidised or reduced.

Suppose you want to balance the following equation, where dark purple manganate(VII) ions react in acid solution with Fe^{2+} ions to produce pale pink Mn^{2+} ions and Fe^{3+} ions.

▲ **Figure 1** *A demonstration of the thermite reaction*

The unbalanced equation is:

$$MnO_4^- + Fe^{2+} + H^+ \rightarrow Mn^{2+} + Fe^{3+} + H_2O$$

1 Write the oxidation state above each element.

$$\overset{+7\ -2}{MnO_4^-} + \overset{+2}{Fe^{2+}} + \overset{+1}{H^+} \rightarrow \overset{+2}{Mn^{2+}} + \overset{+3}{Fe^{3+}} + \overset{+1\ -2}{H_2O}$$

2 Identify the species that has been oxidised and the species that has been reduced.

$\overset{+7}{MnO_4^-} \rightarrow \overset{+2}{Mn^{2+}}$ Manganese has been reduced from +7 to +2 therefore five electrons must be gained.

$MnO_4^- + 5e^- \rightarrow Mn^{2+}$ (this equation is not chemically balanced)
$\overset{+2}{Fe^{2+}} \rightarrow \overset{+3}{Fe^{3+}}$ Fe has been oxidised from +2 to +3 so one electron must be lost.

$$Fe^{2+} \rightarrow Fe^{3+} + e^-$$

In order to balance the number of electrons that are transferred, this step must be multiplied by 5:

$$5Fe^{2+} \rightarrow 5Fe^{3+} + 5e^-$$

So, you know that there are $5Fe^{2+}$ ions to every MnO_4^- ion.

3 Include this information in the unbalanced equation, to balance the redox process.

$$MnO_4^- + 5Fe^{2+} + H^+ \rightarrow Mn^{2+} + 5Fe^{3+} + H_2O$$

(this equation is still not chemically balanced)

4 Balance the remaining atoms, those that are neither oxidised nor reduced. In order to 'use up' the four oxygen atoms on the left-hand side, you need $4H_2O$ on the right-hand side, which will in turn require $8H^+$ on the left-hand side.

$$MnO_4^- + 5Fe^{2+} + 8H^+ \rightarrow Mn^{2+} + 5Fe^{3+} + 4H_2O$$

Notice that this equation is balanced for both atoms and charge.

Disproportionation

In some chemical reactions, atoms of the same element can be both oxidised and reduced. For example, hydrogen peroxide decomposes to oxygen and water.

$$\overset{-1}{2H_2}\overset{}{O_2} \rightarrow \overset{-2}{2H_2}\overset{0}{O} + \overset{}{O_2}$$

Check that you can work out the oxidation state of each oxygen (shown in red) using the rules in Topic 7.2.

Two of the oxygen atoms in the hydrogen peroxide have increased their oxidation state and two have reduced it.

1   Suggest why hairdressers, who use hydrogen peroxide as a bleach, store it in the fridge and in bottles with a small hole in the cap.

2   Here is another disproportionation reaction.
    $Cu_2O \rightarrow Cu + CuO$
    Work out the oxidation states of each atom using the rules in Topic 7.2. Which element disproportionates?

2   Cu (+1 to 0 and +2)

1   Slows down decomposition of H_2O_2. Gases produced from decomposition can escape through small hole.

Half equations from the balanced equation

Example 1

The reaction between copper and *cold dilute* nitric acid produces the gas nitrogen monoxide. The balanced symbol equation is shown:

$$3Cu + 8H^+ + 2NO_3^- \rightarrow 3Cu^{2+} + 2NO + 4H_2O$$

To work out the half equations, you first need to know which elements have been oxidised and which have been reduced.

1   Put in the numbers and look for a change in the oxidation states:

$$\overset{0}{3Cu} + \overset{+1}{8H^+} + \overset{+5\ -2}{2NO_3^-} \rightarrow \overset{+2}{3Cu^{2+}} + \overset{+2\ -2}{2NO} + \overset{+1\ -2}{H_2O}$$

Copper has been oxidised and nitrogen has been reduced.

2   Now work out the half equations.

Each of the three copper atoms loses two electrons, a total of six electrons:

$$3Cu \rightarrow 3Cu^{2+} + 6e^-$$

The two nitrogen atoms NO_3^- have each gained three electrons so the half equation must be based on:

$$2NO_3^- + 6e^- \rightarrow 2NO$$

This half equation is not balanced for atoms or charge. There are six oxygen atoms on the left-hand side and only two on the right-hand side. The total charge on the left is −8 whereas the right-hand side has no charge. Look at the original equation. You need to include the eight H^+ ions on the left-hand side of our half equation (to use up the extra four oxygen atoms that are unaccounted for) and also the four H_2O on the right-hand side. This also accounts for the charge, so the complete half equation is:

$$2NO_3^- + 8H^+ + 6e^- \rightarrow 2NO + 4H_2O$$

This equation is balanced in both atoms and charge.

Example 2

The reaction between copper and *hot concentrated* nitric acid produces the gas nitrogen dioxide.

1   The balanced symbol equation is shown with the oxidation states included:

$$\overset{0}{Cu} + \overset{+1}{4H^+} + \overset{+5\;-2}{2NO_3^-} \rightarrow \overset{+2}{Cu^{2+}} + \overset{+1\;-2}{2H_2O} + \overset{+4\;-2}{2NO_2}$$

Copper has been oxidised and nitrogen has been reduced.

2   Now work out the half equations.
Copper has lost two electrons so the half equation is:

$$Cu \rightarrow Cu^{2+} + 2e^-$$

Nitrogen in NO_3^- has gained an electron so the half equation must be based on:

$$2NO_3^- + 2e^- \rightarrow 2NO_2$$

This is not balanced for charge or atoms. There are an extra two oxygens on the left-hand side and a total charge of −4 whereas the right-hand side is neutral. You need to add the four H^+ ions to the left-hand side to use up the extra oxygen. These will also balance the charge. You then need to add two H_2O to the right-hand side.

The half equation is:

$$2NO_3^- + 4H^+ + 2e^- \rightarrow 2H_2O + 2NO_2$$

Note that if you add the half equations together, the electrons cancel out and you get back to the original balanced equation.

Summary questions

1   The following questions are about the equation:

$$Fe^{2+} + \frac{1}{2}Cl_2 \rightarrow Fe^{3+} + Cl^-$$

   a   Write the oxidation states for each element.

   b   Which element has been oxidised? Explain your answer.

   c   Which element has been reduced? Explain your answer.

   d   Write the half equations for the reaction.

2   a   Use oxidation states to balance the following equations:

   i $Cl_2 + NaOH \rightarrow NaClO_3 + NaCl + H_2O$

   ii $Sn + HNO_3 \rightarrow SnO_2 + NO_2 + H_2O$

   b   Write the half equations for i and ii.

1 **(a)** In terms of electron transfer, what does the reducing agent do in a redox reaction?

(1 mark)

(b) What is the oxidation state of an atom in an uncombined element?

(1 mark)

(c) Deduce the oxidation state of nitrogen in each of the following compounds.
   (i) NCl_3
   (ii) Mg_3N_2
   (iii) NH_2OH

(3 marks)

(d) Lead(IV) oxide, PbO_2, reacts with concentrated hydrochloric acid to produce chlorine, lead(II) ions, Pb^{2+}, and water.
   (i)   Write a half-equation for the formation of Pb^{2+} and water from PbO_2 in the presence of H^+ ions.
   (ii) Write a half-equation for the formation of chlorine from chloride ions.
   (iii) Hence deduce an equation for the reaction which occurs when concentrated hydrochloric acid is added to lead (IV) oxide, PbO_2

(3 marks)
AQA, 2002

2 Chlorine and bromine are both oxidising agents.
  (a) Define an *oxidising agent* in terms of electrons.

(1 mark)

  (b) In aqueous solution, bromine oxidises sulfur dioxide, SO_2, to sulfate ions, SO_4^{2-}
   (i)   Deduce the oxidation state of sulfur in SO_2 and in SO_4^{2-}
   (ii) Deduce a half-equation for the reduction of bromine in aqueous solution.
   (iii) Deduce a half-equation for the oxidation of SO_2 in aqueous solution forming SO_4^{2-} and H^+ ions.
   (iv) Use these two half-equations to construct an overall equation for the reaction between aqueous bromine and sulfur dioxide.

(5 marks)
AQA, 2004

3 **(a)** By referring to electrons, explain the meaning of the term *oxidising agent*.

(1 mark)

  (b) For the element **X** in the ionic compound **MX**, explain the meaning of the term *oxidation state*.

(1 mark)

  (c) Complete the table below by deducing the oxidation state of each of the stated elements in the given ion or compound.

	Oxidation state
carbon in CO_3^{2-}	
phosphorus in PCl_4^+	
nitrogen in Mg_3N_2	

(3 marks)

  (d) In acidified aqueous solution, nitrate ions, NO_3^- react with copper metal forming nitrogen monoxide, NO, and copper(II) ions.
   (i)   Write a half-equation for the oxidation of copper to copper(II) ions.
   (ii) Write a half-equation for the reduction, in an acidified solution, of nitrate ions to nitrogen monoxide.
   (iii) Write an overall equation for this reaction.

(3 marks)
AQA, 2005

4 **(a)** Nitrogen monoxide, NO, is formed when silver metal reduces nitrate ions, NO_3^- in acid solution. Deduce the oxidation state of nitrogen in NO and in NO_3^-.
  (b) Write a half-equation for the reduction of NO_3^- ions in acid solution to form nitrogen monoxide and water.

(c) Write a half-equation for the oxidation of silver metal to $Ag^+(aq)$ ions.

(d) Hence, deduce an overall equation for the reaction between silver metal and nitrate ions in acid solution.

(5 marks)

AQA, 2006

5 Iodine reacts with concentrated nitric acid to produce nitrogen dioxide, NO_2.

(a) (i) Give the oxidation state of iodine in each of the following.

I_2, HIO_3 *(2 marks)*

(ii) Complete the balancing of the following equation.

$I_2 + 10HNO_3 \rightarrow HIO_3 + NO_2 + H_2O$ *(1 mark)*

(b) In industry, iodine is produced from the $NaIO_3$ that remains after sodium nitrate has been crystallised from the mineral Chile saltpetre.

The final stage involves the reaction between $NaIO_3$ and NaI in acidic solution.
Half-equations for the redox processes are given below.

$$IO_3^- + 5e^- + 6H^+ \rightarrow 3H_2O + \frac{1}{2}I_2$$
$$I^- \rightarrow \frac{1}{2}I_2 + e^-$$

Use these half-equations to deduce an overall ionic equation for the production of iodine by this process. Identify the oxidising agent.

Overall ionic equation

The oxidising agent *(2 marks)*

(c) When concentrated sulfuric acid is added to potassium iodide, solid sulfur and a black solid are formed.

(i) Identify the black solid. *(1 mark)*

(ii) Deduce the half-equation for the formation of sulfur from concentrated sulfuric acid.

(1 mark)

(d) When iodide ions react with concentrated sulfuric acid in a different redox reaction, the oxidation state of sulfur changes from +6 to −2. The reduction product of this reaction is a poisonous gas that has an unpleasant smell.
Identify this gas. *(1 mark)*

(e) A yellow precipitate is formed when silver nitrate solution, acidified with dilute nitric acid, is added to an aqueous solution containing iodide ions.

(i) Write the **simplest ionic** equation for the formation of the yellow precipitate.

(1 mark)

(ii) State what is observed when concentrated ammonia solution is added to this precipitate.

(1 mark)

(iii) State why the silver nitrate is acidified when testing for iodide ions.

(1 mark)

(f) Consider the following reaction in which iodide ions behave as reducing agents.

$$Cl_2(aq) + 2I^-(aq) \rightarrow I_2(aq) + 2Cl^-(aq)$$

(i) In terms of electrons, state the meaning of the term *reducing agent*.

(1 mark)

(ii) Write a half-equation for the conversion of chlorine into chloride ions.

(1 mark)

(iii) Explain why iodide ions react differently from chloride ions.

(3 marks)

AQA, 2012

Section 1 practice questions

1   When heated, iron(III) nitrate (M_r = 241.8) is converted into iron(III) oxide, nitrogen dioxide, and oxygen.

$$4Fe(NO_3)_3(s) \rightarrow 2Fe_2O_3(s) \quad + \quad 12NO_2(g) \quad + \quad 3O_2(g)$$

A 2.16 g sample of iron(III) nitrate was completely converted into the products shown.

   (a) (i)   Calculate the amount, in moles, of iron(III) nitrate in the 2.16 g sample. Give your answer to an appropriate number of significant figures.

(1 mark)

      (ii) Calculate the amount, in moles, of oxygen gas produced in this reaction.

(1 mark)

      (iii) Calculate the volume, in m^3, of **nitrogen dioxide** gas at 293 °C and 100 kPa produced from 2.16 g of iron(III) nitrate.

         The gas constant is R = 8.31 J K^{-1} mol^{-1}.

         (If you have been unable to obtain an answer to Question **2 (a) (i)**, you may assume the number of moles of iron(III) nitrate is 0.00642. This is **not** the correct answer.)

(4 marks)

   (b) Suggest a name for this type of reaction that iron(III) nitrate undergoes.

(1 mark)

   (c) Suggest why the iron(III) oxide obtained is pure.

      Assume a complete reaction.

(1 mark)

AQA, 2014

2   Antimony is a solid element that is used in industry. The method used for the extraction of antimony depends on the grade of the ore.

   (a) Antimony can be extracted by reacting scrap iron with low-grade ores that contain antimony sulfide, Sb_2S_3.

      (i)   Write an equation for the reaction of iron with antimony sulfide to form antimony and iron(II) sulfide.

(1 mark)

      (ii) Write a half-equation to show what happens to the iron atoms in this reaction.

(1 mark)

   (b) In the first stage of the extraction of antimony from a high-grade ore, antimony sulfide is roasted in air to convert it into antimony(III) oxide (Sb_2O_3) and sulfur dioxide.

      (i)   Write an equation for this reaction.

(1 mark)

      (ii) Identify **one** substance that is manufactured directly from the sulfur dioxide formed in this reaction.

(1 mark)

   (c) In the second stage of the extraction of antimony from a high-grade ore, antimony(III) oxide is reacted with carbon monoxide at high temperature.

      (i)   Use the standard enthalpies of formation in **Table 1** and the equation given below **Table 1** to calculate a value for the standard enthalpy change for this reaction.

▼ Table 1

	$Sb_2O_3(s)$	$CO(g)$	$Sb(l)$	$CO_2(g)$
$\Delta_f H$ / kJ mol^{-1}	− 705	− 111	+ 20	− 394

$$Sb_2O_3(s) \quad + \quad 3CO(g) \quad \rightarrow \quad 2Sb(l) \quad + \quad 3CO_2(g)$$

(3 marks)

(ii) Suggest why the value for the standard enthalpy of formation of liquid antimony, given in **Table 1**, is **not** zero.

(1 mark)

(iii) State the type of reaction that antimony(III) oxide has undergone in this reaction.

(1 mark)

(d) Deduce **one** reason why the method of extraction of antimony from a low-grade ore, described in part **3 (a)**, is a low-cost process. Do **not** include the cost of the ore.

(1 mark)

AQA, 2014

3 **(a)** Complete the following table.

	Relative mass	Relative charge
proton		
electron		

(2 mark)

(b) An atom has twice as many protons and twice as many neutrons as an atom of ^{19}F

Deduce the symbol, including the mass number, of this atom. *(2 marks)*

(c) The Al^{3+} ion and the Na^+ ion have the same electron arrangement.
   (i) Give the electron arrangement of these ions.
   (ii) Explain why more energy is needed to remove an electron from the Al^{3+} ion than from the Na^+ ion. *(3 marks)*

AQA, 2007

4 Molecules of NH_3, H_2O, and HF contain covalent bonds. The bonds in these molecules are polar.
   (a) (i) Explain why the H–F bond is polar.
      (ii) State which of the molecules NH_3, H_2O, or HF contains the least polar bond.
      (iii) Explain why the bond in your chosen molecule from part (b)(ii) is less polar than the bonds found in the other two molecules. *(4 marks)*
      (iv) Explain why H_2O has a bond angle of 104.5°. *(2 marks)*
   (b) The boiling points of NH_3, H_2O, and HF are all high for molecules of their size. This is due to the type of intermolecular force present in each case.
      (i) Identify the type of intermolecular force responsible.
      (ii) Draw a diagram to show how two molecules of ammonia are attracted to each other by this type of intermolecular force. Include partial charges and all lone pairs of electrons in your diagram. *(4 marks)*
   (c) When an H^+ ion reacts with an NH_3 molecule, an NH_4^+ ion is formed.
      (i) Give the name of the type of bond formed when an H^+ ion reacts with an NH_3 molecule.
      (ii) Draw the shape, including any lone pairs of electrons, of an NH_3 molecule and of an NH_4^+ ion.
      (iii) Name the shape produced by the arrangement of atoms in the NH_3 molecule.
      (iv) Give the bond angle in the NH_4^+ ion. *(7 marks)*

AQA, 2007

Section 1 summary

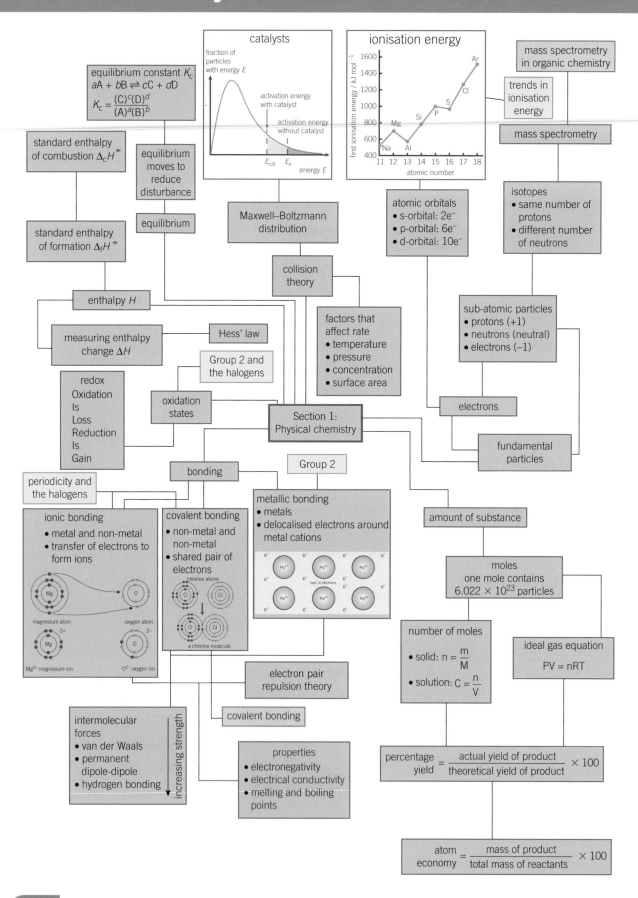

Practical skills

In this section you have met the following ideas:

- Finding the concentration of a solution by titration.

- Finding the yield of a reaction.

- Finding ΔH of reactions using calorimetry and Hess' Law.

- Investigating the effect of temperature, concentration and a catalyst on the rate of reactions.

- Finding out K_c of a reaction.

Maths skills

In this section you have met the following maths skills:

- Using standard form in calculations.

- Carrying out calculations with the Avogradro constant.

- Carrying out calculations using Hess' Law.

- Using appropriate significant figures.

- Calculating weighted means.

- Interpreting mass spectra.

- Working out the shape of molecules using ideas about electron pair repulsion.

Extension

Produce a timeline detailing how our understanding of atoms, atomic structure and chemical bonding has developed.

Suggested resources:

- Atkins, P. (2014), *Physical Chemistry: A very short Introduction*. Oxford University Press, UK. ISBN: 978-0-19-968909-5.

- Dunmar, D., Sluckin, T., (2014), *Soap, Science and Flat-Screen TVs*. Oxford University Press, UK. ISBN: 978-0-19-870083-8.

- Scerri, E. (2013), *The Tale of 7 Elements*. Oxford University Press, UK. ISBN: 978-0-19-539131-2

Section 2
Inorganic chemistry

Chapters in this section

8 Periodicity

9 Group 2, the alkaline earth metals

10 Group 7(17), the halogens

The Periodic Table of elements contains all the elements so far discovered. This unit examines how the properties of the elements are related to their electronic structures and how this determines their position in the Periodic Table.

Periodicity gives an overview of the Periodic Table and classifies blocks of elements in terms of s-, p-, d-, and f-orbitals. It then concentrates on the properties of the elements in Period 3.

Group 2, the alkaline earth metals uses the ideas of electron arrangements to understand the bonding in compounds of these elements and the reactions and trends in reactivity in the group.

Group 7(17) the halogens deals with these reactive non-metal elements, explaining the trends in their reactivity in terms of electronic structure. It includes the reactions of the elements and their compounds using the ideas of redox reactions and oxidation states, and also the uses of chlorine and some of its compounds.

The concepts of the applications of science are found throughout the chapters, where they will provide you with an opportunity to apply your knowledge in a fresh context.

The Periodic Table is a list of all the elements in order of increasing atomic number. You can predict the properties of an element from its position in the table. You can use it to explain the similarities of certain elements and the trends in their properties, in terms of their electronic arrangements.

The structure of the Periodic Table

The Periodic Table has been written in many forms including pyramids and spirals. The one shown below (and at the end of the book) is one common layout. Some areas of the Periodic Table are given names. These are shown in Figure 1.

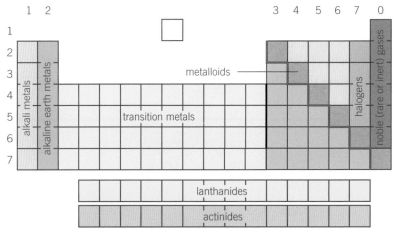

▲ **Figure 1** *Named areas of the Periodic Table*

Metals and non-metals

The red stepped line in Figure 1 (the 'staircase line') divides metals (on its left) from non-metals (on its right). Elements that touch this line, such as silicon, have a combination of metallic and non-metallic properties. They are called metalloids or semi-metals. Silicon, for example, is a non-metal but it looks quite shiny and conducts electricity, although not as well as a metal.

History of the Periodic Table

The development of the Periodic Table is one of the greatest achievements in chemistry. Credit for the final version goes firmly to a Russian, Dmitri Mendeleev, in 1869. He realised that there were undiscovered elements. He left spaces for some unknown elements, and arranged the known elements so that similar elements lined up in columns. Since then, new elements have been discovered that fit into the gaps he left. Mendeleev even accurately

predicted the properties of the missing elements, confirming the success of his Periodic Table.

Many other scientists contributed to the Periodic Table. Research on the internet the parts played by Jons Jacob Berzelius, Robert Bunsen and Gustav Kirchoff, Alexandre Béguyer de Chancourtois, Marie Curie, Sir Humphry Davy, Julius Lothar Meyer, Henry Moseley, John Newlands, Sir William Ramsay, and Glenn T Seaborg.

A common form of the Periodic Table

A version of the Periodic Table is shown in Figure 2. The lanthanides and actinides are omitted and two alternative numbering schemes for groups are shown.

key

relative atomic mass
atomic symbol
name
atomic (proton) number

1.0 H hydrogen 1

1 (1)	2 (2)	(3)	(4)	(5)	(6)	(7)	(8)	(9)	(10)	(11)	(12)	3 (13)	4 (14)	5 (15)	6 (16)	7 (17)	0 (18)
																	4.0 He helium 2
6.9 Li lithium 3	9.0 Be beryllium 4											10.8 B boron 5	12.0 C carbon 6	14.0 N nitrogen 7	16.0 O oxygen 8	19.0 F fluorine 9	20.2 Ne neon 10
23.0 Na sodium 11	24.3 Mg magnesium 12											27.0 Al aluminium 13	28.1 Si silicon 14	31.0 P phosphorus 15	32.1 S sulfur 16	35.5 Cl chlorine 17	39.9 Ar argon 18
39.1 K potassium 19	40.1 Ca calcium 20	45.0 Sc scandium 21	47.9 Ti titanium 22	50.9 V vanadium 23	52.0 Cr chromium 24	54.9 Mn manganese 25	55.8 Fe iron 26	58.9 Co cobalt 27	58.7 Ni nickel 28	63.5 Cu copper 29	65.4 Zn zinc 30	69.7 Ga gallium 31	72.6 Ge germanium 32	74.9 As arsenic 33	79.0 Se selenium 34	79.9 Br bromine 35	83.8 Kr krypton 36
85.5 Rb rubidium 37	87.6 Sr strontium 38	88.9 Y yttrium 39	91.2 Zr zirconium 40	92.9 Nb niobium 41	95.9 Mo molybdenum 42	[98] Tc technetium 43	101.1 Ru ruthenium 44	102.9 Rh rhodium 45	106.4 Pd palladium 46	107.9 Ag silver 47	112.4 Cd cadmium 48	114.8 In indium 49	118.7 Sn tin 50	121.8 Sb antimony 51	127.6 Te tellurium 52	126.9 I iodine 53	131.3 Xe xenon 54
132.9 Cs caesium 55	137.3 Ba barium 56	138.9 La* lanthanum 57	178.5 Hf hafnium 72	180.9 Ta tantalum 73	183.8 W tungsten 74	186.2 Re rhenium 75	190.2 Os osmium 76	192.2 Ir iridium 77	195.1 Pt platinum 78	197.0 Au gold 79	200.6 Hg mercury 80	204.4 Tl thallium 81	207.2 Pb lead 82	209.0 Bi bismuth 83	[209] Po polonium 84	[210] At astatine 85	[222] Rn radon 86
[223] Fr francium 87	[226] Ra radium 88	[227] Ac* actinium 89	[261] Rf rutherfordium 104	[262] Db dubnium 105	[266] Sg seaborgium 106	[264] Bh bohrium 107	[277] Hs hassium 108	[268] Mt meitnerium 109	[271] Ds darmstadtium 110	[272] Rg roentgenium 111							

elements with atomic numbers 112–116 have been reported but are not fully authenticated

*the lanthanides (atomic numbers 58–71) and the actinides (atomic numbers 90–103) have been omitted

▲ **Figure 2** *The full form of the Periodic Table*

The s-, p-, d-, and f-blocks of the Periodic Table

Figure 3 shows the elements described in terms of their electronic arrangement.

Areas of the table are labelled s-block, p-block, d-block, and f-block.

- All the elements that have their highest energy electrons in s-orbitals are in the s-block, for example, sodium, Na ($1s^2\ 2s^2\ 2p^6\ 3s^1$).

- All the elements that have their highest energy electrons in p-orbitals are called p-block, for example, carbon, C ($1s^2\ 2s^2\ 2p^2$).

- All the elements that have their highest energy electrons in d-orbitals are called d-block, for example, iron, Fe ($1s^2\ 2s^2\ 2p^6\ 3s^2\ 3p^6\ 4s^2\ 3d^6$) and so on.

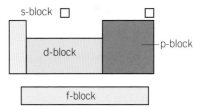

▲ **Figure 3** *The s-, p-, d-, and f-block areas of the Periodic Table*

Strictly speaking the transition metals and the d-block elements are not exactly the same. Scandium and zinc are not transition metals because they do not form any compounds in which they have partly filled d-orbitals, which is the characteristic of transition metals.

The origin of the terms s, p, d, and f is historical. When elements are heated they give out light energy at certain wavelengths, as excited electrons fall back from one energy level to a lower one. This causes lines to appear in the spectrum of light they give out. The letters s, p, d, and f stand for words that were used first to describe the lines – s for sharp, p for principal, d for diffuse, and f for fine.

Groups

A **group** is a vertical column of elements. The elements in the same group form a chemical 'family' – they have similar properties. Elements in the same group have the same number of electrons in the outer main level. The groups were traditionally numbered I–VII in Roman numerals plus zero for the noble gases, missing out the transition elements. It is now common to number them in ordinary numbers as 1–7 and 0 (or 1–8) and sometimes as 1–18 including the transition metals.

Reactivity

In the s-block, elements (metals) get more reactive going down a group. To the right (non-metals), elements tend to get more reactive going up a group.

Transition elements are a block of rather unreactive metals. This is where most of the useful metals are found.

Lanthanides are metals which are not often encountered. They all tend to form +3 ions in their compounds and have broadly similar reactivity.

Actinides are radioactive metals. Only thorium and uranium occur naturally in the Earth's crust in anything more than trace quantities.

Periods

Horizontal rows of elements in the Periodic Table are called **periods**. The periods are numbered starting from Period 1, which contains only hydrogen and helium. Period 2 contains the elements lithium to neon, and so on. There are trends in physical properties and chemical behaviour as you go across a period, see Topic 8.2.

Placing hydrogen and helium

The positions of hydrogen and helium vary in different versions of the table. Helium is usually placed above the noble gases (Group 0) because of its properties. But, it is not a p-block element– its electronic arrangement is $1s^2$.

Hydrogen is sometimes placed above Group 1 but is often placed on its own. It usually forms singly charged +1 (H^+) ions like the Group 1 elements but otherwise is not similar to them since they are all reactive metals and hydrogen is a gas. It is sometimes placed above the halogens because it can form H^- ions and also bond covalently.

Summary questions

1 From the elements, Br, Cl, Fe, K, Cs, and Sb, pick out:

   a two elements

      i in the same period

      ii in the same group

      iii that are non-metals

   b one element

      i that is in the d-block

      ii that is in the s-block.

2 From the elements Tl, Ge, Xe, Sr, and W, pick out:

   a a noble gas

   b the element described by Group 4, Period 4

   c an s-block element

   d a p-block element

   e a d-block element.

Trends in the properties of elements of Period 3

The Periodic Table reveals patterns in the properties of elements. For example, every time you go across a period you go from metals on the left to non-metals on the right. This is an example of **periodicity**. The word periodic means recurring regularly.

Periodicity and properties of elements in Period 3

Periodicity is explained by the electron arrangements of the elements.

- The elements in Groups 1, 2, and 3 (sodium, magnesium, and aluminium) are metals. They have giant structures. They lose their outer electrons to form ionic compounds.

- Silicon in Group 4 has four electrons in its outer shell with which it forms four covalent bonds. The element has some metallic properties and is classed as a semi-metal.

- The elements in Groups 5, 6, and 7 (phosphorus, sulfur, and chlorine) are non-metals. They either accept electrons to form ionic compounds, or share their outer electrons to form covalent compounds.

- Argon in Group 0 is a noble gas– it has a full outer shell and is unreactive.

Table 1 shows some trends across Period 3. Similar trends are found in other periods.

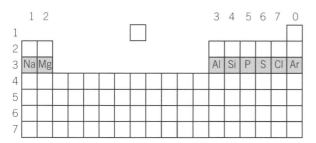

▲ **Figure 1** *The Periodic Table with Period 3 highlighted*

Learning objectives:
→ Describe the trends in melting and boiling temperatures of the elements in Period 3.
→ Explain these trends in terms of bonding and structure.

Specification reference: 3.2.1

Study tip
- Remember that when a molecular substance melts, the covalent bonds remain intact but the van der Waals forces break.
- Learn the formulae P_4, S_8, Cl_2.

▼ **Table 1** *Some trends across Period 3*

Group	1	2	3	4	5	6	7	0
Element	sodium	magnesium	aluminium	silicon	phosphorus	sulfur	chlorine	argon
Electron arrangement	[Ne] $3s^1$	[Ne] $3s^2$	[Ne] $3s^2 3p^1$	[Ne] $3s^2 3p^2$	[Ne] $3s^2 3p^3$	[Ne] $3s^2 3p^4$	[Ne] $3s^2 3p^5$	[Ne] $3s^2 3p^6$
	s-block			p-block				
	metals			semi-metal	non-metals			noble gas
Structure of element	giant metallic			macromolecular (giant covalent)	molecular			atomic
					P_4	S_8	Cl_2	Ar
Melting point, T_m / K	371	922	933	1683	317 (white)	392 (monoclinic)	172	84
Boiling point, T_b / K	1156	1380	2740	2628	553 (white)	718	238	87

Hint

The melting temperature of a substance is also the freezing temperature.

Synoptic link

To revisie metallic bonding, look back at Topic 3.3, Metallic bonding.

Trends in melting and boiling points

The trends in melting and boiling points are shown in Figure 2.

There is a clear break in the middle of the figure between elements with high melting points (on the left, with sodium, Na, in Group 1 as the exception) and those with low melting points (on the right). These trends are due to their structures.

- Giant structures (found on the left) tend to have high melting points and boiling points.
- Molecular or atomic structures (found on the right) tend to have low melting points and boiling points.

The melting points and boiling points of the metals increase from sodium to aluminium because of the strength of metallic bonding. As you go from left to right the charge on the ion increases so more electrons join the delocalised electron 'sea' that holds the giant metallic lattice together.

The melting points of the non-metals with molecular structures depend on the sizes of the van der Waals forces between the molecules. This in turn depends on the number of electrons in the molecule and how closely the molecules can pack together. As a result the melting points of these non-metals are ordered: $S_8 > P_4 > Cl_2$. Silicon with its giant structure has a much higher melting point. Boiling points follow a similar pattern.

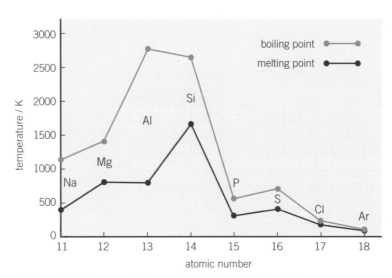

▲ **Figure 2** *Melting and boiling points of elements in Period 3*

Summary questions

1 Whereabouts in a period do you find the following? Choose from left, right, middle.

   a Elements that lose electrons when forming compounds

   b Elements that accept electrons when forming compounds

2 In what group do you find an element that exists as the following?

   a separate atoms

   b a macromolecule

3 A and B are both elements. Both conduct electricity— A well, B slightly. A melts at a low temperature, B at a much higher temperature. Suggest the identity of A and B and explain how their bonding and structure account for their properties.

Some key properties of atoms, such as size and ionisation energy, are periodic, that is, there are similar trends as you go across each period in the Periodic Table.

Atomic radii

These tell us about the sizes of atoms. You cannot measure the radius of an isolated atom because there is no clear point at which the electron cloud density around it drops to zero. Instead half the distance between the centres of a pair of atoms is used, see Figure 1.

The atomic radius of an element can differ as it is a general term. It depends on the type of bond that it is forming – covalent, ionic, metallic, van der Waals, and so on. The covalent radius is most commonly used as a measure of the size of the atom. Figure 2 shows a plot of covalent radius against atomic number.

(Even metals can form covalent molecules such as Na_2 in the gas phase. Since noble gases do not bond covalently with one another, they do not have covalent radii and so they are often left out of comparisons of atomic sizes.)

The graph shows that:

- atomic radius is a periodic property because it decreases across each period and there is a jump when starting the next period
- atoms get larger down any group.

Why the radii of atoms decrease across a period

You can explain this trend by looking at the electronic structures of the elements in a period, for example, sodium to chlorine in, Period 3, as shown in Figure 3.

As you move from sodium to chlorine you are adding protons to the nucleus and electrons to the outer main level, the third shell. The charge on the nucleus increases from +11 to +17. This increased charge pulls the electrons in closer to the nucleus. There are no additional electron shells to provide more shielding. So the size of the atom *decreases* as you go across the period.

atom	Na	Mg	Al	Si	P	S	Cl
size of atom							
	2,8,1	2,8,2	2,8,3	2,8,4	2,8,5	2,8,6	2,8,7
atomic (covalent) radius / nm	0.156	0.136	0.125	0.117	0.110	0.104	0.099
nuclear charge	11+	12+	13+	14+	15+	16+	17+

▲ **Figure 3** *The sizes and electronic structures of the elements sodium to chlorine*

Learning objectives:
→ Describe the trends in atomic radius and first ionisation energy of the elements in Period 3.
→ Explain these trends.
Specification reference: 3.2.1

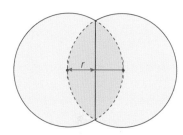

▲ **Figure 1** *Atomic radii are taken to be half the distance between the centres of a pair of atoms*

Hint

1 nm is 1×10^{-9} m

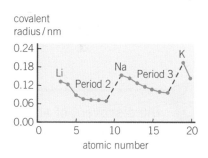

▲ **Figure 2** *The periodicity of covalent radii. The noble gases are not included because they do not form covalent bonds with one another*

Study tip

It is a common mistake to think that atoms increase in size as you cross a period. While the nuclei have more protons (and neutrons) the radius of the atom depends on the size of the electron shells.

Why the radii of atoms increase down a group

Going down a group in the Periodic Table, the atoms of each element have one extra complete main level of electrons compared with the one before. So, for example, in Group 1 the outer electron in potassium is in main level 4, whereas in sodium it is in main level 3. So going down the group, the outer electron main level is further from the nucleus and the atomic radii increase.

Synoptic link

You learnt about ionisation energy in Topic 1.6, Electron arrangements and ionistion energy.

First ionisation energy

The first ionisation energy is the energy required to convert a mole of isolated gaseous atoms into a mole of singly positively charged gaseous ions, that is, to remove one electron from each atom.

$$E(g) \rightarrow E^+(g) + e^-(g) \qquad \text{where E stands for any element}$$

The first ionisation energies also have periodic patterns. These are shown in Figure 4.

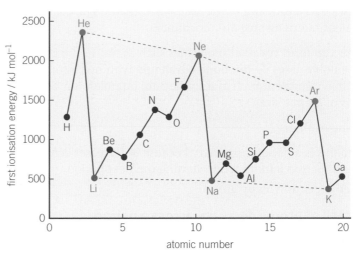

▲ **Figure 4** *The periodicity of first ionisation energies*

✚ The discovery of argon

When the Periodic Table was first put forward, none of the noble gases had been discovered. The first, argon, was discovered by Scottish chemist William Ramsay. He noticed that the density of nitrogen prepared by a chemical reaction was 1.2505 g dm^{-3} while nitrogen prepared by removing oxygen and carbon dioxide from air had a density of 1.2572 g dm^{-3}. He reasoned that there must be a denser impurity in the second sample, which he showed to be a previously unknown and very unreactive gas – argon. He later went on to discover the whole group of noble gases for which he won the Nobel (not noble!) Prize.

Chemists at the time had difficulty placing an unreactive gas of A$_r$ approximately 40 in the Periodic Table.

A suggestion was made that argon might be an allotrope of nitrogen, N$_3$, analogous to the O$_3$ allotrope of oxygen.

1. To how many significant figures were the two densities measured?
2. Using relative atomic masses from the Periodic Table, explain why argon is denser than nitrogen.
3. Suggest how oxygen could be removed from a sample of air.
4. What would be the M_r of N$_3$ (to the nearest whole number)?
5. Suggest why chemists were reluctant to regard argon as an element.

The first ionisation energy generally increases across a period– alkali metals like sodium, Na, and lithium, Li, have the lowest values and the noble gases (helium, He, neon, Ne, and argon, Ar) have the highest values.

The first ionisation energy decreases going down any group. The trends for Group 1 and Group 0 are shown dotted in red and green, respectively on the graph.

You can explain these patterns by looking at electronic arrangements (Figure 5).

Outer electrons are harder to remove as nuclear charge increases

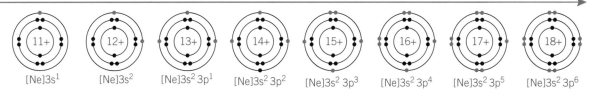

$[Ne]3s^1$ $[Ne]3s^2$ $[Ne]3s^2 3p^1$ $[Ne]3s^2 3p^2$ $[Ne]3s^2 3p^3$ $[Ne]3s^2 3p^4$ $[Ne]3s^2 3p^5$ $[Ne]3s^2 3p^6$

▲ **Figure 5** *The electronic structures of the elements sodium to argon*

Why the first ionisation energy increases across a period
As you go across a period from left to right, the number of protons in the nucleus increases but the electrons enter the same main level, see Figure 5. The increased charge on the nucleus means that it gets increasingly difficult to remove an electron.

Why the first ionisation energy decreases going down a group
The number of filled inner levels increases down the group. This results in an increase in shielding. Also, the electron to be removed is at an increasing distance from the nucleus and is therefore held less strongly. Thus the outer electrons get easier to remove going down a group because they are further away from the nucleus.

Why there is a drop in ionisation energy from one period to the next
Moving from neon in Period 0 (far right) with electron arrangement 2,8 to sodium, 2,8,1 (Period 1, far left) there is a sharp drop in the first ionisation energy. This is because at sodium a new main level starts and so there is an increase in atomic radius, the outer electron is further from the nucleus, less strongly attracted and easier to remove.

Summary questions

1 What happens to the size of atoms as you go from left to right across a period? Choose from increase, decrease, no change.

2 What happens to the first ionisation energy as you go from left to right across a period? Choose from increase, decrease, no change.

3 What happens to the nuclear charge of the atoms as you go left to right across a period?

4 Why do the noble gases have the highest first ionisation energy of all the elements in their period?

This chapter revisits the trends in ionisation energies first dealt with in Topic 1.6, in the context of periodicity. The graph of first ionisation energy against atomic number across a period is not smooth. Figure 1 below shows the plot for Period 3.

> **Hint**
>
> Ionisation energies are sometimes called ionisation enthalpies.

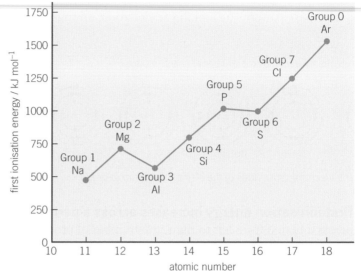

▲ **Figure 1** *Graph of first ionisation energy against atomic number for the elements of Period 3*

It shows that:

• the first ionisation energy actually drops between Group 2 and Group 3, so that aluminium has a lower ionisation energy than magnesium

• the ionisation energy drops again slightly between Group 5 (phosphorus) and Group 6 (sulfur).

Similar patterns occur in other periods. You can explain this if you look at the electron arrangements of these elements.

The drop in first ionisation energy between Groups 2 and 3

For the first ionisation energy:

• magnesium, $1s^2 2s^2 2p^6 3s^2$, loses a 3s electron

• aluminium, $1s^2 2s^2 2p^6 3s^2 3p^1$, loses the 3p electron.

The p-electron is already in a higher energy level than the s-electron, so it takes less energy to remove it, see Figure 2.

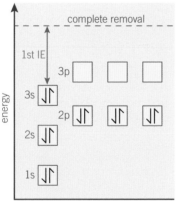

magnesium $1s^2 2s^2 2p^6 3s^2$

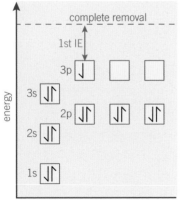

aluminium $1s^2 2s^2 2p^6 3s^2 3p^1$

▲ **Figure 2** *The first ionisation energies of magnesium and aluminium (not to scale)*

The drop in first ionisation energy between Groups 5 and 6

An electron in a pair will be easier to remove that one in an orbital on its own because it is already being repelled by the other electron. As shown in Figure 3:

- phosphorus, $1s^2 2s^2 2p^6 3s^2 3p^3$, has no paired electrons in a p-orbital because each p-electron is in a different orbital
- sulfur, $1s^2 2s^2 2p^6 3s^2 3p^4$, has two of its p-electrons paired in a p-orbital so one of these will be easier to remove than an unpaired one due to the repulsion of the other electron in the same orbital.

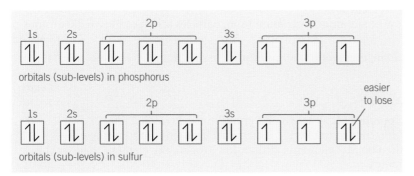

▲ **Figure 3** *Electron arrangements of phosphorus and sulfur*

Successive ionisation energies

If you remove the electrons from an atom one at a time, each one is harder to remove than the one before. Figure 4 is a graph of ionisation energy against number of electrons removed for sodium, electron arrangement 2,8,1.

You can see that there is a sharp increase in ionisation energy between the first and second electrons. This is followed by a gradual increase over the next eight electrons and then another jump before the final two electrons. Sodium, in Group 1 of the Periodic Table, has one electron in its outer main level (the easiest one to remove), eight in the next main level and two (very hard to remove) in the innermost main level.

Figure 5 is a graph of successive ionisation energies against number of electrons removed for aluminium, electron arrangement 2,8,3.

It shows three electrons that are relatively easy to remove – those in the outer main level – and then a similar pattern to that for sodium.

If you plotted a graph for chlorine, the first seven electrons would be relatively easier to remove than the next eight.

This means that the number of electrons that are relatively easy to remove tells us the group number in the Periodic Table. For example, the values of 906, 1763, 14 855, and 21 013 kJ mol⁻¹ for the first five ionisation energies of an element, tell us that the element is in Group 2. This is because the big jump occurs after two electrons have been removed.

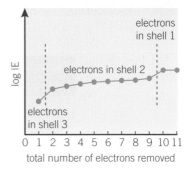

▲ **Figure 4** *Graph of successive ionisation energies against number of electrons removed for sodium. Note that the log of the ionisation energies is plotted in order to fit the large range of values onto the scale*

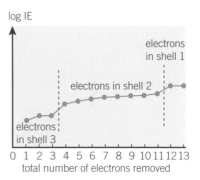

▲ **Figure 5** *Graph of successive ionisation energy against number of electrons removed for aluminium*

Summary questions

1   Write the electron arrangement in the form $1s^2$... for:

   a beryllium

   b boron.

2   If one electron is lost from for the following atoms, from what main level does it come?

   a beryllium

   b boron

3   Why is the first ionisation energy of boron less than that of beryllium?

Practice questions

1   The following table gives the melting points of some elements in Period 3.

Element	Na	Al	Si	P	S
Melting point / K	371	933	1680	317	392

   (a) State the type of structure shown by a crystal of silicon.
        Explain why the melting point of silicon is very high.

(3 marks)

   (b) State the type of structure shown by crystals of sulfur and phosphorus.
        Explain why the melting point of sulfur is higher than the melting point of phosphorus.

(3 marks)

   (c) Draw a diagram to show how the particles are arranged in aluminium and explain
        why aluminium is malleable.
        (You should show a minimum of six aluminium particles arranged in two dimensions.)

(3 marks)

   (d) Explain why the melting point of aluminium is higher than the melting point of
        sodium.

(3 marks)

AQA, 2011

2   Trends in physical properties occur across all Periods in the Periodic Table.
    This question is about trends in the Period 2 elements from lithium to nitrogen.
    (a) Identify, from the Period 2 elements lithium to nitrogen, the element that has the
         largest atomic radius.

(1 mark)

    (b) (i)   State the general trend in first ionisation energies for the Period 2 elements
              lithium to nitrogen.

(1 mark)

    (b) (ii) Identify the element that deviates from this general trend, from lithium to
              nitrogen, and explain your answer.

(3 marks)

    (c) Identify the Period 2 element that has the following successive ionisation energies.

	First	Second	Third	Fourth	Fifth	Sixth
Ionisation energy / kJ mol^{-1}	1090	2350	4610	6220	37 800	47 000

(1 mark)

AQA, 2012

    (d) Draw a cross on the diagram to show the melting point of nitrogen.

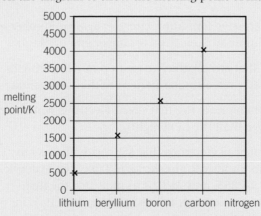

(1 mark)

    (e) Explain, in terms of structure and bonding, why the melting point of carbon is high.

(3 marks)

3 **(a)** Use your knowledge of electron configuration and ionisation energies to answer this question.
The following diagram shows the **second** ionisation energies of some Period 3 elements.

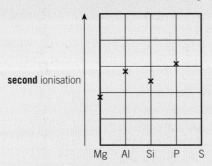

(i) Draw an '**X**' on the diagram to show the **second** ionisation energy of sulfur.

(1 mark)

(ii) Write the full electron configuration of the Al^{2+} ion.

(1 mark)

(iii) Write an equation to show the process that occurs when the **second** ionisation energy of aluminium is measured.

(1 mark)

(iv) Give **one** reason why the **second** ionisation energy of silicon is lower than the **second** ionisation energy of aluminium.

(1 mark)

(b) Predict the element in Period 3 that has the highest **second** ionisation energy.
Give a reason for your answer.

(2 marks)

(c) The following table gives the successive ionisation energies of an element in Period 3.

	First	Second	Third	Fourth	Fifth	Sixth
Ionisation energy / kJ mol^{-1}	786	1580	3230	4360	16 100	19 800

Identify this element.

(1 mark)

(d) Explain why the ionisation energy of every element is endothermic.

(1 mark)

AQA, 2013

4 The elements in Period 2 show periodic trends.
(a) Identify the Period 2 element, from carbon to fluorine, that has the largest atomic radius. Explain your answer.

(3 marks)

(b) State the general trend in first ionisation energies from carbon to neon.
Deduce the element that deviates from this trend and explain why this element deviates from the trend.

(4 marks)

(c) Write an equation, including state symbols, for the reaction that occurs when the first ionisation energy of carbon is measured.

(1 mark)

(d) Explain why the second ionisation energy of carbon is higher than the first ionisation energy of carbon.

(1 mark)

(e) Deduce the element in Period 2, from lithium to neon, that has the highest second ionisation energy.

(1 mark)

AQA, 2013

Learning objectives

→ Explain the changes in the atomic radius of the Group 2 elements from Mg to Ba.

→ Explain the changes in the first ionisation energy of the Group 2 elements from Mg to Ba.

→ Explain the changes in the melting point of the Group 2 elements from Mg to Ba.

→ State the trend in reactivity of the group.

→ State the trend in solubilities of a) the hydroxides b) the sulfates.

Specification reference: 3.2.2

The elements in Group 2 are sometimes called the **alkaline earth metals**. This is because their oxides and hydroxides are alkaline. Like Group 1, they are s-block elements. They are similar in many ways to Group 1 but they are less reactive. Beryllium is not typical of the group and is not considered here.

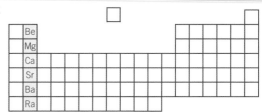

The physical properties of the Group 2 elements, magnesium to barium

A summary of some of the physical properties of the elements from magnesium to barium is given in Table 1. Trends in properties are shown by the arrows, which show the direction of increase.

Electron arrangement
The elements all have two electrons in an outer s-orbital. This s-orbital becomes further away from the nucleus going down the group.

The sizes of the atoms
The atoms get bigger going down the group. The atomic (metallic) radii increase because each element has an extra filled main level of electrons compared with the one above it.

Melting points
Group 2 elements are metals with high melting points, typical of a giant metallic structure. Going down the group, the electrons in the 'sea' of delocalised electrons are further away from the positive nuclei. As a result, the strength of the metallic bonds decreases going down the group. For this reason the melting points of Group 2 elements decrease slightly going down the group, starting with calcium.

Magnesium, with the lowest melting point, does not fit this trend but there is no simple explanation for this anomaly.

Ionisation energies
In *all* their reactions, atoms of elements in Group 2 lose their two outer electrons to form ions with two positive charges.

$$M(g) \rightarrow M(g)^{2+} + 2e^-$$

So, an amount of energy equal to the sum of the first and the second ionisation energies is needed for complete ionisation.

$$M(g) \rightarrow M^+(g) + e^- \quad \text{plus} \quad M^+(g) \rightarrow M^{2+}(g) + e^-$$

Both the first ionisation energy and the second ionisation energy decrease going down the group – it takes less energy to remove the electrons as they become further and further away from the positive nucleus. The nucleus is more effectively shielded by more inner shells of electrons.

> ## Synoptic link
>
> Look back at Topic 3.3, Metallic bonding for more on the giant metallic structure.

> ## Hint
>
> Remember the shorthand for writing electron arrangements using the previous inert gas. So $[Ne]3s^2$ is shorthand for $1s^2, 2s^2, 2p^6, 3s^2$.

▼ **Table 1** *The physical properties of Group 2, magnesium to barium*

	Atomic number Z	Electron arrangement	Metallic radius / nm	First + second IEs / kJ mol^{-1}	Melting point T_m / K	Boiling point T_b / K	Density ρ / g cm^{-3}
magnesium, Mg	12	[Ne]$3s^2$	0.160	738 + 1451 = 2189 ▲	922	1380	1.74
calcium, Ca	20	[Ar]$4s^2$	0.197	590 + 1145 = 1735	1112	1757	1.54
strontium, Sr	38	[Kr]$5s^2$	0.215	550 + 1064 = 1614	1042	1657	2.60
barium, Ba	56	[Xe]$6s^2$	0.224 ▼	503 + 965 = 1468	998	1913	3.51 ▼

In all their reactions, the metals get more reactive going down the group.

Lime kilns

Disused lime kilns can be found in many areas where there is limestone rock. Limestone is mainly calcium carbonate, $CaCO_3$, and it was heated in kilns fired by wood or coal to produce quicklime (calcium oxide, CaO) which was used to make building mortar, to treat acidic soils, and in making glass.

In the kiln, heat decomposes the limestone:

$$CaCO_3(s) \rightarrow CaO(s) + CO_2(g)$$

A typical kiln contained around 25 tonnes of limestone.

▲ **Figure 1** *A disused lime kiln*

1　⊞ Calculate how many tonnes of lime this would produce.

2　Give two reasons why lime kilns were significant emitters of carbon dioxide.

3　The limestone was broken into fist-sized lumps before firing. Suggest why.

In practice limestone is unlikely to be found 100% pure. One contaminant is silicon dioxide (sand), which is unaffected by heat. Imagine limestone that contains 15% sand, so 100 tonnes of limestone would contain 85 tonnes of calcium carbonate.

4　Rework the calculation above to calculate how much lime would actually be produced.

Both lime (calcium oxide, CaO) and slaked lime (calcium hydroxide, $Ca(OH)_2$) may be used to neutralise acids in soil.

The equations for their reactions with hydrochloric acid (for simplicity) are given below.

$$CaO(s) + 2HCl(aq) \rightarrow CaCl_2(aq) + H_2O(l)$$

$$Ca(OH)_2(s) + 2HCl(aq) \rightarrow CaCl_2(aq) + 2H_2O(l)$$

5　How many *moles* of **a** lime and **b** quicklime are needed to neutralise 2 mol HCl?

6　How many *grams* of **a** lime and **b** slaked lime are needed to neutralise 2 mol HCl?

7　What implications does this have for the farmer or gardener?

8　Suggest other factors to be considered when deciding which compound to use.

Hint

Remember *lower* pH means *more* acidic.

The chemical reactions of the Group 2 elements, magnesium to barium

Oxidation is loss of electrons so in all their reactions the Group 2 metals are oxidised. The metals go from oxidation state 0 to oxidation state +2. These are redox reactions.

Reaction with water

With water you see a trend in reactivity – the metals get more reactive going down the group. These are also redox reactions.

The basic reaction is as follows, where M is any Group 2 metal:

$$\overset{0}{M}(s) + 2H_2O(l) \rightarrow \overset{+1}{M}\overset{+2}{(OH)_2}(aq) + \overset{0}{H_2}(g)$$

Magnesium hydroxide is milk of magnesia and is used in indigestion remedies to neutralise excess stomach acid which causes heartburn, indigestion, and wind.

Magnesium reacts very slowly with cold water but rapidly with steam to form an alkaline oxide and hydrogen.

$$Mg(s) + H_2O(g) \rightarrow MgO(s) + H_2(g)$$

Calcium reacts in the same way but more vigorously, even with cold water. Strontium and barium react more vigorously still. Calcium hydroxide is sometimes called slaked lime and is used to treat acidic soil. Most plants have an optimum level of acidity or alkalinity in which they thrive. For example, grass prefers a pH of around 6 so if the soil has a pH much below this, then it will not grow as well as it could. Crops such as wheat, corn, oats, and barley prefer soil that is nearly neutral.

▲ **Figure 2** *Two applications of Group 2 hydroxides.*

The extraction of titanium

Titanium is a metal with very useful properties – it is strong, low density, and has a high melting point. It is used in the aerospace industry and also for making replacement hip joints. It is a relatively common metal in the Earth's crust but it is not easy to extract.

Most metals are found in the Earth as oxides, and the metal is extracted by reacting the oxide with carbon:

Metal oxide + carbon → metal + carbon dioxide

This method cannot be used for titanium as the metal reacts with carbon to form titanium carbide, TiC, which makes the metal brittle. So the titanium oxide is first reacted with chlorine and carbon (coke) to form titanium chloride, $TiCl_4$, and carbon monoxide.

The titanium chloride is then reduced to titanium by reaction with magnesium:

$$TiCl_4(l) + 2Mg(s) \rightarrow 2MgCl_2(s) + Ti(s)$$

1   Write a balanced symbol equation for the reaction of iron oxide, Fe_2O_3, with carbon.
2   What is unusual about titanium chloride as a metal compound and what does this suggest about the bonding in it?
3   Write a balanced symbol equation for the reaction of titanium oxide with chlorine and carbon.
4   Work out the oxidation state of each element in this equation before and after reaction. What has been oxidised and what has been reduced.

The solubilities of the hydroxides and sulfates

Hydroxides

There is a clear trend in the solubilities of the hydroxides – going down the group they become more soluble. The hydroxides are all white solids.

Magnesium hydroxide, $Mg(OH)_2$ (milk of magnesia), is almost insoluble. It is sold as a suspension in water, rather than a solution.

- Calcium hydroxide, $Ca(OH)_2$, is sparingly soluble and a solution is used as lime water.
- Strontium hydroxide, $Sr(OH)_2$, is more soluble.
- Barium hydroxide, $Ba(OH)_2$, dissolves to produce a strongly alkaline solution:

   $Ba(OH)_2(s) + aq \rightarrow Ba^{2+}(aq) + 2OH^-(aq)$

Sulfates

The solubility trend in the sulfates is exactly the opposite – they become less soluble going down the group. So, barium sulfate is virtually insoluble. This means that it can be taken by mouth as a barium meal to outline the gut in medical X-rays. (The heavy barium atom is very good at absorbing X-rays.) This test is safe, despite the fact that barium compounds are highly toxic, because barium sulfate is so insoluble.

The insolubility of barium sulfate is also used in a simple test for sulfate ions in solution. The solution is first acidified with nitric or hydrochloric acid. Then barium chloride solution is added to the solution under test and if a sulfate is present a white precipitate of barium sulfate is formed.

$Ba^{2+}(aq) + SO_4^{2-}(aq) \rightarrow BaSO_4(s)$

The addition of acid removes carbonate ions as carbon dioxide. (Barium carbonate is also a white insoluble solid, which would be indistinguishable from barium sulfate).

Hint

The trends in solubilities of the hydroxide and sulfates can be used as the basis of a test for Ca^{2+}, Sr^{2+}, and Ba^{2+} ions in compounds.

Hint

The symbol aq is used to represent an unspecified amount of water.

Summary questions

1 a What is the oxidation number of all Group 2 elements in their compounds?

   b Explain your answer.

2 Why does it become easier to form +2 ions going down Group 2?

3 Explain why this is a redox reaction.
   $Ca + Cl_2 \rightarrow CaCl_2$

4 Write the equation for the reaction of calcium with water. Include the oxidation state of each element.

5 How would you expect the reaction of strontium with water to compare with those of the following? Explain your answers.

   a calcium

   b barium

6 Radium is below strontium in Group 2. How would you predict the solubilities of the following compounds would compare with the other members of the group? Explain your answers.

   a radium hydroxide

   b radium sulfate

1   State and explain the trend in melting point of the Group 2 elements Ca to Ba.

(3 marks)
AQA, 2006

2   State the trends in solubility of the hydroxides and of the sulfates of the Group 2 elements Mg to Ba.

Describe a chemical test you could perform to distinguish between separate aqueous solutions of sodium sulfate and sodium nitrate. State the observation you would make with each solution. Write an equation for any reaction which occurs.

(6 marks)
AQA, 2006

3   **(a)**   For the elements Mg to Ba, state how the solubilities of the hydroxides and the solubilities of the sulfates change down Group 2.

(b)   Describe a test to show the presence of sulfate ions in an aqueous solution. Give the results of this test when performed on separate aqueous solutions of magnesium chloride and magnesium sulfate. Write equations for any reactions occurring.

(c)   State the trend in the reactivity of the Group 2 elements Mg to Ba with water. Write an ionic equation with state symbols to show the reaction of barium with an excess of water.

(9 marks)
AQA, 2005

4   Group 2 metals and their compounds are used commercially in a variety of processes and applications.
(a)   State a use of magnesium hydroxide in medicine.

(1 mark)

(b)   Calcium carbonate is an insoluble solid that can be used in a reaction to lower the acidity of the water in a lake.

Explain why the rate of this reaction decreases when the temperature of the water in the lake falls.

(3 marks)

(c)   Strontium metal is used in the manufacture of alloys.
   (i)   Explain why strontium has a higher melting point than barium.

(2 marks)

   (ii)   Write an equation for the reaction of strontium with water.

(1 mark)

(d)   Magnesium can be used in the extraction of titanium.
   (i)   Write an equation for the reaction of magnesium with titanium(IV) chloride.

(1 mark)

   (ii)   The excess of magnesium used in this extraction can be removed by reacting it with dilute sulfuric acid to form magnesium sulfate.

Use your knowledge of Group 2 sulfates to explain why the magnesium sulfate formed is easy to separate from the titanium.

(1 mark)
AQA, 2010

5   Group 2 elements and their compounds have a wide range of uses.
(a)   For parts **(a)(i)** to **(a)(iii)**, choose the correct answer to complete each sentence.
   (i)   From $Mg(OH)_2$ to $Ba(OH)_2$, the solubility in water

decreases	increases	stays the same

(1 mark)

   (ii)   From Mg to Ba, the first ionisation energy

decreases	increases	stays the same

(1 mark)

   (iii)   From Mg to Ba, the atomic radius

decreases	increases	stays the same

(1 mark)

(b) Explain why calcium has a higher melting point than strontium.

(2 marks)

(c) Acidified barium chloride solution is used as a reagent to test for sulfate ions.

(i) State why sulfuric acid should **not** be used to acidify the barium chloride.

(1 mark)

(ii) Write the **simplest ionic** equation with state symbols for the reaction that occurs when acidified barium chloride solution is added to a solution containing sulfate ions.

(1 mark)

AQA, 2012

6 The following diagram shows the first ionisation energies of some Period 3 elements.

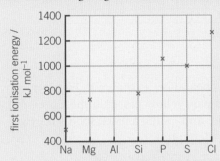

(a) Draw a cross on the diagram to show the first ionisation energy of aluminium.

(1 mark)

(b) Write an equation to show the process that occurs when the first ionisation energy of aluminium is measured.

(2 marks)

(c) State which of the first, second or third ionisations of aluminium would produce an ion with the electron configuration $1s^2 2s^2 2p^6 3s^1$

(1 mark)

(d) Explain why the value of the first ionisation energy of sulfur is less than the value of the first ionisation energy of phosphorus.

(2 marks)

(e) Identify the element in Period 2 that has the highest first ionisation energy and give its electron configuration.

(2 marks)

(f) State the trend in first ionisation energies in Group 2 from beryllium to barium. Explain your answer in terms of a suitable model of atomic structure.

(3 marks)

AQA, 2010

7 **(a)** There are many uses for compounds of barium.

(i) Write an equation for the reaction of barium with water.

(1 mark)

(ii) State the trend in reactivity with water of the Group 2 metals from Mg to Ba

(1 mark)

(b) Give the formula of the **least** soluble hydroxide of the Group 2 metals from Mg to Ba

(1 mark)

(c) State how barium sulfate is used in medicine. Explain why this use is possible, given that solutions containing barium ions are poisonous.

(2 marks)

AQA, 2012

Learning objectives:

→ Explain how and why the atomic radius changes in Group 7 of the Periodic Table.

→ Explain how and why electronegativity changes in Group 7 of the Periodic Table.

Specification reference: 3.2.3

Group 7, on the right-hand side of the Periodic Table, is made up of non-metals. As elements they exist as diatomic molecules, F_2, Cl_2, Br_2, and I_2, called the halogens. (Astatine is rare and radioactive.)

Hint

The word halogen means 'salt former'. The halogens readily react with many metals to form fluoride, chloride, bromide, and iodide salts.

▲ **Figure 1** *Fluorine chlorine, bromine and iodine in their gaseous states*

Physical properties

The gaseous halogens vary in appearance, as shown in Figure 1. At room temperature, fluorine is a pale yellow gas, chlorine a greenish gas, bromine a red-brown liquid, and iodine a black solid – they get darker and denser going down the group.

They all have a characteristic 'swimming-bath' smell.

A number of the properties of fluorine are untypical. Many of these untypical properties stem from the fact that the F—F bond is unexpectedly weak, compared with the trend for the rest of the halogens, see Table 1. The small size of the fluorine atom leads to repulsion between non-bonding electrons because they are so close together:

▼ **Table 1** *Bond energies for fluorine, chlorine, bromine, and iodine*

Bond	Bond energy / kJ mol^{-1}
F–F	158
Cl–Cl	243
Br–Br	193
I–I	151

The physical properties of fluorine, chlorine, bromine, and iodine are shown in Table 2.

There are some clear trends shown by the red arrows. These can be explained as follows.

Size of atoms

The atoms get bigger going down the group because each element has one extra filled main level of electrons compared with the one above it, see Figure 2.

▼ **Table 2** *The physical properties of Group 7, fluorine to iodine*

Halogen	Atomic number, Z	Electron arrangement	Electronegativity	Atomic (covalent) radius / nm	Melting point T_m / K	Boiling point T_b / K
fluorine	9	[He] $2s^2\,2p^5$	4.0	0.071	53	85
chlorine	17	[Ne] $3s^2\,3p^5$	3.0	0.099	172	238
bromine	35	[Ar] $3d^{10}\,4s^2\,4p^5$	2.8	0.114	266	332
iodine	53	[Kr] $4d^{10}\,5s^2\,5p^5$	2.5	0.133	387	457

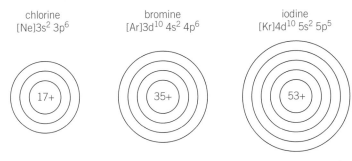

chlorine
[Ne]$3s^2 3p^6$

bromine
[Ar]$3d^{10} 4s^2 4p^6$

iodine
[Kr]$4d^{10} 5s^2 5p^5$

▲ **Figure 2** *The outer shell gets further from the nucleus going down the group*

Electronegativity

Electronegativity is a measure of the ability of an atom to attract electrons, or electron density, towards itself within a covalent bond.

Electronegativity depends on the attraction between the nucleus and bonding electrons in the outer shell. This, in turn, depends on a balance between the number of protons in the nucleus (nuclear charge) and the distance between the nucleus and the bonding electrons, plus the shielding effect of inner shells of electrons.

For example, consider the hydrogen halides, HX. The shared electrons in the H—X bond get further away from the nucleus as the atoms get larger going down the group. This makes the shared electrons further from the halogen nucleus and increases the shielding by more inner shells of electrons. These factors are more important than the increasing nuclear charge, so the electronegativity decreases going down the group.

Melting and boiling points

These increase as going down the group. This is because larger atoms have more electrons and this makes the van der Waals forces between the molecules stronger. The lower the boiling point, the more volatile the element. So chlorine, which is a gas at room temperature, is more volatile than iodine, which is a solid.

Study tip

Remember that melting and boiling points involve weakening and breaking van der Waals forces only. The covalent bonds in the halogen molecules stay intact.

Study tip

Remember to write a halogen element as a diatomic molecule, for example, as F_2 not F.

Synoptic link

Look back at Topic 3.2, Covalent bonding.

Summary questions

1 Predict the properties of astatine compared with the other halogens in terms of:

   a physical state at room temperature, including colour

   b size of atom

   c electronegativity.

2 Explain your answers to question **1**.

3 a Use the data in Table 2 to make a rough estimate of the boiling point of astatine.

   b Why would you expect the boiling point of astatine to be the largest?

Learning objectives:

→ State the trend in oxidising ability of the halogens.

→ Describe the experimental evidence that confirms this trend.

Specification reference: 3.2.3

Hint

Remember OIL RIG: Oxidation Is Loss of electrons. Reduction Is Gain of electrons.

Trends in oxidising ability

Halogens usually react by gaining electrons to become negative ions, with a charge of -1. These reactions are redox reactions – halogens are oxidising agents and are themselves reduced. For example:

$$Cl_2 + 2e^- \xrightarrow{\text{gain of electrons}} 2Cl^-$$

The oxidising ability of the halogens increases going up the group.

Fluorine is one of the most powerful oxidising agents known.

fluorine   chlorine   bromine   iodine

$\longleftarrow$ increasing oxidising power $\longrightarrow$

Displacement reactions

Halogens will react with metal halides in solution in such a way that the halide in the compound will be displaced by a more reactive halogen but not by a less reactive one. This is called a **displacement reaction**.

For example, chlorine will displace bromide ions, but iodine will not.

$$\overset{0}{Cl_2}(aq) + 2Na\overset{-1}{Br}(aq) \rightarrow \overset{0}{Br_2}(aq) + 2Na\overset{-1}{Cl}(aq)$$

The ionic equation for this reaction is:

$$Cl_2(aq) + \cancel{2Na^+(aq)} + 2Br^-(aq) \rightarrow Br_2(aq) + \cancel{2Na^+(aq)} + 2Cl^-(aq)$$

The sodium ions are spectator ions.

The two colourless starting materials react to produce the red-brown colour of bromine.

In this redox reaction the chlorine is acting as an oxidising agent, by removing electrons from Br^- and so oxidising $2Br^-$ to Br_2 (the oxidation number of the bromine increases from -1 to 0). In general, a halogen will always displace the ion of a halogen below it in the Periodic Table, see Table 1.

▼ **Table 1** *The oxidation of a halide by a halogen*

	F^-	Cl^-	Br^-	I^-
F_2	–	yes	yes	yes
Cl_2	no	–	yes	yes
Br_2	no	no	–	yes
I_2	no	no	no	–

You cannot investigate fluorine in an aqueous solution because it reacts with water.

The extraction of bromine from sea water

The oxidation of a halide by a halogen is the basis of a method for extracting bromine from sea water. Sea water contains small amounts of bromide ions which can be oxidised by chlorine to produce bromine:

$$Cl_2(aq) + 2Br^-(aq) \rightarrow Br_2(aq) + 2Cl^-(aq)$$

Extraction of iodine from kelp

Iodine was discovered in 1811. It was extracted from kelp, which is obtained by burning seaweed. Some iodine is still produced in this way. Salts such as sodium chloride, potassium chloride, and potassium sulfate are removed from the kelp by washing with water. The residue is then heated with manganese dioxide and concentrated sulfuric acid and iodine is liberated.

$$2I^- + MnO_2 + 4H^+ \rightarrow Mn^{2+} + 2H_2O + I_2$$

1 Is the reaction an oxidation or a reduction of the iodide ion? Explain your answer.
2 Find out why our table salt often has potassium iodide added to it.

Summary questions

1 **a** Which of the following mixtures would react?

    i $Br_2(aq) + 2NaCl(aq)$

    ii $Cl_2(aq) + 2NaI(aq)$

  b Explain your answers.

  c Complete the equation for the mixture that reacts.

Halide ions as reducing agents

Halide ions can act as reducing agents. In these reactions the halide ions lose (give away) electrons and become halogen molecules. There is a definite trend in their reducing ability. This is linked to the size of the ions. The larger the ion, the more easily it loses an electron. This is because the electron is lost from the outer shell which is further from the nucleus as the ion gets larger so the attraction to the outer electron is less.

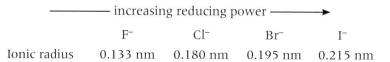

	increasing reducing power $\longrightarrow$		
	F^-	Cl^-	Br^-	I^-
Ionic radius	0.133 nm	0.180 nm	0.195 nm	0.215 nm

This trend can be seen in the reactions of solid sodium halides with concentrated sulfuric acid.

The reactions of sodium halides with concentrated sulfuric acid

Solid sodium halides react with concentrated sulfuric acid. The products are different and reflect the reducing powers of the halide ions shown above.

Sodium chloride (solid)

In this reaction, drops of concentrated sulfuric acid are added to solid sodium chloride. Steamy fumes of hydrogen chloride are seen. The solid product is sodium hydrogensulfate.

The reaction is:

$$NaCl(s) + H_2SO_4(l) \rightarrow NaHSO_4(s) + HCl(g)$$

This is not a redox reaction because no oxidation state has changed. The chloride ion is too weak a reducing agent to reduce the sulfur (oxidation state = +6) in sulfuric acid. It is an acid–base reaction.

$$\overset{+1\ -1}{Na\,Cl(s)} + \overset{+1\ +6\ -2}{H_2\,S\,O_4(l)} \rightarrow \overset{+1\ +1\ +6\ -2}{Na\,H\,S\,O_4(s)} + \overset{+1\ -1}{H\,Cl(g)}$$

This reaction can be used to prepare hydrogen chloride gas which, because of this reaction, was once called salt gas.

A similar reaction occurs with sodium fluoride to produce hydrogen fluoride, an extremely dangerous gas that will etch glass. The fluoride ion is an even weaker reducing agent than the chloride ion.

Sodium bromide (solid)

In this case you will see steamy fumes of hydrogen bromide *and* brown fumes of bromine. Colourless sulfur dioxide is also formed.

Two reactions occur.

First sodium hydrogensulfate and hydrogen bromide are produced (in a similar acid–base reaction to sodium chloride).

$$NaBr(s) + H_2SO_4(l) \rightarrow NaHSO_4(s) + HBr(g)$$

However, bromide ions are strong enough reducing agents to reduce the sulfuric acid to sulfur dioxide. The oxidation state of the sulfur is reduced from +6 to +4 and that of the bromine increases from –1 to 0.

$$\overset{-1}{2H^+ + 2Br^-} + \overset{+6}{H_2SO_4(l)} \rightarrow \overset{+4}{SO_2(g)} + 2H_2O(l) + \overset{0}{Br_2(l)}$$

This is a redox reaction. The reactions are exothermic and some of the bromine vaporises.

Sodium iodide (solid)

In this case you see steamy fumes of hydrogen iodide, the black solid of iodine, and the bad egg smell of hydrogen sulfide gas is present. Yellow solid sulfur may also be seen. Colourless sulfur dioxide is also evolved.

Several reactions occur. Hydrogen iodide is produced in an acid–base reaction as before.

$$NaI(s) + H_2SO_4(l) \rightarrow NaHSO_4(s) + HI(g)$$

Iodide ions are better reducing agents than bromide ions so they reduce the sulfur in sulfuric acid even further (from +6 to zero and –2) so that sulfur dioxide, sulfur, and hydrogen sulfide gas are produced. For example:

$$\overset{-1}{8H^+ + 8I^-} + \overset{+6}{H_2SO_4(l)} \rightarrow \overset{-2}{H_2S(g)} + 4H_2O(l) + \overset{0}{4I_2(s)}$$

During the reduction from +6 to –2, the sulfur passes through oxidation state 0 and some yellow, solid sulfur may be seen.

Study tip

Remember that the reactions take place between *solid* halide salts and *concentrated* sulfuric acid.

Further redox equations

As the sulfur in the sulfuric acid is reduced by the iodide ions from oxidation state +6 to –2, it passes through oxidation states +4 (sulfur dioxide, SO_2) and 0 (uncombined sulfur, S).

Use the oxidation state technique to help you to write equations for these two processes.

$$\overset{+6}{H_2SO_4} + \overset{-1}{6I^-} + 6H^+ \rightarrow \overset{0}{S} + \overset{0}{3I_2} + 4H_2O$$

$$\overset{+6}{H_2SO_4} + \overset{-1}{2I^-} + 2H^+ \rightarrow \overset{+4}{SO_2} + \overset{0}{I_2} + 2H_2O$$

Identifying metal halides with silver ions

All metal halides (except fluorides) react with silver ions in aqueous solution, for example, in silver nitrate, to form a precipitate of the insoluble silver halide. For example:

$$Cl^-(aq) + Ag^+(aq) \rightarrow AgCl(s)$$

Silver fluoride does not form a precipitate because it is soluble in water.

1 Dilute nitric acid HNO_3 or $(H^+(aq) + NO_3^-(aq))$ is first added to the halide solution to get rid of any soluble carbonate, $CO_3^{2-}(aq)$, or hydroxide, $OH^-(aq)$ impurities:

$$CO_3^{2-}(aq) + 2H^+(aq) + 2NO_3^-(aq) \rightarrow CO_2(g) + H_2O(l) + 2NO_3^-(aq)$$

$$OH^-(aq) + H^+(aq) + NO_3^-(aq) \rightarrow H_2O(l) + NO_3^-(aq)$$

These would interfere with the test by forming insoluble silver carbonate:

$$2Ag^+(aq) + CO_3^{2-}(aq) \rightarrow Ag_2CO_3(s)$$

or insoluble silver hydroxide:

$$Ag^+(aq) + OH^-(aq) \rightarrow AgOH(s)$$

Hint

Sulfuric acid or hydrochloric acid cannot be used. Sulfuric acid would give a precipitate of silver sulfate and hydrochloric acid a precipitate of silver chloride. Either of these would invalidate the test.

Hint

Silver hydroxide in fact is converted into silver oxide – a brown precipitate:

$$2AgOH \rightarrow Ag_2O + H_2O$$

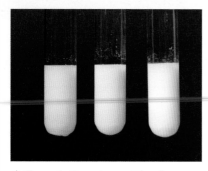

▲ **Figure 1** *The colours of the silver halides: (from left to right) AgCl, AgBr, AgI*

2 Then a few drops of silver nitrate solution are added and the halide precipitate forms.

The reaction can be used as a test for halides because you can tell from the colour of the precipitate which halide has formed, see Table 1. The colours of silver bromide and silver iodide are similar but if you add a few drops of concentrated ammonia solution, silver bromide dissolves but silver iodide does not.

▼ **Table 1** *Tests for halides*

Halide	silver fluoride	silver chloride	silver bromide	silver iodide
Colour	no precipitate	white ppt	cream ppt	pale yellow ppt
Further test		dissolves in dilute ammonia	dissolves in concentrated ammonia	insoluble in concentrated ammonia

Study tip

You need to learn the colours of the silver halides, especially AgBr and AgI.

Summary questions

1 The reaction between concentrated sulfuric acid and solid sodium fluoride is not usually carried out in the laboratory.

   a How does the reducing power of the fluoride ion compare with the other halide ions?

   b Explain why you would predict this.

   c Write a balanced symbol equation for the reaction between concentrated sulfuric acid and sodium fluoride.

   d Is this a redox reaction? Explain your answer.

2 🧪 A few drops of silver nitrate were added to an acidified solution, to show the presence of sodium bromide.

   a What would you see?

   b Write the equation for the reaction.

   c What would happen if you now added a few drops of concentrated ammonia solution?

   d Why is an acid added to sodium bromide solution initially?

   e Neither hydrochloric nor sulfuric acid may be used to acidify the solution. Explain why this is so.

   f Why can't this test be used to find out if fluoride ions are present?

10.4 Uses of chlorine

Chlorine is a poisonous gas and was notoriously used as such in the First World War. However, it is soluble in water and in this form has become an essential part of our life in the treatment of water both for drinking and in swimming pools.

Reaction with water

Chlorine reacts with water in a reversible reaction to form chloric(I) acid, HClO, and hydrochloric acid, HCl:

$$\underset{Cl_2(g)}{0} + H_2O(l) \rightleftharpoons \underset{HClO(aq)}{+1} + \underset{HCl(aq)}{-1}$$

In this reaction, the oxidation number of one of the chlorine atoms increases from 0 to +1 and that of the other decreases from 0 to −1. This type of redox reaction, where the oxidation state of some atoms of the same element increase and others decrease, is called **disproportionation**.

This reaction takes place when chlorine is used to purify water for drinking and in swimming baths, to prevent life-threatening diseases. Chloric(I) acid is an oxidising agent and kills bacteria by oxidation. It is also a bleach.

The other halogens react similarly, but much more slowly going down the group.

In sunlight, a different reaction occurs:

$$\underset{\text{pale green}}{2Cl_2(g)} + 2H_2O(l) \rightarrow \underset{\text{colourless}}{4HCl(aq)} + O_2(g)$$

Chlorine is rapidly lost from pool water in sunlight so that shallow pools need frequent addition of chlorine.

An alternative to the direct chlorination of swimming pools is to add solid sodium (or calcium) chlorate(I). This dissolves in water to form chloric(I) acid, HClO(aq,) in a reversible reaction:

$$NaClO(s) + H_2O \rightleftharpoons Na^+(aq) + OH^-(aq) + HClO(aq)$$

In alkaline solution, this equilibrium moves to the left and the HClO is removed as ClO⁻ ions. To prevent this happening, swimming pools need to be kept slightly acidic. However, this is carefully monitored and the water never gets acidic enough to corrode metal components and affect swimmers.

Reaction with alkali

Chlorine reacts with cold, dilute sodium hydroxide to form sodium chlorate(I), NaClO. This is an oxidising agent and the active ingredient in household bleach. This is also a disproportionation reaction – see the oxidation numbers above the relevant species.

$$\underset{Cl_2(g)}{0} + 2NaOH(aq) \rightarrow \underset{NaClO(aq)}{+1} + \underset{NaCl(aq)}{-1} + H_2O(l)$$

The other halogens behave similarly.

Learning objectives:

→ Describe how chlorine reacts with water.

→ Describe how chlorine reacts with alkali.

Specification reference: 3.2.3

Summary questions

1 Write the equations for bromine reacting with:

   a water

   b alkali.

2 Why is chlorine added to the domestic water supply?

3 a What products are obtained when an aqueous solution of chlorine is left in the sunlight?

   b Write the equation for the reaction, giving the oxidation states of every atom before and after reaction.

   c What has been oxidised?

   d What has been reduced?

   e What is the oxidising agent?

   f What is the reducing agent?

Practice questions

1 (a) State the trend in electronegativity of the elements down Group 7. Explain this trend.

(3 marks)

   (b) (i) State the trend in reducing ability of the halide ions down Group 7.
        (ii) Give an example of a reagent which could be used to show that the reducing ability of bromide ions is different from that of chloride ions.

(2 marks)

   (c) The addition of silver nitrate solution followed by dilute aqueous ammonia can be used as a test to distinguish between chloride and bromide ions. For each ion, state what you would observe if an aqueous solution containing the ion was tested in this way.

(4 marks)

   (d) Write an equation for the reaction between chlorine and cold, dilute aqueous sodium hydroxide. Give two uses of the resulting solution.

(3 marks)
AQA, 2006

2 (a) Explain, by referring to electrons, the meaning of the terms *reduction* and *reducing agent*.

(2 marks)

   (b) Iodide ions can reduce sulfuric acid to three different products.
        (i) Name the **three** reduction products and give the oxidation state of sulfur in each of these products.
        (ii) Describe how observations of the reaction between solid potassium iodide and concentrated sulfuric acid can be used to indicate the presence of any two of these reduction products.
        (iii) Write half-equations to show how two of these products are formed by reduction of sulfuric acid.

(10 marks)

   (c) Write an ionic equation for the reaction that occurs when chlorine is added to cold water. State whether or not the water is oxidised and explain your answer.

(3 marks)
AQA, 2006

3 (a) State the trend in the boiling points of the halogens from fluorine to iodine and explain this trend.

(4 marks)

   (b) Each of the following reactions may be used to identify bromide ions. For each reaction, state what you would observe and, where indicated, write an appropriate equation.
        (i) The reaction of aqueous bromide ions with chlorine gas
        (ii) The reaction of aqueous bromide ions with aqueous silver nitrate followed by the addition of concentrated aqueous ammonia
        (iii) The reaction of solid potassium bromide with concentrated sulfuric acid

(7 marks)

   (c) Write an equation for the redox reaction that occurs when potassium bromide reacts with concentrated sulfuric acid.

(2 marks)
AQA, 2005

4 (a) State and explain the trend in electronegativity down Group 7 from fluorine to iodine.

(3 marks)

   (b) State what you would observe when chlorine gas is bubbled into an aqueous solution of potassium iodide. Write an equation for the reaction that occurs.

(2 marks)

   (c) Identify **two** sulfur-containing reduction products formed when concentrated sulfuric acid oxidises iodide ions. For each reduction product, write a half-equation to illustrate its formation from sulfuric acid.

(4 marks)

   (d) Write an equation for the reaction between chlorine gas and dilute aqueous sodium hydroxide. Name the **two** chlorine-containing products of this reaction and give the oxidation state of chlorine in each of these products.

(5 marks)
AQA, 2005

5   A student investigated the chemistry of the halogens and the halide ions.

(a)   In the first two tests, the student made the following observations.

Test	Observation
1 Add chlorine water to aqueous potassium iodide solution.	The colourless solution turned a brown colour.
2 Add silver nitrate solution to aqueous potassium chloride solution.	The colourless solution produced a white precipitate.

(i)   Identify the species responsible for the brown colour in Test **1**.
Write the **simplest ionic** equation for the reaction that has taken place in Test **1**.
State the type of reaction that has taken place in Test **1**.

(3 marks)

(ii)   Name the species responsible for the white precipitate in Test **2**.
Write the **simplest ionic** equation for the reaction that has taken place in Test **2**.
State what would be observed when an excess of dilute ammonia solution is added to the white precipitate obtained in Test **2**.

(3 marks)

(b)   In two further tests, the student made the following observations.

Test	Observation
3 Add concentrated sulfuric acid to solid potassium chloride.	The white solid produced misty white fumes which turned blue litmus paper to red.
4 Add concentrated sulfuric acid to solid potassium iodide.	The white solid turned black. A gas was released that smelled of rotten eggs. A yellow solid was formed.

(i)   Write the **simplest ionic** equation for the reaction that has taken place in Test **3**.
Identify the species responsible for the misty white fumes produced in Test **3**.

(2 marks)

(ii)   The student had read in a textbook that the equation for one of the reactions in Test **4** is as follows.

$$8H^+ + 8I^- + H_2SO_4 \rightarrow 4I_2 + H_2S + 4H_2O$$

Write the **two** half-equations for this reaction.
State the role of the sulfuric acid and identify the yellow solid that is also observed in Test **4**.

(4 marks)

(iii)   The student knew that bromine can be used for killing microorganisms in swimming pool water.
The following equilibrium is established when bromine is added to cold water.

$$Br_2(l) + H_2O(l) \rightleftharpoons HBrO(aq) + H^+(aq) + Br^-(aq)$$

Use Le Chatelier's principle to explain why this equilibrium moves to the right when sodium hydroxide solution is added to a solution containing dissolved bromine.
Deduce why bromine can be used for killing microorganisms in swimming pool water, even though bromine is toxic.

(3 marks)
AQA, 2012

Section 2 practice questions

1   For each of the following reactions, select from the list below, the **formula** of a sodium halide that would react as described.

NaF   NaCl   NaBr   NaI

Each **formula** may be selected once, more than once or not at all.

(a) This sodium halide is a white solid that reacts with concentrated sulfuric acid to give a brown gas.

(1 mark)

(b) When a solution of this sodium halide is mixed with silver nitrate solution, no precipitate is formed.

(1 mark)

(c) When this solid sodium halide reacts with concentrated sulfuric acid, the reaction mixture remains white and steamy fumes are given off.

(1 mark)

(d) A colourless aqueous solution of this sodium halide reacts with orange bromine water to give a dark brown solution.

(1 mark)

AQA, 2010

2   There are many uses for Group 2 metals and their compounds.

(a) State a medical use of barium sulfate.

State why this use of barium sulfate is safe, given that solutions containing barium ions are poisonous.

(2 marks)

(b) Magnesium hydroxide is used in antacid preparations to neutralise excess stomach acid.

Write an equation for the reaction of magnesium hydroxide with hydrochloric acid.

(1 mark)

(c) Solutions of barium hydroxide are used in the titration of weak acids.

State why magnesium hydroxide solution could **not** be used for this purpose.

(1 mark)

(d) Magnesium metal is used to make titanium from titanium(IV) chloride.

Write an equation for this reaction of magnesium with titanium(IV) chloride.

(1 mark)

(e) Magnesium burns with a bright white light and is used in flares and fireworks.

Use your knowledge of the reactions of Group 2 metals with water to explain why water should **not** be used to put out a fire in which magnesium metal is burning.

(2 marks)

AQA, 2014

3   This question is about Group 7 chemistry.

(a) Sea water is a major source of iodine.

The iodine extracted from sea water is impure. It is purified in a two-stage process.

Stage **1** $I_2 + 2H_2O + SO_2 \rightarrow 2HI + H_2SO_4$

Stage **2** $2HI + Cl_2 \rightarrow I_2 + 2HCl$

(i) State the initial oxidation state and the final oxidation state of sulfur in Stage **1**.

(2 marks)

(ii) State, in terms of electrons, what has happened to chlorine in Stage **2**.

(1 mark)

(b) When concentrated sulfuric acid is added to potassium iodide, iodine is formed in the following redox equations.

......KI + ...H_2SO_4 $\rightarrow$ ......$KHSO_4$ + ...I_2 + S   + ...H_2O

$8KI$ + $9H_2SO_4$ $\rightarrow$ $8KHSO_4$ + $4I_2$ + H_2S + $4H_2O$

(i) Balance the equation for the reaction that forms sulfur.

(1 mark)

(ii) Deduce the half-equation for the formation of iodine from iodide ions.

(1 mark)

(iii) Deduce the half-equation for the formation of hydrogen sulfide from concentrated sulfuric acid.

(1 mark)

(c) A yellow precipitate is formed when silver nitrate solution, acidified with dilute nitric acid, is added to an aqueous solution containing iodide ions.

(i) Write the **simplest ionic** equation including state symbols for the formation of the yellow precipitate.

(1 mark)

(ii) State what is observed when concentrated ammonia solution is added to this yellow precipitate.

(1 mark)

(iii) State why the silver nitrate solution is acidified when testing for iodide ions.

(1 mark)

(iv) Explain why dilute hydrochloric acid is **not** used to acidify the silver nitrate solution in this test for iodide ions.

(1 mark)

(d) Chlorine is toxic to humans. This toxicity does not prevent the large-scale use of chlorine in water treatment.

(i) Give one reason why water is treated with chlorine.

(1 mark)

(ii) Explain why the toxicity of chlorine does **not** prevent this use.

(1 mark)

(iii) Write an equation for the reaction of chlorine with cold water.

(1 mark)

(e) Give the formulas of the **two** different chlorine-containing compounds that are formed when chlorine reacts with cold, dilute, aqueous sodium hydroxide.
State the oxidation state of chlorine in each of the chlorine-containing ions formed.

(1 mark)
AQA, 2014

4 (a) The diagram below shows the melting points of some of the elements in Period 3.

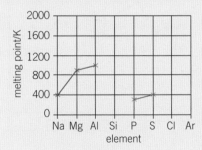

(i) On a copy of the diagram, use crosses to mark the approximate positions of the melting points for the elements silicon, chlorine and argon. Complete the diagram by joining the crosses.

(ii) By referring to its structure and bonding, explain your choice of position for the melting point of silicon.

(iii) Explain why the melting point of sulfur, S_8, is higher than that of phosphorus, P_4.

(8 marks)

(b) State and explain the trend in melting point of the Group 2 elements Ca–Ba.

(3 marks)
AQA, 2006

Section 2 summary

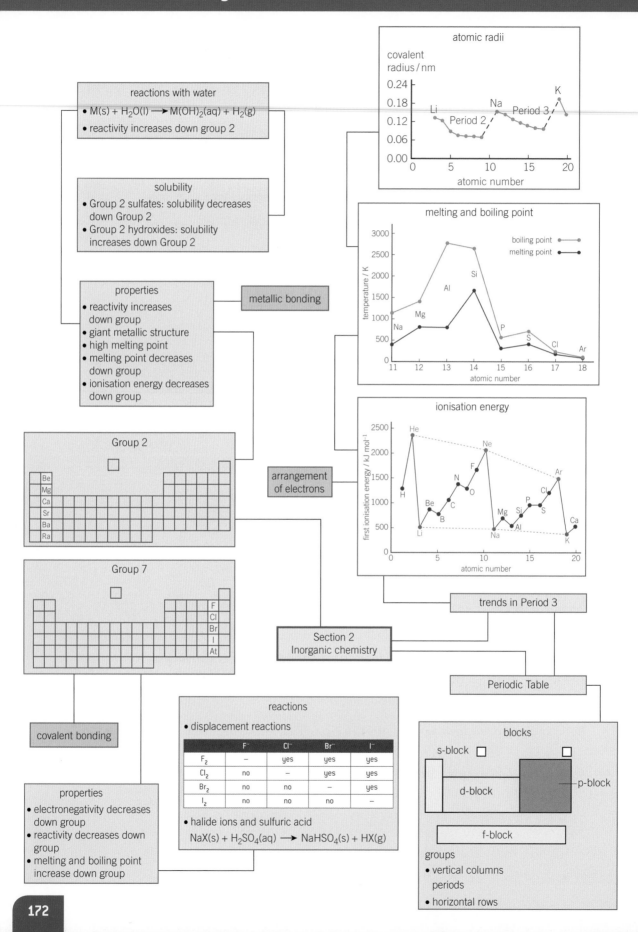

atomic radii

reactions with water
- $M(s) + H_2O(l) \longrightarrow M(OH)_2(aq) + H_2(g)$
- reactivity increases down group 2

solubility
- Group 2 sulfates: solubility decreases down Group 2
- Group 2 hydroxides: solubility increases down Group 2

properties
- reactivity increases down group
- giant metallic structure
- high melting point
- melting point decreases down group
- ionisation energy decreases down group

metallic bonding

Group 2

| Be |
| Mg |
| Ca |
| Sr |
| Ba |
| Ra |

arrangement of electrons

Group 7

| F |
| Cl |
| Br |
| I |
| At |

covalent bonding

properties
- electronegativity decreases down group
- reactivity decreases down group
- melting and boiling point increase down group

reactions
- displacement reactions

	F⁻	Cl⁻	Br⁻	I⁻
F_2	–	yes	yes	yes
Cl_2	no	–	yes	yes
Br_2	no	no	–	yes
I_2	no	no	no	–

- halide ions and sulfuric acid
$NaX(s) + H_2SO_4(aq) \longrightarrow NaHSO_4(s) + HX(g)$

Section 2
Inorganic chemistry

trends in Period 3

Periodic Table

blocks
s-block
d-block
p-block
f-block

groups
- vertical columns
periods
- horizontal rows

Practical skills

In this section you have met the following ideas:

- Testing reactions of Group 2 metals with water.
- Testing solubility of Group 2 hydroxides and sulfates.
- Testing for sulfate ions using acidified barium chloride.
- Investigating the reactions of halogens and their halide ions.
- Testing for halide ions using acidified silver nitrate and ammonia.

Maths skills

In this section you have met the following maths skills:

- Identify trends and patterns in data.
- Balancing symbol equations.

Extension

Although Dimitri Mendeleev is credited with establishing the Periodic Table, it was really developed over time by a number of different scientists. One of the key achievements of Mendeleev's work was that he was able to predict the properties of elements that were yet to be discovered and was even able to predict the properties of those missing elements accurately.

1 Suggest how Mendeleev could have correctly predicted the properties of gallium, almost five years before it had been discovered.

2 Investigate the research on X-ray spectra by Henry Moseley and its impact on Mendeleev's Periodic Table.

3 Research the reaction between chlorine and sodium hydroxide. Using your knowledge and understanding of redox reactions explain why this reaction is particularly interesting.

Section 3
Organic chemistry

Chapters in this section

11 Introduction to organic chemistry

12 Alkanes

13 Halogenoalkanes

14 Alkenes

15 Alcohols

16 Organic analysis

Carbon atoms have the ability to bond in chains which may be straight, branched, or in rings, forming millions of compounds. Organic chemistry is the study of compounds based on carbon chains. Hydrogen is almost always present.

Introduction to organic chemistry looks at the nature of carbon compounds and explains the different types of formulae that can be used to represent a compound, and also the IUPAC naming system, used to describe organic compounds. It looks at the different sorts of isomers that are possible in some organic compounds. (Isomers have the same formula but a different arrangement of atoms.)

Alkanes is about crude oil and its fractional distillation. It also looks at the different ways that large alkane molecules can be cracked into smaller, more useful molecules. It deals with the combustion of carbon compounds.

Halogenoalkanes looks at how these compounds are formed, how they react and their role in the problem of depletion of the ozone layer.

Alkenes describes the reactions of these compounds which have one or more double bonds.

Alcohols shows the importance of ethanol and describes the primary, secondary, and tertiary structures of alcohols and their reactions.

Organic analysis revisits the mass spectrometer and describes its use in determining the relative molecular masses of compounds and also their molecular formulae. Infra-red spectroscopy is introduced as a vital tool for identifying the functional groups in organic compounds. Some test tube reactions that may be used to help identify organic compounds are described.

The concepts of the applications of science are found throughout the chapters, where they will provide you with an opportunity to apply your knowledge in a fresh context.

What you already know

The material in this section builds upon knowledge and understanding that you will have developed at GCSE, in particular the following:

- ☐ Carbon atoms have four outer electrons and can form four single bonds.
- ☐ Organic compounds are based on chains of carbon atoms.
- ☐ Double covalent bonds can be formed by sharing four electrons between a pair of atoms.

Learning objectives:

→ State what is meant by the terms empirical formula, molecular formula, skeletal formula, and structural formula.

Specification reference: 3.3.1

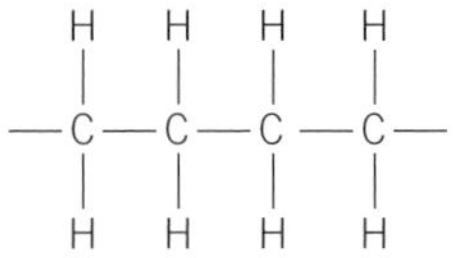

▲ **Figure 1** *Part of a hydrocarbon chain*

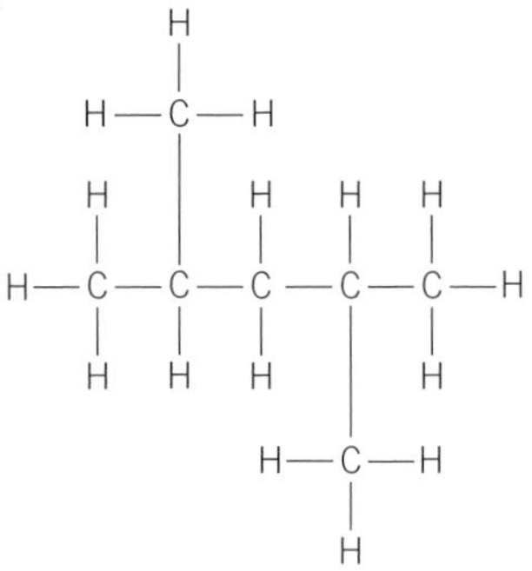

▲ **Figure 2** *A branched hydrocarbon chain*

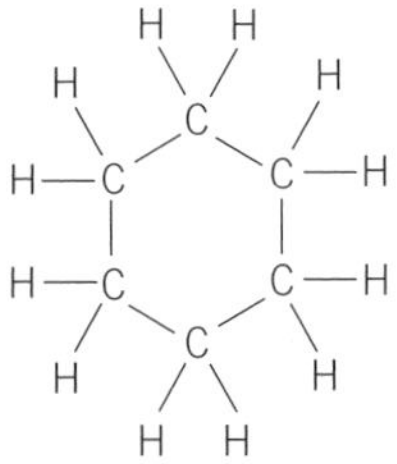

▲ **Figure 3** *A hydrocarbon ring*

Organic chemistry is the chemistry of carbon compounds. Life on our planet is based on carbon, and organic means to do with living beings. Nowadays, many carbon-based materials, like plastics and drugs, are made synthetically and there are large industries based on synthetic materials. There are far more compounds of carbon known than those of all the other elements put together, well over 10 million.

What is special about carbon?

Carbon can form rings and very long chains, which may be branched. This is because:

- a carbon atom has four electrons in its outer shell, so it forms four covalent bonds
- carbon–carbon bonds are relatively strong (347 kJ mol^{-1}) and non-polar.

The carbon–hydrogen bond is also strong (413 kJ mol^{-1}) and relatively non-polar. Hydrocarbon chains form the skeleton of most organic compounds, see Figures 1, 2, and 3.

➕ From inorganic to organic

Organic compounds were originally thought to be produced by living things only. This was disproved by Friedrich Wöhler in 1828. He made urea (an organic compound found in urine) from ammonium cyanate (an inorganic compound).

$$NH_4^+(NCO)^- \longrightarrow (NH_2)_2CO$$

ammonium cyanate   urea

He reported to a fellow chemist

"I cannot, so to say, hold my chemical water and must tell you that I can make urea without thereby needing to have kidneys, or anyhow, an animal, be it human or dog".

This reaction is an isomerism reaction.

1   What is meant by an isomerism reaction?
2   What is the atom economy of this reaction? Explain your answer.

Bonding in carbon compounds

In *all* stable carbon compounds, carbon forms four covalent bonds and has eight electrons in its outer shell. It can do this by forming bonds in different ways.

- By forming four single bonds as in methane:

- By forming two single bonds and one double bond as in ethene:

- By forming one single bond and one triple bond as in ethyne:

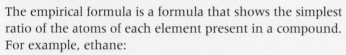

Formulae of carbon compounds

The empirical formula

Worked example: Empirical formula

The empirical formula is a formula that shows the simplest ratio of the atoms of each element present in a compound. For example, ethane:

3.00 g of ethane contains 2.40 g of carbon, $A_r = 12.0$, and 0.60 g of hydrogen, $A_r = 1.0$. What is its empirical formula?

$$\text{number of moles of carbon} = \frac{2.40}{12.0} = 0.20 \text{ mol of carbon}$$

$$\text{number of moles of hydrogen} = \frac{0.60}{1} = 0.60 \text{ mol of hydrogen}$$

Dividing through by the smaller number (0.20) gives C : H as 1 : 3

So the empirical formula of ethane is CH_3.

Hint

The mass of carbon plus the mass of hydrogen adds up to the mass of ethane so no other element is present.

The molecular formula

The molecular formula is the formula that shows the actual number of atoms of each element in the molecule. It is found from:

- the empirical formula
- the relative molecular mass of the empirical formula
- the relative molecular mass of the molecule.

Synoptic link

Look back at Topic 2.4, Empirical and molecular formula, if you are not sure about this calculation.

Worked example: Molecular formula

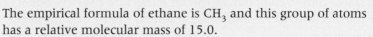

The empirical formula of ethane is CH_3 and this group of atoms has a relative molecular mass of 15.0.

The relative molecular mass of ethane is 30.0, which is 2×15.0. So, there must be two units of the empirical formula in every molecule of ethane.

The molecular formula is therefore $(CH_3)_2$ or C_2H_6.

Other formulae

Other, different types of formulae are used in organic chemistry because, compared with inorganic compounds, organic molecules are more varied. The type of formula required depends on the information that you are dealing with. You may want to know about the way the atoms are arranged within the molecule, as well as just the number of each atom present. There are different ways of doing this.

The displayed formula

This shows every atom and every bond in the molecule:

— is a single bond

= is a double bond

≡ is a triple bond

For ethene, C_2H_4, the displayed formula is:

$$\underset{H}{\overset{H}{\diagdown}} C = C \underset{H}{\overset{H}{\diagup}}$$

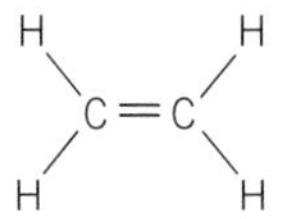

For ethanol, C_2H_6O, the displayed formula is:

$$H - \overset{\overset{\displaystyle H}{|}}{C} - \overset{\overset{\displaystyle H}{|}}{\underset{\underset{\displaystyle H}{|}}{C}} - O - H$$

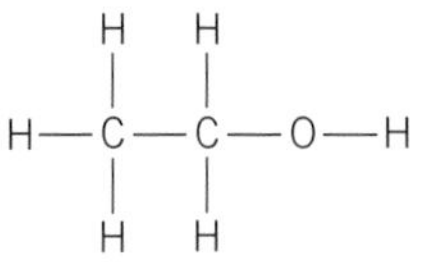

The structural formula

This shows the unique arrangement of atoms in a molecule in a simplified form, without showing all the bonds.

Each carbon is written separately, with the atoms or groups that are attached to it.

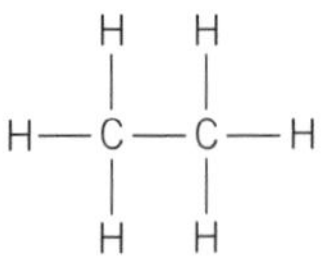

$$H - \overset{\overset{\displaystyle H}{|}}{\underset{\underset{\displaystyle H}{|}}{C}} - \overset{\overset{\displaystyle H}{|}}{\underset{\underset{\displaystyle H}{|}}{C}} - H \qquad \text{is written } CH_3CH_3$$

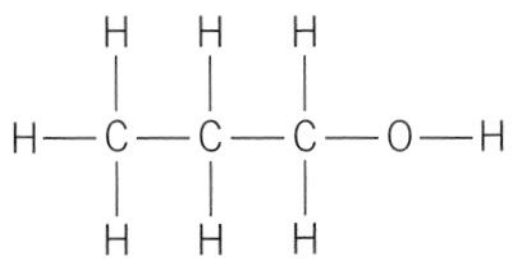

H—C—C—C—O—H   is written $CH_3CH_2CH_2OH$

Branches in the carbon chains are shown in brackets:

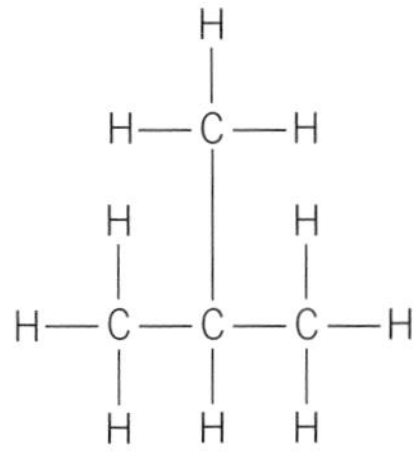

is written $CH_3CH(CH_3)CH_3$

Skeletal formulae

With more complex molecules, displayed structural formulae become time-consuming to draw. In skeletal notation the carbon atoms are not drawn at all. Straight lines represent carbon–carbon bonds and carbon atoms are assumed to be where the bonds meet. Neither hydrogen atoms nor C—H bonds are drawn. Each carbon is assumed to form enough C—H bonds to make a total of four bonds (counting double bonds as two).

Synoptic link

You will learn more about reaction mechanisms and free radicals in Topic 13.2, Nucleophilic substitution in halogenoalkanes, and Topic 12.3, Industrial cracking.

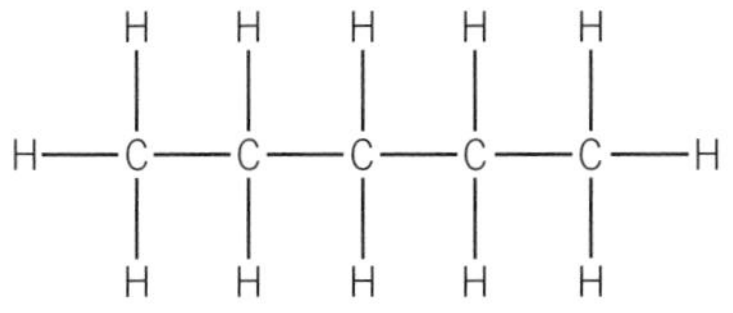

would be written

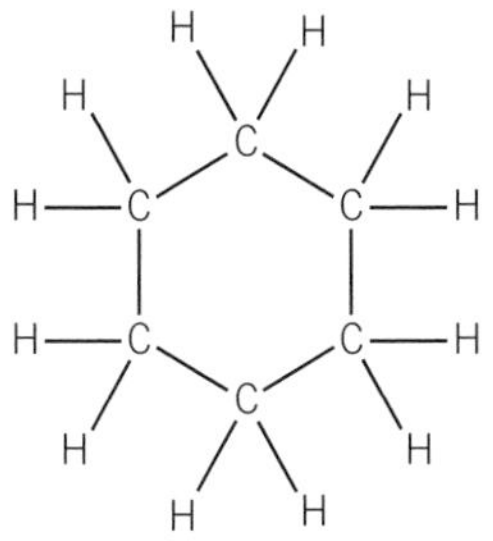

would be written

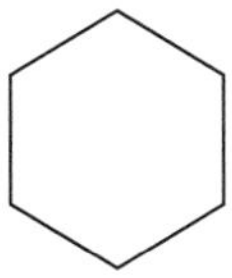

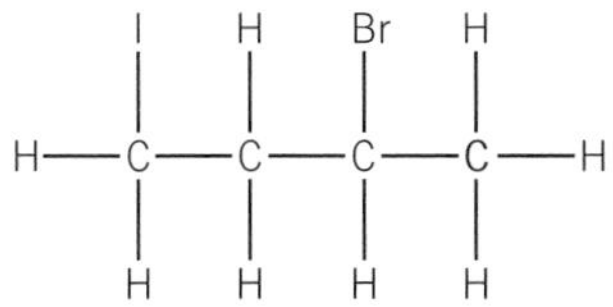

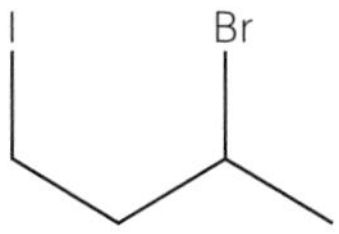

would be written

The choice of type of formula to use depends on the circumstances and the type of information you need to give. Notice that skeletal formulae give a rough idea of the bond angles. In an unbranched alkane chain these are 109.5°.

Three-dimensional structural formulae

These attempt to show the three-dimensional structure of the molecule. Bonds coming out of the paper are shown by wedges  and bonds going into the paper by dotted lines .

Some examples of different types of formulae are given in the Table 1.

▼ **Table 1** *Different types of formulae*

Empirical formula	Molecular formula/name	Structural formula				
		Shorthand	Displayed	Skeletal	Three-dimensional	
CH_2	C_6H_{12} hex-1-ene	$CH_2CHCH_2CH_2CH_2CH_3$				
C_2H_6O	C_2H_6O ethanol	CH_3CH_2OH				
C_3H_7Cl	C_3H_7Cl 2-chloropropane	$CH_3CHClCH_3$				
C_3H_6O	C_3H_6O propanone	CH_3COCH_3				

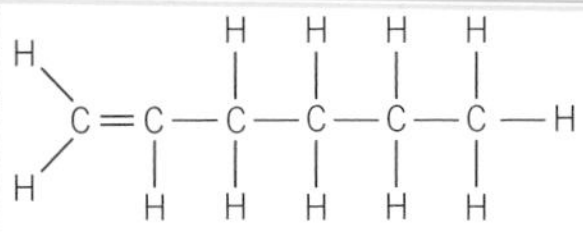

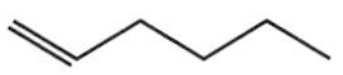

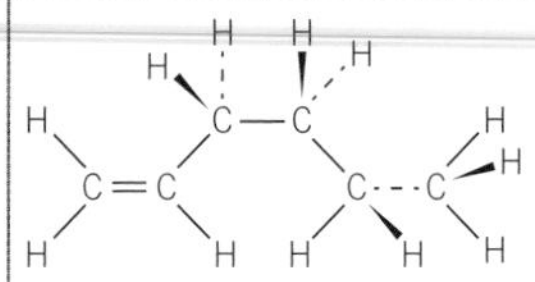

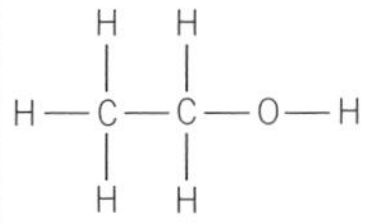

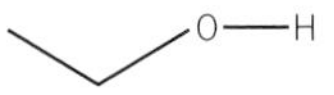

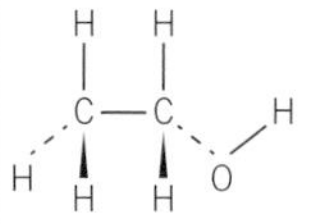

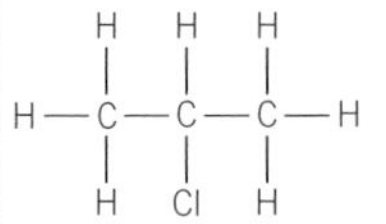

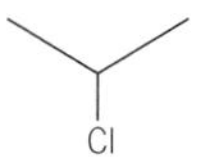

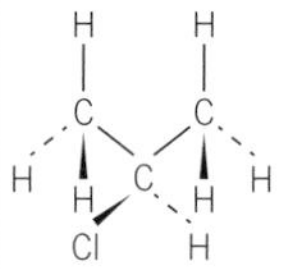

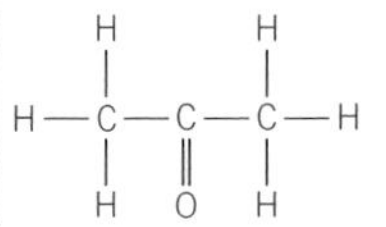

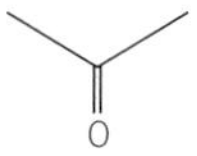

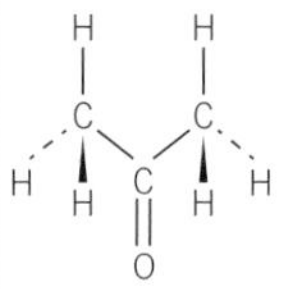

Summary questions

1 A compound comprising only carbon and hydrogen, in which 4.8 g of carbon combine with 1.0 g of hydrogen, has a relative molecular mass of 58.

a How many moles of carbon are there in 4.8 g?

b How many moles of hydrogen are there in 1.0 g?

c What is the empirical formula of this compound?

d What is the molecular formula of this compound?

e Draw the structural formula and the skeletal formula of the compound that has a straight chain.

f Draw the displayed formula and the skeletal formula of the compound that has a branched chain.

Reaction mechanisms

Curly arrows

You can often explain what happens in organic reactions by considering the movement of elections. As electrons are negatively charged, they tend to move from areas of high electron density to more positively charged areas. For example, a lone pair of electrons will be attracted to the positive end of a polar bond, written as $C^{\delta+}$. The movement of a pair of electrons is shown by a curly arrow that starts from a lone pair of electrons or from a covalent bond and moves towards a positively charged area of a molecule to form a new bond.

Free radicals

Sometimes a covalent bond (which consists of a pair of electrons shared between two atoms) may break in such a way that one electron goes to each atom that originally formed the bond. These fragments of the original molecule have an unpaired electron and are called free radicals. They are usually extremely reactive.

11.2 Nomenclature – naming organic compounds

Learning objectives:

→ Explain the IUPAC rules for naming alkanes and alkenes.

→ State what is meant by a functional group.

→ State what is meant by a homologous series.

Specification reference: 3.3.1

The system used for naming compounds was developed by the International Union of Pure and Applied Chemistry or **IUPAC**. This is an international organisation of chemists that draws up standards so that chemists throughout the world use the same conventions – rather like a universal language of chemistry. Systematic names tell us about the structures of the compounds rather than just the formula. Only the basic principles are covered here.

Roots

A systematic name has a *root* that tells us the longest unbranched hydrocarbon chain or ring, see Table 1.

The syllable after the root tells us whether there are any double bonds.

-ane means no double bonds. For example, ethane

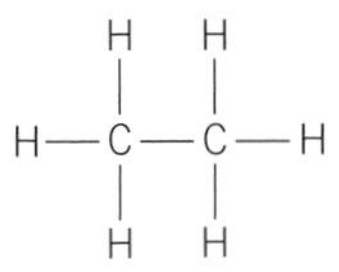

has two carbon atoms and no double bond.

-ene means there is a double bond. For example, ethene

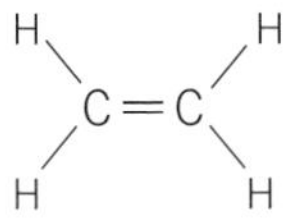

has two carbon atoms and one double bond.

▼ **Table 1** *The first six roots used in naming organic compounds*

Number of carbons	Root
1	meth
2	eth
3	prop
4	but
5	pent
6	hex

Prefixes and suffixes

Prefixes and suffixes describe the changes that have been made to the root molecule.

- Prefixes are added to the beginning of the root.

For example, side chains are shown by a prefix, whose name tells us the number of carbons:

methyl	CH_3-	ethyl	C_2H_5-
propyl	C_3H_7-	butyl	C_4H_9-

For example:

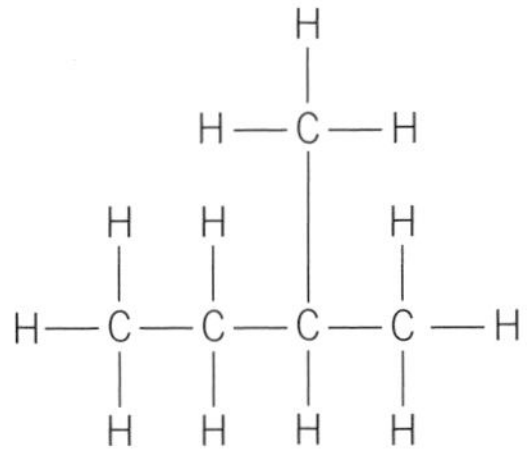

is called *methyl*butane. The longest unbranched chain is four carbons long, which gives us butane (as there are no double bonds) and there is a side chain of one carbon, a methyl group.

Study tip

It is vital to learn the root names from C1 to C6.

Hydrocarbon ring molecules have the additional prefix cyclo. So the compound below would be named cyclohexane:

- Suffixes are added to the end of the root.

For example, alcohols, —OH, have the suffix -ol, as in methanol, CH_3OH.

Functional groups

Most organic compounds are made up of a hydrocarbon chain that has one or more reactive groups attached to it. These reactive groups are called **functional groups**. The functional group reacts in the same way, whatever the length of the hydrocarbon chain. So, for example, if you learn the reactions of one alkene, such as ethene, you can apply this knowledge to any alkene.

Functional groups are named by using a suffix or prefix as shown in Table 2.

▼ **Table 2** *The suffixes of some functional groups*

Family	General functional group	Suffix	Example
alkanes	C_nH_{2n+2}	-ane	ethane
alkenes	R—CH=CH—R	-ene	propene
halogenoalkanes	R—X (X = F, Cl, Br, or I)	none	chloromethane CH_3Cl
alcohols	R—OH	-ol	ethanol C_2H_5OH
aldehydes	R—C⬩O H	-al	ethanal CH_3CHO
ketones	R—C—R' ‖ O	-one	propanone CH_3COCH_3
carboxylic acids	R—C⬩O OH	-oic acid	ethanoic acid CH_3COOH

Study tip

At this stage, you will not need to learn how to name all the functional groups in the table but they are useful to illustrate the principles of naming.

Note that the halogenoalkanes are named using a prefix (fluoro-, chloro-, bromo-, iodo-) rather than a suffix. R is often used to represent a hydrocarbon chain (of any length). Think of it as representing the rest of the molecule.

Examples

eth indicates that the molecule has a chain of two carbon atoms, *ane* that it is has no double or triple bonds and *bromo* that one of the hydrogen atoms of ethane is replaced by a bromine atom.

bromoethane

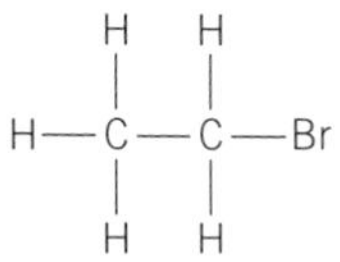

Prop indicates a chain of three carbon atoms and *ene* that there is one C=C (double bond).

propene

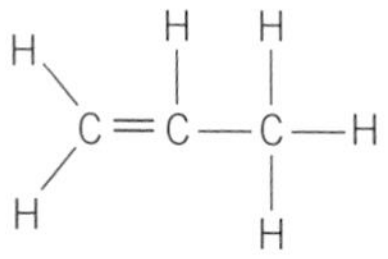

Meth indicates a single carbon, *an* that there are no double bonds and *ol* that there is an OH group (an alcohol).

methanol

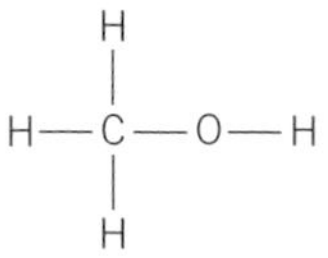

Chain and position isomers

With longer chains, you need to say where a side chain or a functional group is located on the main chain. For example, methylpentane could refer to:

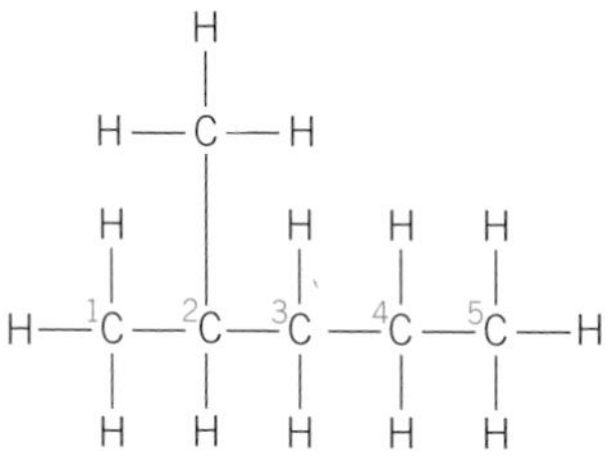

2–methylpentane   3–methylpentane

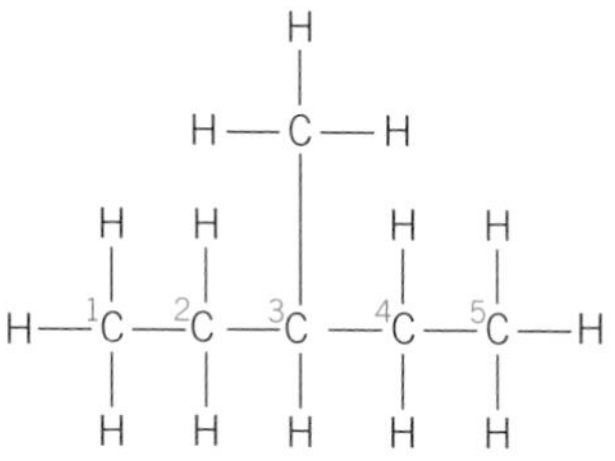

A number (sometimes called a locant) is used to tell us the position of any branching in a chain and the position of any functional group. The examples above are structural isomers. Structural isomers have the same molecular formula but different structural formulae, see Topic 11.3.

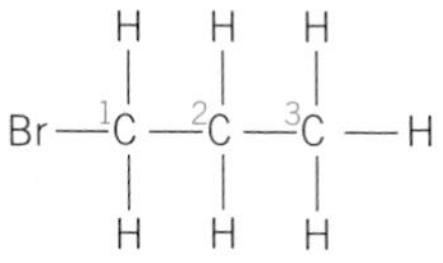

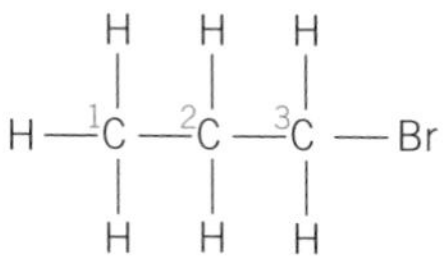

Both molecules are 1-bromopropane. The right-hand one is not 3-bromopropane because the smallest possible number is always used.

1-bromopropane may also be represented by either of the structural formulae below because all the hydrogens on carbon 1 are equivalent.

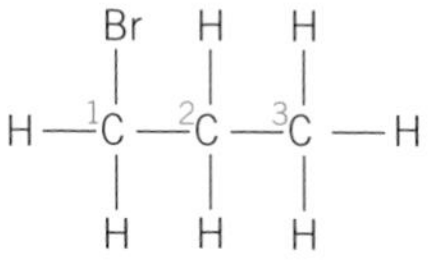

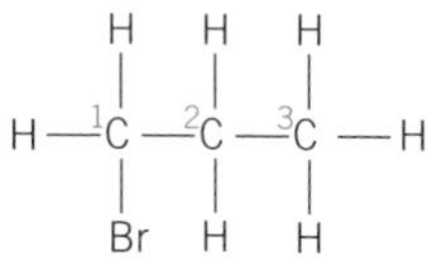

Hint

Take care. Don't get confused by the way the formula is drawn. These are the same molecule.

What's in a name?

Before the advent of systematic naming, chemicals could be given any old name. For example methanoic acid was called formic acid because it is produced by ants as a defence against predators. The Latin for ant is *formica*. However the name tells us nothing about the chemical structure of formic acid (except that it is acidic), whereas *meth*anoic acid tells us that it has one carbon atom. Acetic acid's name comes from the Latin *acetum*, meaning vinegar but the systematic name *eth*anoic acid tells us that it has two carbon atoms.

Sometimes, though, systematic names are just too long. Buckminsterfullerene was named from the similarity of its structure to the geodesic domes designed by the American architect Richard Buckminster Fuller (Figure 1).

Housane is named after the resemblance of its structural formula to a house – its systematic name is Bicyclo[2.1.0]pentane.

If we replace one of the hydrogen atoms with a methyl group —CH_3, as shown we get roof-methylhousane.

> Draw the formulae of the two positional isomers of roof-methylhousane which might be called eave-methylhousane and floor-methylhousane.

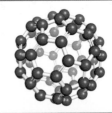

▲ **Figure 1** *Buckminster fullerene (top) and a geodesic dome (bottom)*

▲ **Figure 2** *Housane (left) and roof-methylhousane (right)*

Molecules with more than one functional group or side chain

A molecule may have more than one functional group. For example:

2-bromo-1-iodopropane

Even though iodine is on carbon 1 and bromine is on carbon 2, *bromo* is written before *iodo* because the substituting groups are put in *alphabetical order* rather than in the numerical order of the functional groups.

You can show that you have more than one of the same substituting group by adding prefixes as well as functional groups. di-, tri-, and tetra- mean two, three, and four, respectively.

So, is called 1,1-dichloroethane

and is called 1,2-dichloroethane.

> **Hint**
>
> In chemical names, strings of numbers are separated by commas. A hyphen is placed between words and numbers.

Homologous series

A **homologous series** is a family of organic compounds with the same functional group, but different carbon chain length.

- Members of a homologous series have a general formula. For example, the alkanes are C_nH_{2n+2} and alkenes, with one double bond, are C_nH_{2n}.
- Each member of the series differs from the next by CH_2.
- The length of the carbon chain has little effect on the *chemical* reactivity of the functional group.
- The length of the carbon chain affects physical properties, like melting point, boiling point, and solubility. Melting points and boiling points increase by a small amount as the number of carbon atoms in the chain increases. This is because the intermolecular forces increase. In general, small molecules are gases, larger ones liquids or solids.
- Chain branching generally reduces melting points because the molecules pack together less well.

▼ **Table 3** *Examples of systematic naming of organic compounds. Try covering up the name or structure to test yourself.*

Structural formula	Name
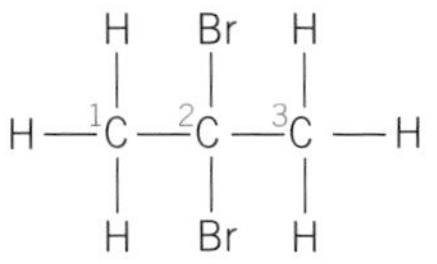	2,2-dibromopropane
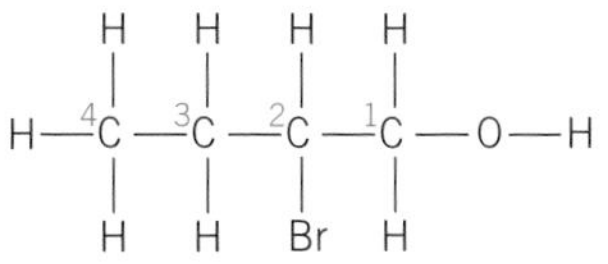	2-bromobutan-1-ol The suffix -ol defines the end of the chain you count from
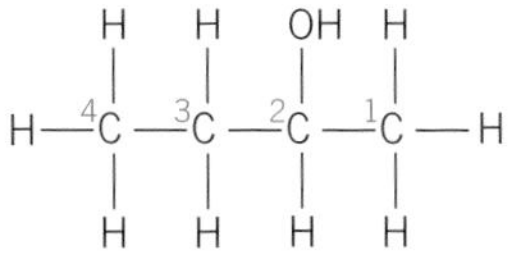	butan-2-ol
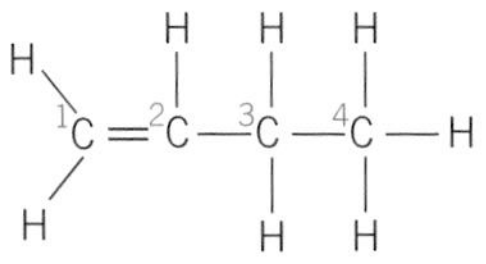	but-1-ene *Not* but-2-ene, but-3-ene, or but-4-ene as we use the smallest locant possible

Summary questions

1 What is the name of each of the following?

   a $CH_3CH_2CH_2Cl$

   b $CH_3CH_2CH_2CH_2CH_3$

   c $CH_3CH_2CH{=}CHCH_3$

   d $CH_3CH_2CH_2CH(CH_3)CH_3$

2 Draw the displayed formulae for:

   a methylbutanone

   b but-2-ene

   c 2-chlorohexane

   d but-1-ene.

Learning objectives:

→ State what is meant by structural isomers.

→ Describe the three ways in which structural isomerism can occur.

Specification reference: 3.3.1

Isomers

Isomers are molecules that have the same molecular formula but whose atoms are arranged differently. There are two basic types of isomerism in organic chemistry – structural isomerism and stereoisomerism.

Structural isomerism

Structural isomers are defined as having the same molecular formula but different structural formulae, see Topic 11.2. There are three sub-divisions. Structural isomers can have:

1   the same functional groups attached to the main chain at different points – this is called positional isomerism
2   functional groups that are different – this is called functional group isomerism
3   a different arrangement of the hydrocarbon chain (such as branching) – this is called chain isomerism.

Positional isomerism

The functional group is attached to the main chain at different points. For example, the molecular formula C_3H_7Cl could represent:

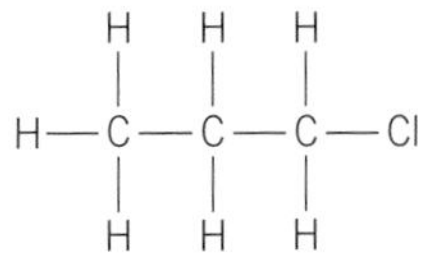

1-chloropropane   or   2-chloropropane

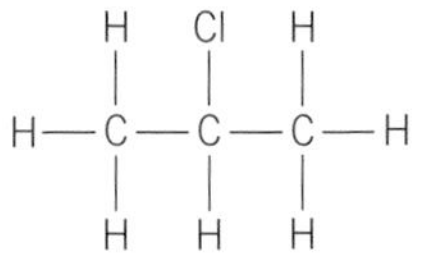

Functional group isomerism

There are different functional groups. For example, the molecular formula C_2H_6O could represent:

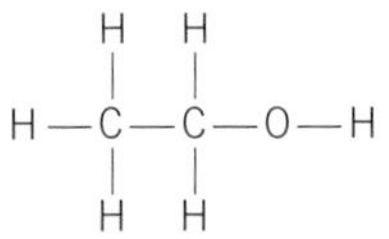

ethanol (an alcohol)   or   methoxymethane (an ether)

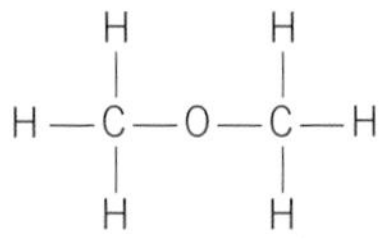

Chain isomerism

The hydrocarbon chain is arranged differently. For example, the molecular formula C_4H_9OH could represent:

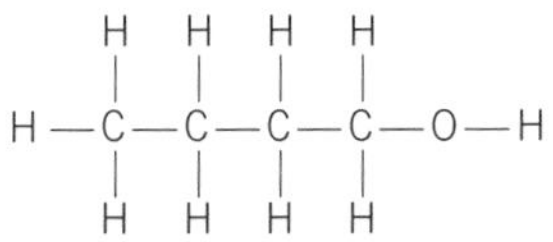

butan-1-ol   or   2-methylpropan-1-ol

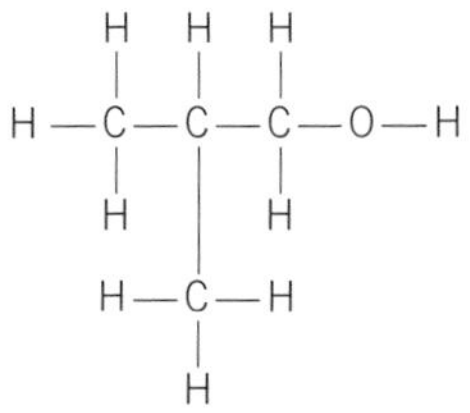

These isomers are called chain-branching isomers.

The existence of isomers makes the task of identifying an unknown organic compound more difficult. This is because there may be a number of compounds with different structures that all have the same molecular formula. So, you have to use analytical methods that tell you about the structure.

Stereoisomerism

Stereoisomerism is where two (or more) compounds have the same structural formula. They differ in the *arrangement* of the bonds in space.

There are two types:

- *E-Z* isomerism and
- optical isomerism.

E-Z isomerism

E-Z isomerism tells us about the positions of substituents at either side of a carbon–carbon double bond. Two substituents may either be on the same side of the bond *Z* (*cis*) or on opposite sides *E* (*trans*).

Z-1,2-dichloroethene *E*-1,2-dichloroethene

Substituted groups joined by a single bond can rotate around the single bond, so there are no isomers (Figure 1) but there is no rotation around a double bond. So, *Z*- and *E*-isomers are separate compounds and are not easily converted from one to the other.

E-Z is from the German *Entgegen* (opposite – *trans*) and *Zusammen* (together – *cis*).

Hint

You will also see the terms *cis* and *trans* used to describe the positions of groups attached to a double bond.

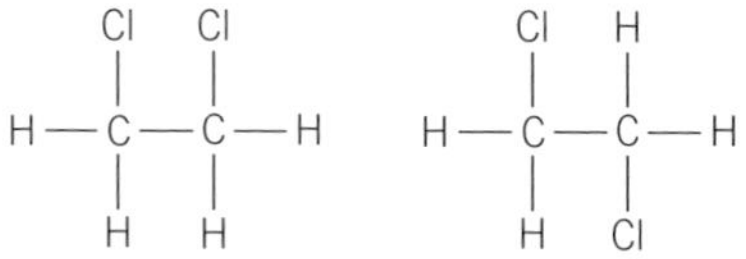

▲ **Figure 1** *Groups can rotate around a single bond. These are representations of the same molecule and are not isomers*

Synoptic link

E-Z isomerism is discussed further in Topic 14.2, Reactions of alkenes.

Summary questions

1 What type of structural isomerism is shown by the following pairs of molecules? Choose from: A = functional groups at different points, B = different functional groups, C = chain branching.

  a $CH_3CH_2OCH_3$ and $CH_3CH_2CH_2OH$

  b $CH_3CH_2CH_2OH$ and $CH_3CH(OH)CH_3$

  c $CH_3CH_2CH_2CH_2CH_3$ and $CH_3CH(CH_3)CH_2CH_3$

2 a Write the displayed and structural formulae for all the five isomers of hexane, C_6H_{14}.

  b Name these isomers.

3 Which of these molecules can show *E-Z* (*cis-trans*) isomerism?

  A $CH_2{=}CH_2$

  B $CH_3{-}CH_3$

  C $RCH{=}CH_2$

  D $RCH{=}CHR$

4 a Give the name of this:

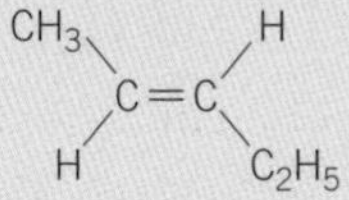

  b What is the name of its geometrical isomer?

Practice questions

1 The alkanes form a homologous series of hydrocarbons. The first four straight-chain alkanes are shown below.

methane CH_4
ethane CH_3CH_3
propane $CH_3CH_2CH_3$
butane $CH_3CH_2CH_2CH_3$

   (a) (i)   State what is meant by the term *hydrocarbon*.
        (ii) Give the general formula for the alkanes.
        (iii) Give the molecular formula for hexane, the sixth member of the series.

   (3 marks)

   (b) Each homologous series has its own general formula. State **two** other characteristics of an homologous series.

   (2 marks)

   (c) Branched-chain structural isomers are possible for alkanes which have more than three carbon atoms.
        (i)   State what is meant by the term *structural isomers*.
        (ii) Name the **two** isomers of hexane shown below.

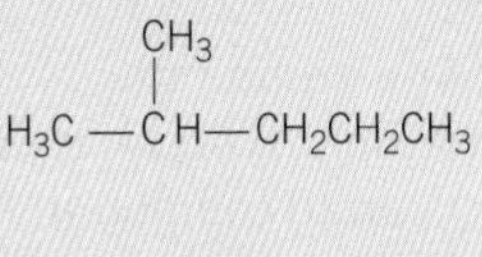

isomer 1 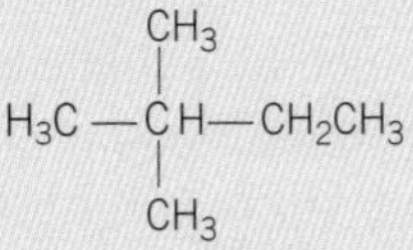   *isomer 2*

        (iii) Give the structures of **two** other branched-chain isomers of hexane.

   (6 marks)

   (d) A hydrocarbon, **W**, contains 92.3% carbon by mass. The relative molecular mass of **W** is 78.0
        (i)   Calculate the empirical formula of **W**.
        (ii) Calculate the molecular formula of **W**.

   (4 marks)
   AQA, 2003

2 (a) Give the systematic chemical name of CCl_2F_2.   *(1 mark)*
   (b) Give the systematic chemical name of CCl_4.   *(1 mark)*
   (c) Give the systematic chemical name of $CHCl_2\,CHCl_2$.   *(1 mark)*
   AQA, 2001

3 There are five structural isomers of the molecular formula C_5H_{10} which are alkenes. The displayed formulae of two of these isomers are given.

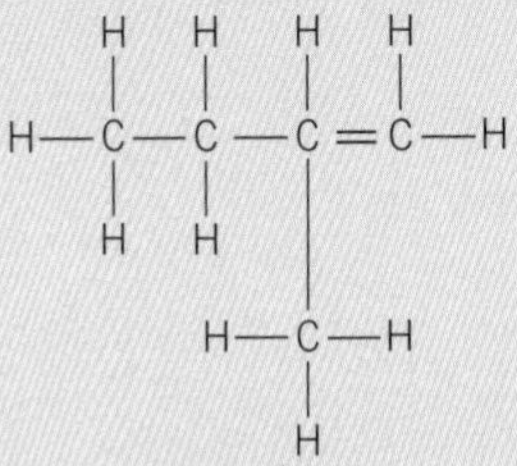

isomer 1 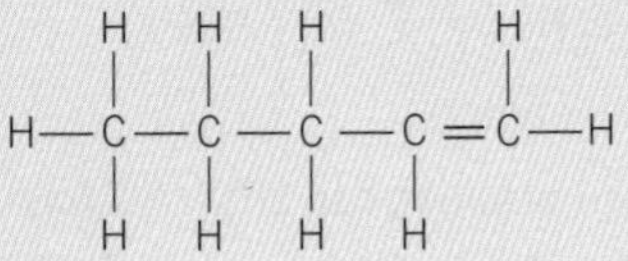   *isomer 2*

   (a) Draw the displayed formulae of two of the remaining alkene structural isomers.

   (2 marks)

(b) Consider the reaction scheme shown below and answer the question that follows.

$$\text{Isomer 1} \xrightarrow{\text{HBr}} CH_3CH_2CBr(CH_3)_2$$

$$\mathbf{Y}$$

Give the name of compound **Y**.

(1 mark)
AQA, 2000

4 There are four structural isomers of molecular formula C_4H_9Br. The structural formulae of two of these isomers are given below.

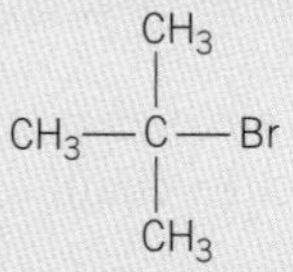

isomer 1   isomer 2

    (i)   Draw the structural formulae of the remaining two isomers.   *(3 marks)*
    (ii) Name isomer 1.

AQA, 2001

5 **(a)** The structure of the bromoalkane **Z** is

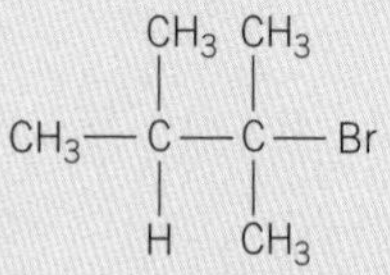

Give the IUPAC name for **Z**.
Give the general formula of the homologous series of straight-chain bromoalkanes that contains one bromine atom per molecule.
Suggest **one** reason why 1-bromohexane has a higher boiling point than **Z**.

(3 marks)

(b) Draw the displayed formula of 1,2-dichloro-2-methylpropane.
State its empirical formula.

(2 marks)
AQA, 2013

12 Alkanes
12.1 Alkanes

Learning objectives:
- → State the definition of an alkane.
- → Explain how alkanes are named.
- → Describe their properties.

Specification reference: 3.3.2

Alkanes are **saturated hydrocarbons** – they contain only carbon–carbon and carbon–hydrogen *single* bonds. They are among the least reactive organic compounds. They are used as fuels and lubricants and as starting materials for a range of other compounds. This means that they are very important to industry. The main source of alkanes is crude oil.

The general formula

The general formula for all chain alkanes is C_nH_{2n+2}. Hydrocarbons may be unbranched chains, branched chains, or rings.

Unbranched chains

Unbranched chains are often called straight chains but the C—C—C angle is 109.5°. This means that the chains are not actually straight. In an unbranched alkane, each carbon atom has two hydrogen atoms except the end carbons which have one extra.

For example, pentane, C_5H_{12}:

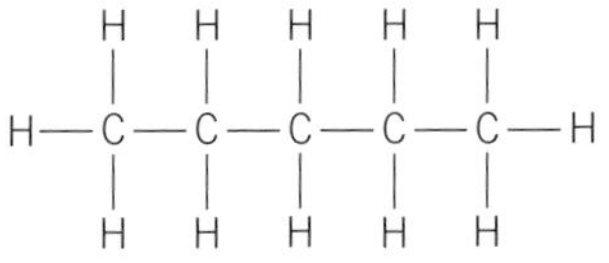

$CH_3CH_2CH_2CH_2CH_3$

displayed structural

Synoptic link

You learnt about bond angles in alkanes in Topic 3.6, The shapes of molecules and ions.

Branched chains

For example, methylbutane, C_5H_{12}, which is an isomer of pentane:

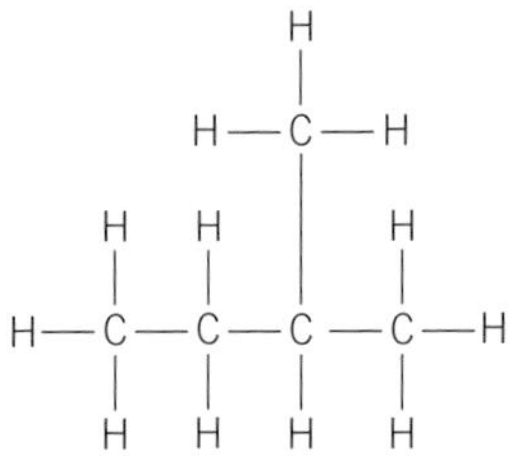

$CH_3CH_2CH(CH_3)CH_3$

displayed structural

Ring alkanes

Ring alkanes have the general molecular formula C_nH_{2n} because the end hydrogens are not required.

How to name alkanes

Straight chains

Alkanes are named from the root, which tells us the number of carbon atoms, and the suffix -ane, denoting an alkane, see Table 1.

▼ **Table 1** *Names of the first six alkanes*

methane	CH_4
ethane	C_2H_6
propane	C_3H_8
butane	C_4H_{10}
pentane	C_5H_{12}
hexane	C_6H_{14}

Branched chains

When you are naming a hydrocarbon with a branched chain, you must first find the longest unbranched chain which can sometimes be a bit tricky, see the example below. This gives the root name. Then name the branches or *side chains* as prefixes – methyl-, ethyl-, propyl-, and so on. Finally, add numbers to say which carbon atoms the side chains are attached to.

Example

Both the hydrocarbons below are the same, though they seem different at first sight.

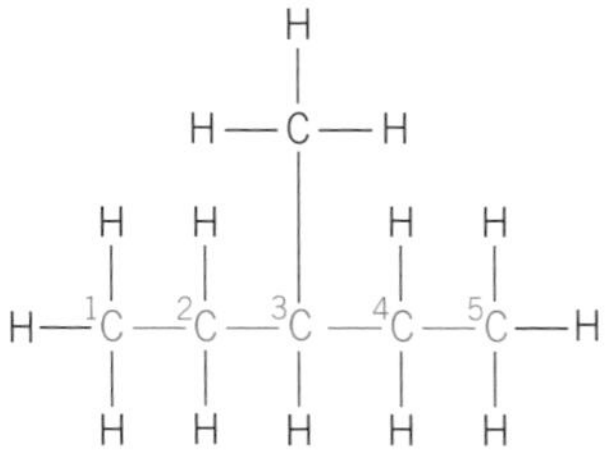

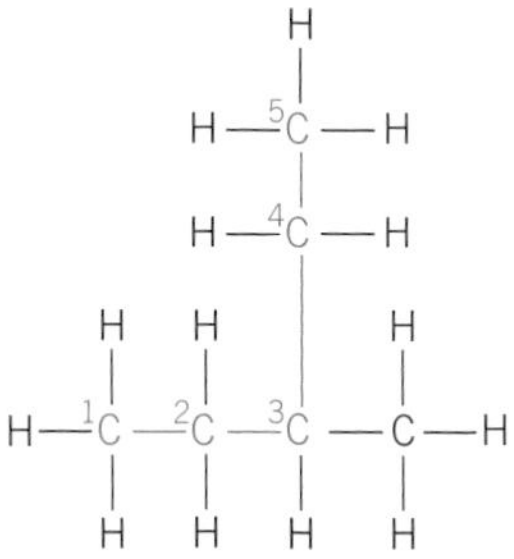

The skeletal formulae for 3-methylpentane is:

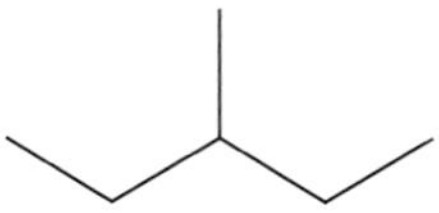

In both representations, the longest unbranched chain (in red) is five carbons, so the root is pentane. The only side chain has one carbon so it is methyl-. It is attached at carbon 3 so the full name is 3-methylpentane.

Structure

Isomerism

Methane, ethane, and propane have no isomers but after that, the number of possible isomers increases with the number of carbons in the alkane. For example, butane, with four carbons, has two isomers whilst pentane has three:

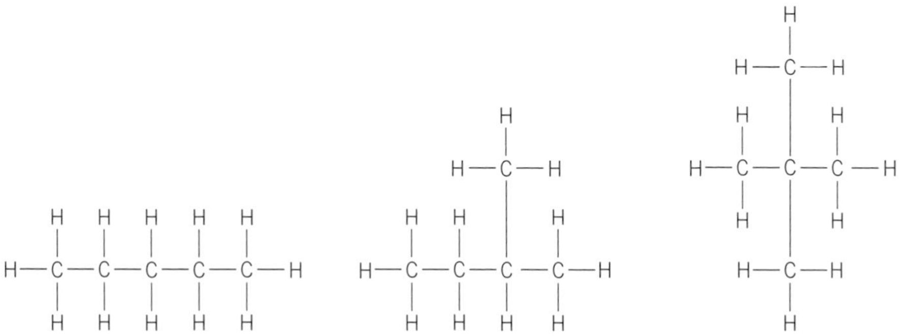

pentane   methylbutane   2,2-dimethylpropane

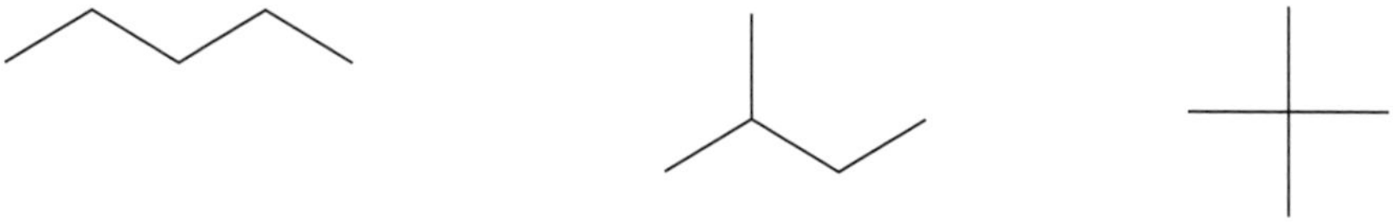

The number of isomers rises rapidly with chain length. Decane, $C_{10}H_{22}$, has 75 and $C_{30}H_{62}$ has over 4 billion.

Synoptic link

You learnt about electronegitivty in Topic 3.4, Electronegatvity – bond polarity in covalent bonds.

▲ **Figure 1** *Camping Gaz is a mixture of propane and butane. Polar expeditions use special gas mixtures with a higher proportion of propane, because butane is liquid below 272 K (−1 °C)*

increasing chain length →

▲ **Figure 2** *The effect of increasing chain length on the physical properties of alkanes*

Physical properties

Polarity

Alkanes are almost non-polar because the electronegativities of carbon (2.5) and hydrogen (2.1) are so similar. As a result, the only intermolecular forces between their molecules are weak van der Waals forces, and the larger the molecule, the stronger the van der Waals forces.

Boiling points

This increasing intermolecular force is why the boiling points of alkanes increase as the chain length increases. The shorter chains are gases at room temperature. Pentane, with five carbons, is a liquid with a low boiling point of 309 K (36 °C). At a chain length of about 18 carbons, the alkanes become solids at room temperature. The solids have a waxy feel.

Alkanes with branched chains have lower melting points than straight chain alkanes with the same number of carbon atoms. This is because they cannot pack together as closely as unbranched chains and so the van der Waals forces are not so effective.

Solubility

Alkanes are insoluble in water. This is because water molecules are held together by hydrogen bonds which are much stronger than the van der Waal's forces that act between alkane molecules. However, alkanes do mix with other relatively non-polar liquids.

How alkanes react

Alkanes are relatively unreactive. They have strong carbon–carbon and carbon–hydrogen bonds. They do not react with acids, bases, oxidising agents, and reducing agents. However, they do burn and they will react with halogens under suitable conditions. They burn in a plentiful supply of oxygen to form carbon dioxide and water (or, in a restricted supply of oxygen, to form carbon monoxide or carbon).

Summary questions

1 Name the alkane $CH_3CH_2CH(CH_3)CH_3$ and draw its displayed formula.

2 Draw the displayed formula and structural formula of 2-methylhexane.

3 Name an isomer of 2-methylhexane that has a straight chain.

4 Which of the two isomers in question **3** will have the higher melting point? Explain your answer.

Crude oil is at present the world's main source of organic chemicals. It is called a fossil fuel because it was formed millions of years ago by the breakdown of plant and animal remains at the high pressures and temperatures deep below the Earth's surface. Because it forms very slowly, it is effectively non-renewable.

Crude oil is a mixture mostly of alkanes, both unbranched and branched. Crude oils from different sources have different compositions. The composition of a typical North Sea oil is given in Table 1.

Learning objectives:
→ State the origin of crude oil.
→ Explain how crude oil is separated into useful fractions on an industrial scale.

Specification reference: 3.3.2

▼ **Table 1** *The composition of a typical North Sea crude oil*

	Gases	Petrol	Naphtha	Kerosene	Gas oil	Fuel oil and wax
Approximate boiling temperature / K	310	310–450	400–490	430–523	590–620	above 620
Chain length	1–5	5–10	8–12	11–16	16–24	25+
Percentage present	2	8	10	14	21	45

Crude oil contains small amounts of other compounds dissolved in it. These come from other elements in the original plants and animals the oil was formed from, for example, some contain sulfur. These produce sulfur dioxide, SO_2, when they are burnt. This is one of the causes of acid rain – sulfur dioxide reacts with oxygen high in the atmosphere to form sulfur trioxide. This reacts with water in the atmosphere to form sulfuric acid.

Hint

Crude oil is being produced now but accumulation of a deposit of this oil is a very slow process.

Fractional distillation of crude oil

To convert crude oil into useful products you have to separate the mixture. This is done by heating it and collecting the **fractions** that boil over different ranges of temperatures. Each fraction is a mixture of hydrocarbons of similar chain length and therefore similar properties, see Figure 1. The process is called **fractional distillation** and it is done in a **fractionating tower**.

- The crude oil is first heated in a furnace.
- A mixture of liquid and vapour passes into a tower that is cooler at the top than at the bottom.
- The vapours pass up the tower via a series of trays containing bubble caps until they arrive at a tray that is sufficiently cool (at a lower temperature than their boiling point). Then they condense to liquid.
- The mixture of liquids that condenses on each tray is piped off.
- The shorter chain hydrocarbons condense in the trays nearer to the top of the tower, where it is cooler, because they have lower boiling points.
- The thick residue that collects at the base of the tower is called tar or bitumen. It can be used for road surfacing but, as supply often exceeds demand, this fraction is often further processed to give more valuable products

Combustion of sulfur

Write balanced equations for the three steps in which sulfur is converted into sulfuric acid.

$$SO_3 + H_2O \rightarrow H_2SO_4$$
$$2SO_2 + O_2 \rightarrow 2SO_3$$
$$S + O_2 \rightarrow SO_2$$

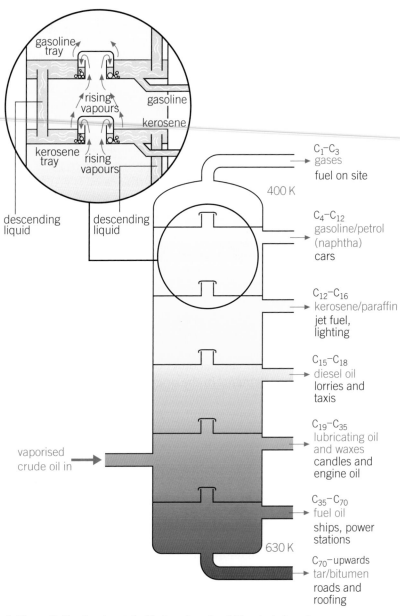

▲ **Figure 2** *The fractional distillation of crude oil. The chain length ranges are approximate*

▲ **Figure 3** *Crude oil is separated into fractions by distillation in cylindrical towers typically 8 m in diameter and 40 m high. Oil refineries vary but a typical one might process 3.5 million tonnes of crude oil per year.*

Fracking

Almost half the UK's electricity is generated from natural gas (largely methane). Around half of this comes from the North Sea but this percentage is decreasing as these wells become depleted and more and more gas is being imported via pipeline from Europe.

However, many areas of the UK have resources of natural gas – not caught under impervious rock layers as in the North Sea but trapped within shale rock rather like water in a sponge. This gas can be extracted by drilling into the shale and forcing pressurised water mixed with sand into the shale. This causes the rather soft shale rock to break up or fracture (giving the term fracking, short for hydraulic fracturing) releasing the trapped gas which flows to the surface. A number of chemicals are added to the water such as hydrochloric acid to help break up the shale and methanol to prevent corrosion in the system.

Many people are opposed to fracking for a variety of reasons:

- they do not like the infrastructure of wells and the associated traffic in their 'backyard'
- there is concern about the amount of water used
- they worry about the chemical additives polluting water supplies
- occasionally fracking appears to have caused small earthquakes
- burning natural gas produces carbon dioxide – a cause of global warming.

Set against these objections is the appeal of gas supplies for many years which are not subject to control by other countries.

Balance the equation for the combustion of methane.

$$CH_4 + O_2 \rightarrow CO_2 + H_2O$$

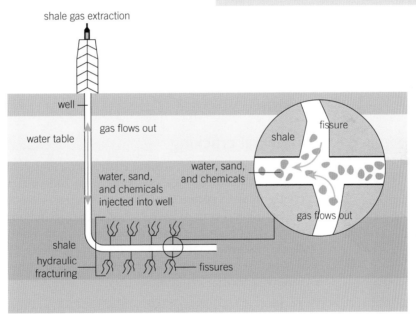

▲ **Figure 1** *The process of fracking*

$$CH_4 + 2O_2 \rightarrow CO_2 + 2H_2O$$

Summary questions

1 Draw the displayed formula and structural formula of hexane.

2 In which of the crude oil fractions named in Table 1 is hexane most likely to be found?

3 What is fractional distillation and how is it different from distillation?

4 Give the names of two gases produced in fractional distillation.

Learning objectives:

→ Describe what cracking is.

→ Describe what the conditions and products of thermal cracking are.

→ Describe the conditions and products of catalytic cracking.

→ Explain the economic reasons for cracking.

Specification reference: 3.3.2

> ### Study tip
>
> You should understand the commercial benefits of cracking.

▲ **Figure 1** *A range of products obtained from crude oil*

The naphtha fraction from the fractional distillation of crude oil is in huge demand, for petrol and by the chemical industry. The longer chain fractions are not as useful and therefore of lower value economically. Most crude oil has more of the longer chain fractions than is wanted and not enough of the naphtha fraction.

The shorter chain products are economically more valuable than the longer chain material. To meet the demand for the shorter chain hydrocarbons, many of the longer chain fractions are broken into shorter lengths (cracked). This has two useful results:

- shorter, more useful chains are produced, especially petrol
- some of the products are alkenes, which are more reactive than alkanes.

Note that petrol is a mixture of mainly alkanes containing between four and twelve carbon atoms.

Alkenes are used as chemical feedstock (which means they supply industries with the starting materials to make different products) and are converted into a huge range of other compounds including polymers and a variety of products from paints to drugs. Perhaps the most important alkene is ethene, which is the starting material for poly(ethene) (also called polythene) and a wide range of other everyday materials.

Alkanes are very unreactive and harsh conditions are required to break them down. There are a number of different ways of carrying out cracking.

Thermal cracking

This reaction involves heating alkanes to a high temperature, 700–1200 K, under high pressure, up to 7000 kPa. Carbon–carbon bonds break in such a way that one electron from the pair in the covalent bond goes to each carbon atom. So initially two shorter chains are produced, each ending in a carbon atom with an unpaired electron. These fragments are called free radicals. Free radicals are highly reactive intermediates and react in a number of ways to form a variety of shorter chain molecules.

As there are not enough hydrogen atoms to produce two alkanes, one of the new chains must have a C=C, and is therefore an alkene:

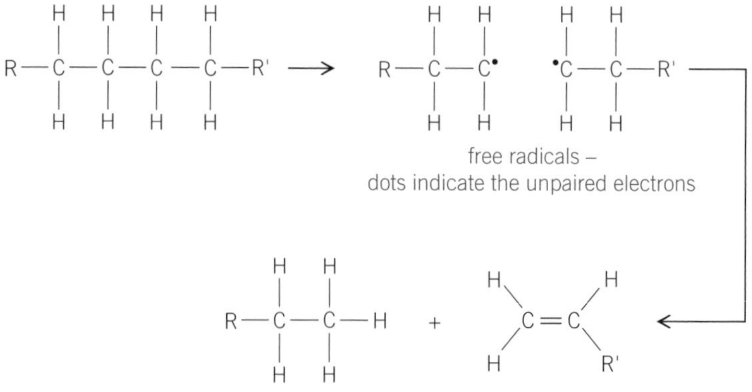

▲ **Figure 2** *Thermal cracking*

Any number of carbon–carbon bonds may break and the chain does not necessarily break in the middle. Hydrogen may also be produced. Thermal cracking tends to produce a high proportion of alkenes. To avoid too much decomposition (ultimately to carbon and hydrogen) the alkanes are kept in these conditions for a very short time, typically one second. The equation in Figure 2 shows cracking of a long chain alkane to give a shorter chain alkane and an alkene. The chain could break at any point.

Catalytic cracking

Catalytic cracking takes place at a lower temperature (approximately 720 K) and lower pressure (but more than atmospheric), using a zeolite catalyst, consisting of silicon dioxide and aluminium oxide (aluminosilicates). Zeolites have a honeycomb structure with an enormous surface area. They are also acidic. This form of cracking is used mainly to produce motor fuels. The products are mostly branched alkanes, cycloalkanes (rings), and aromatic compounds, see Figure 3.

The products obtained from cracking are separated by fractional distillation.

In the laboratory, catalytic cracking may be carried out in the apparatus shown in Figure 4, using lumps of aluminium oxide as a catalyst.

a
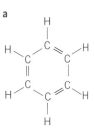

b

▲ **Figure 3** *Aromatic compounds are based on the benzene ring C_6H_6. Although it appears to have three double bonds as in **a**, the electrons are spread around the ring, making it more stable than expected. It is usually represented as in **b***

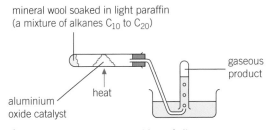

mineral wool soaked in light paraffin
(a mixture of alkanes C_{10} to C_{20})

gaseous product

heat

aluminium oxide catalyst

▲ **Figure 4** *Laboratory cracking of alkanes*

The products are mostly gases, showing that they have chain lengths of less than C_5 and the mixture decolourises bromine solution. This is a test for a carbon–carbon double bond showing that the product contains alkenes.

Study tip

You should be able to predict the products and write equations for typical thermal and catalytic cracking reactions.

Summary questions

1 Complete the word equation for one possibility for the thermal cracking of decane.

decane → octane + ...

2 🔬 In the laboratory cracking of alkanes, how can you tell that the products have shorter chains than the starting materials?

3 Why would we not crack octane industrially?

4 How can the temperature required for cracking be reduced?

5 Give two economic reasons for cracking long chain alkanes.

▲ **Figure 1** *Incomplete combustion is potentially dangerous because of carbon monoxide formation. Carbon monoxide detectors in kitchens can warn of dangerous levels of this gas.*

Alkanes are quite unreactive. They do not react with acids, bases, oxidising agents, or reducing agents. However, they do burn and they will react with halogens under suitable conditions.

Combustion

The shorter chain alkanes burn completely in a plentiful supply of oxygen to give carbon dioxide and water.

For example, methane:

$$CH_4(g) + 2O_2(g) \rightarrow CO_2(g) + 2H_2O(l) \qquad \Delta H = -890 \text{ kJ mol}^{-1}$$

Or ethane:

$$C_2H_6(g) + 3\frac{1}{2} O_2(g) \rightarrow 2CO_2(g) + 3H_2O(l) \qquad \Delta H = -1559.7 \text{ kJ mol}^{-1}$$

Combustion reactions give out heat and have large negative enthalpies of combustion. The more carbons present, the greater the heat output. For this reason they are important as fuels. Fuels are substances that release heat energy when they undergo combustion. They also store a large amount of energy for a small amount of weight. For example, octane produces approximately 48 kJ of energy per gram when burnt, which is about twice the energy output per gram of coal. Examples of alkane fuels include:

- methane (the main component of natural or North Sea gas)
- propane (camping gas)
- butane (Calor gas)
- petrol (a mixture of hydrocarbons of approximate chain length C_8)
- paraffin (a mixture of hydrocarbons of chain lengths C_{10} to C_{18}).

Incomplete combustion

In a limited supply of oxygen, the poisonous gas carbon monoxide, CO, is formed. For example, with propane:

$$C_3H_8(g) + 3\frac{1}{2} O_2(g) \rightarrow 3CO(g) + 4H_2O(l)$$

This is called **incomplete combustion**.

With even less oxygen, carbon (soot) is produced. For example, when a Bunsen burner is used with a closed air hole, the flame is yellow and a black sooty deposit appears on the apparatus. Incomplete combustion often happens with longer chain hydrocarbons, which need more oxygen to burn compared with shorter chains.

Polluting the atmosphere

All hydrocarbon-based fuels derived from crude oil may produce polluting products when they burn. They include the following:

- carbon monoxide, CO, a poisonous gas produced by incomplete combustion

- nitrogen oxides, NO, NO_2, and N_2O_4 (often abbreviated to NO_x) produced when there is enough energy for nitrogen and oxygen in the air to combine, for example:

$$N_2(g) + O_2(g) \rightarrow 2NO(g)$$

This happens in a petrol engine at the high temperatures present, when the sparks ignite the fuel. These oxides may react with water vapour and oxygen in the air to form nitric acid. They are therefore contributors to acid rain and photochemical smog

- sulfur dioxide is another contributor to acid rain. It is produced from sulfur-containing impurities present in crude oil. This oxide combines with water vapour and oxygen in the air to form sulfuric acid

- carbon particles, called particulates, which can exacerbate asthma and cause cancer

- unburnt hydrocarbons may also enter the atmosphere and these are significant greenhouse gases. They contribute to photochemical smog which can cause a variety of health problems (Figure 2)

- carbon dioxide, CO_2, is a greenhouse gas. It is always produced when hydrocarbons burn. Although carbon dioxide is necessary in the atmosphere, its level is rising and this is a cause of the increase in the Earth's temperature and consequent climate change.

- water vapour which is also a greenhouse gas.

▲ **Figure 2** *Photochemical smog is the chemical reaction of sunlight, nitrogen oxides, NO_x and volatile organic compounds in the atmosphere, which leaves airborne particles (called particulate matter) and ground-level ozone.*

Flue gas desulfurisation

Large numbers of power stations generate electricity by burning fossil fuels such as coal or natural gas. These fuels contain sulfur compounds and one of the products of their combustion is sulfur dioxide, SO_2, a gas that causes acid rain by combining with oxygen and water in the atmosphere to form sulfuric acid.

$$SO_2(g) + \frac{1}{2} O_2(g) + H_2O(l) \rightarrow H_2SO_4(l)$$

The gases given out by power stations are called flue gases so the process of removing the sulfur dioxide is called flue gas desulfurisation. In one method, a slurry of calcium oxide (lime) and water is sprayed into the flue gas which reacts with the calcium oxide and water to form calcium sulfite, which can be further oxidised to calcium sulfate, also called gypsum. The overall reaction is:

$$CaO(s) + 2H_2O(l) + SO_2(g) + \frac{1}{2} O_2 \rightarrow CaSO_4 \cdot 2H_2O(s)$$

Gypsum is a saleable product as it is used to make builders' plaster and plasterboard.

An alternative process uses calcium carbonate (limestone) rather than calcium oxide:

$$CaCO_3(s) + \frac{1}{2} O_2(g) + SO_2(g) \rightarrow CaSO_4(s) + CO_2(g)$$

Catalytic converters

The internal combustion engine produces most of the pollutants listed above, though sulfur is now removed from petrol so that sulfur dioxide has become less of a problem.

▲ **Figure 3** *A catalytic converter*

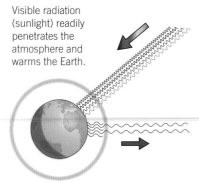

Visible radiation (sunlight) readily penetrates the atmosphere and warms the Earth.

Invisible infra-red radiation is emitted by the Earth and cools it down. But some of this infrared is trapped by greenhouse gases in the atmosphere which act as a blanket, keeping the heat in.

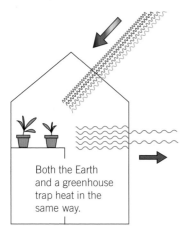

Both the Earth and a greenhouse trap heat in the same way.

▲ **Figure 4** *The greenhouse effect*

Summary questions

1   Write word and balanced symbol equations for:

    a the complete combustion of propane

    b the incomplete combustion of butane to produce carbon monoxide and water.

2   **a** What are the problems with using carbon-based fuels?

    b What steps are taken to reduce these problems?

3   Pick two alternative sources of power which do not use carbon-based fuels and discuss the pros and cons of each.

All new cars with petrol engines are now equipped with catalytic converters in their exhaust systems (Figure 3). These reduce the output of carbon monoxide, nitrogen oxides, and unburnt hydrocarbons in the exhaust gas mixture.

The catalytic converter is a honeycomb made of a ceramic material coated with platinum and rhodium metals. These are the catalysts. The honeycomb shape provides an enormous surface area, so a little of these expensive metals goes a long way. As the polluting gases pass over the catalyst, they react with each other to form less harmful products by the following reactions:

- $$2CO(g) \;+\; 2NO(g) \;\rightarrow\; N_2(g) \;+\; 2CO_2(g)$$
  carbon monoxide   nitrogen oxide   nitrogen   carbon dioxide

- hydrocarbons + nitrogen oxide → nitrogen + carbon dioxide + water

For example, $C_8H_{18} \;+\; 25NO \;\rightarrow\; 12\frac{1}{2}N_2 \;+\; 8CO_2 \;+\; 9H_2O$

The reactions take place on the surface of the catalyst, on the layer of platinum and rhodium metals.

Global warming and the greenhouse effect

Greenhouses become very warm inside. This is because the visible rays from the sun pass through the glass. Rather than escaping, their energy is absorbed by everything inside the greenhouse and re-radiated as infrared energy, which is heat. Infrared energy has a longer wavelength and cannot pass back out through the glass.

Carbon dioxide behaves rather like glass. It traps infrared radiation so that the Earth's atmosphere heats up. This is important for life because without carbon dioxide and other greenhouse gases, the Earth would be too cold to sustain life. Other greenhouse gases are water vapour and methane. These are even more effective than carbon dioxide, but there has not been as much change in the level of these gases in the atmosphere in recent years. However, since the industrial revolution fossil fuels have been used to fuel industrial plants and the level of carbon dioxide has been rising. Gradually, the Earth's temperature has been rising too and the majority of scientists believe that the increasing level of carbon dioxide is the cause of global warming.

The concentration of water vapour, the most abundant greenhouse gas, in the atmosphere tends to stay roughly the same (except locally – by waterfalls, for example) because of the equilibrium that exists between water vapour and liquid water. However, if the temperature of the atmosphere rises, there will be more water vapour in the air and therefore more greenhouse warming. This may be offset by greater cloud formation and clouds reflect solar radiation. The role of water is therefore recognised as very important but as yet not fully understood.

Carbon neutral activities

Many people are concerned about activities, such as airline flights, that produce large amounts of carbon dioxide. A flight from London to Paris produces about 350 kg of carbon dioxide per passenger (from burning hydrocarbon fuels). Activities that produce no carbon dioxide emissions overall are referred to as carbon neutral.

12.5 The formation of halogenoalkanes

When you put a mixture of an alkane and a halogen into bright sunlight, or shine a photoflood lamp onto the mixture, the alkane and the halogen will react to form a halogenoalkane. The ultraviolet component of the light starts the reaction. Alkanes do not react with halogens in the dark at room temperature.

For example, if you put a mixture of hexane and a little liquid bromine into a test tube and leave it in the dark, it stays red-brown (the colour of bromine). However, if you shine ultraviolet light onto it, the mixture becomes colourless and misty fumes of hydrogen bromide appear.

A substitution reaction has taken place. One or more of the hydrogen atoms in the alkane has been replaced by a bromine atom and hydrogen bromide is given off as a gas. The main reaction is:

$$C_6H_{14}(g) \; + \; Br_2(l) \; \rightarrow \; C_6H_{13}Br(l) \; + \; HBr(g)$$

hexane   bromine   bromohexane   hydrogen bromide

Bromohexane is a halogenoalkane.

Chain reactions

The reaction above is called a free-radical substitution. It starts off a **chain reaction** which takes place in three stages – **initiation**, **propagation**, and **termination**.

The reaction between any alkane and a halogen goes by the same mechanism.

For example, methane and chlorine:

$$CH_4(g) + Cl_2(g) \rightarrow CH_3Cl(g) + HCl(g)$$

Initiation
- The first, or initiation, step of the reaction is breaking the Cl—Cl bond to form two chlorine atoms.
- The chlorine molecule absorbs the energy of a single quantum of ultraviolet (UV) light. The energy of one quantum of UV light is greater than the Cl—Cl bond energy, so the bond will break.
- Since both atoms are the same, the Cl—Cl bond breaks homolytically, that is, one electron going to each chlorine atom.
- This results in two separate chlorine atoms, written Cl•. They are called **free radicals**. The dot is used to show the unpaired electron.

UV light

Cl—Cl $\longrightarrow$ 2Cl•

- Free radicals are highly reactive.
- The C—H bond in the alkane needs more energy to break than is available in a quantum of ultraviolet radiation, so this bond does not break.

Learning objectives:
→ Define what a free radical is.
→ Describe the reaction mechanism for the free-radical substitution of methane.

Specification reference: 3.3.2

Hint

In fact a mixture of many bromoalkanes is formed.

Hint

You can test for hydrogen bromide by wafting the fumes from a bottle of ammonia over the mouth of the test tube; white fumes of ammonium bromide are formed. This test will also give a positive result for other hydrogen halides.

▲ **Figure 1** *Slip-Slop-Slap is the name for the health campaign in Australia exhorting people to 'slip on a shirt, slop on sunscreen, and slap on a hat' to prevent skin damage caused by UV in sunlight*

Propagation

This takes place in two stages:

1 The chlorine free radical takes a hydrogen atom from methane to form hydrogen chloride, a stable compound. This leaves a methyl free radical, $\bullet CH_3$.

$$Cl\bullet + CH_4 \rightarrow HCl + \bullet CH_3$$

2 The methyl free radical is also very reactive and reacts with a chlorine molecule. This produces another chlorine free radical and a molecule of chloromethane – a stable compound.

$$\bullet CH_3 + Cl_2 \rightarrow CH_3Cl + Cl\bullet$$

The effect of these two steps is to produce hydrogen chloride, chloromethane, and a new $Cl\bullet$ free radical. This is ready to react with more methane and repeat the two steps. This is the chain part of the chain reaction. These steps may take place thousands of times before the radicals are destroyed in the termination step.

Termination

Termination is the step in which the free radicals are removed. This can happen in any of the following three ways:

$Cl\bullet + Cl\bullet \rightarrow Cl_2$	Two chlorine free radicals react together to give chlorine.
$\bullet CH_3 + \bullet CH_3 \rightarrow C_2H_6$	Two methyl free radicals react together to give ethane.
$Cl\bullet + \bullet CH_3 \rightarrow CH_3Cl$	A chlorine free radical and a methyl free radical react together to give chloromethane.

Notice that in every case, two free radicals react to form a stable compound with no unpaired electrons.

Other products of the chain reaction

Other products are formed as well as the main ones, chloromethane and hydrogen chloride.

- Some ethane is produced at the termination stage, as shown above.
- Dichloromethane may be made at the propagation stage, if a chlorine radical reacts with some chloromethane that has already formed.

$$CH_3Cl + Cl\bullet \rightarrow \bullet CH_2Cl + HCl$$

followed by $\bullet CH_2Cl + Cl_2 \rightarrow CH_2Cl_2 + Cl\bullet$

- With longer-chain alkanes there will be many isomers formed because the $Cl\bullet$ can replace *any* of the hydrogen atoms.
- Chain reactions are not very useful because they produce such a mixture of products. They will also occur without light at high temperatures.

Why are chain reactions important?

It is believed that chlorofluorocarbons (CFCs) in the stratosphere are destroying the ozone layer.

Ozone is a molecule made from three oxygen atoms, O_3. It decomposes to oxygen. Too much ozone at ground level causes lung irritation and degradation of paints and plastics, but high in the atmosphere it has a vital role.

The ozone layer is important because it protects the Earth from the harmful exposure to too many ultraviolet (UV) rays. Without this protective layer, life on Earth would be very different. For example, plankton in the sea, which are at the very bottom of the food chain of the oceans, need protection from too much UV radiation. Also, too much UV radiation causes skin cancer in people by damaging DNA.

Chlorine free radicals are formed from CFCs because the C—Cl bond breaks homolytically in the presence of UV radiation to produce chlorine free radicals, Cl•. Ozone molecules are then attacked by these:

$$Cl\bullet + O_3 \rightarrow \underset{\text{free radical}}{ClO\bullet} + O_2$$

The resulting free radicals also attack ozone and regenerate Cl•:

$$ClO\bullet + O_3 \rightarrow 2O_2 + Cl\bullet$$

Adding the two equations, you can see that the chlorine free radical is not destroyed in this process. It acts as a catalyst in the breakdown of ozone to oxygen.

$$2O_3 \rightarrow 3O_2$$

Summary questions

1   What stage of a free-radical reaction of bromine with methane is represented by the following?

   a $Br\bullet + Br\bullet \rightarrow Br_2$

   b $CH_4 + Br\bullet \rightarrow CH_3\bullet + HBr$

   c $\bullet CH_3 + Br_2 \rightarrow CH_3Br + Br\bullet$

   d $Br_2 \rightarrow 2Br\bullet$

2   Look at the equations for the destruction of ozone in the last section of this topic.

   a   Which two are propagation steps?

   b   Suggest three possible termination steps.

Practice questions

1 Octane is the eighth member of the alkane homologous series.
   (a) State **two** characteristics of a homologous series. *(2 marks)*
   (b) Name a process used to separate octane from a mixture containing several
       different alkanes. *(1 mark)*
   (c) The structure shown below is one of several structural isomers of octane.

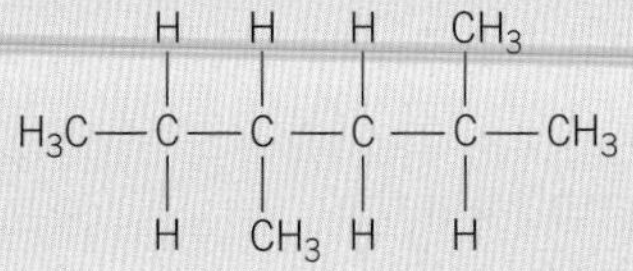

   $$H_3C-\underset{\underset{H}{|}}{\overset{\overset{H}{|}}{C}}-\underset{\underset{CH_3}{|}}{\overset{\overset{H}{|}}{C}}-\underset{\underset{H}{|}}{\overset{\overset{H}{|}}{C}}-\underset{\underset{H}{|}}{\overset{\overset{CH_3}{|}}{C}}-CH_3$$

   Give the meaning of the term structural isomerism. Name this isomer and
   state its empirical formula. *(4 marks)*
   (d) Suggest why the branched chain isomer shown above has a lower boiling
       point than octane. *(2 marks)*

   AQA, 2011

2 Cetane, $C_{16}H_{34}$, is a major component of diesel fuel.
   (a) Write an equation to show the complete combustion of cetane. *(1 mark)*
   (b) Cetane has a melting point of 18°C and a boiling point of 287°C.
       In polar regions vehicles that use diesel fuel may have ignition problems.
       Suggest **one** possible cause of this problem with the diesel fuel. *(1 mark)*
   (c) The pollutant gases NO and NO_2 are sometimes present in the exhaust gases of
       vehicles that use petrol fuel.
       (i) Write an equation to show how NO is formed and give a condition
           needed for its formation. *(2 marks)*
       (ii) Write an equation to show how NO is removed from the exhaust gases
           in a catalytic converter. Identify a catalyst used in the converter. *(2 marks)*
       (iii) Deduce an equation to show how NO_2 reacts with water and oxygen
           to form nitric acid, HNO_3. *(1 mark)*
   (d) Cetane, $C_{16}H_{34}$, can be cracked to produce hexane, butene and ethene.
       (i) State **one** condition that is used in this cracking reaction. *(1 mark)*
       (ii) Write an equation to show how one molecule of cetane can be
           cracked to form hexane, butene and ethene. *(1 mark)*
       (iii) State **one** type of useful solid material that could be formed from alkenes.
           (1 mark)

   AQA, 2011

3 Hexane, C_6H_{14}, is a member of the homologous series of alkanes.
   (a) (i) Name the raw material from which hexane is obtained. *(1 mark)*
       (ii) Name the process used to obtain hexane from this raw material. *(1 mark)*
   (b) C_6H_{14} has structural isomers.
       (i) Deduce the number of structural isomers with molecular formula C_6H_{14} *(1 mark)*
       (ii) State **one** type of structural isomerism shown by the isomers of C_6H_{14} *(1 mark)*
   (c) One molecule of an alkane **X** can be cracked to form one molecule of hexane
       and two molecules of propene.
       (i) Deduce the molecular formula of **X**. *(1 mark)*
       (ii) State the type of cracking that produces a high percentage of alkenes.
           State the conditions needed for this type of cracking. *(2 marks)*
       (iii) Explain the main economic reason why alkanes are cracked. *(1 mark)*
   (d) Hexane can react with chlorine under certain conditions as shown in the
       following equation.

   $$C_6H_{14} + Cl_2 \rightarrow C_6H_{13}Cl + HCl$$

       (i) Both the products are hazardous. The organic product would be labelled
           'flammable'.
           Suggest the most suitable hazard warning for the other product. *(1 mark)*
       (ii) Calculate the percentage atom economy for the formation of
           $C_6H_{13}Cl$ ($M_r = 120.5$) in this reaction. *(1 mark)*

(e) A different chlorinated compound is shown below. Name this compound and state its empirical formula.

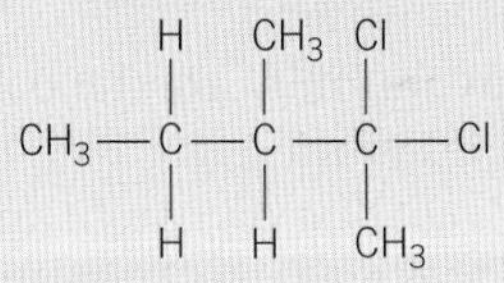

$$CH_3-\underset{\underset{\displaystyle H}{|}}{\overset{\overset{\displaystyle H}{|}}{C}}-\underset{\underset{\displaystyle H}{|}}{\overset{\overset{\displaystyle CH_3}{|}}{C}}-\underset{\underset{\displaystyle CH_3}{|}}{\overset{\overset{\displaystyle Cl}{|}}{C}}-Cl$$

(2 marks)
AQA, 2012

4 Alkanes are used as fuels. A student burned some octane, C_8H_{18}, in air and found that the combustion was incomplete.

(a) **(i)** Write an equation for the incomplete combustion of octane to produce carbon monoxide as the only carbon-containing product. *(1 mark)*

(ii) Suggest **one** reason why the combustion was incomplete. *(1 mark)*

(b) Catalytic converters are used to remove the toxic gases NO and CO that are produced when alkane fuels are burned in petrol engines.

(i) Write an equation for a reaction between these two toxic gases that occurs in a catalytic converter when these gases are removed. *(1 mark)*

(ii) Identify a metal used as a catalyst in a catalytic converter.
Suggest **one** reason, other than cost, why the catalyst is coated on a ceramic honeycomb. *(2 marks)*

(c) If a sample of fuel for a power station is contaminated with an organic sulfur compound, a toxic gas is formed by complete combustion of this sulfur compound.

(i) State **one** environmental problem that can be caused by the release of this gas. *(1 mark)*

(ii) Identify **one** substance that could be used to remove this gas.
Suggest **one** reason, other than cost, why this substance is used. *(2 marks)*
AQA, 2012

5 Chlorine can be used to make chlorinated alkanes such as dichloromethane.

(a) Write an equation for each of the following steps in the mechanism for the reaction of chloromethane, CH_3Cl, with chlorine to form dichloromethane, CH_2Cl_2.
Initiation step
First propagation step
Second propagation step
The termination step that forms a compound with empirical formula CH_2Cl *(4 marks)*

(b) When chlorinated alkanes enter the upper atmosphere, carbon–chlorine bonds are broken. This process produces a reactive intermediate that catalyses the decomposition of ozone. The overall equation for this decomposition is $2O_3 \rightarrow 3O_2$

(i) Name the type of reactive intermediate that acts as a catalyst in this reaction. *(1 mark)*

(ii) Write **two** equations to show how this intermediate is involved as a catalyst in the decomposition of ozone. *(2 marks)*
AQA, 2013

Learning objectives:

→ Explain why halogenoalkanes are more reactive than alkanes.

→ Explain why carbon–halogen bonds are polar.

→ Explain the trends in bond enthalpy and bond polarity of the carbon–halogen bond.

Specification reference: 3.3.3

Hint

Halogenoalkanes are sometimes called haloalkanes.

▲ **Figure 1** *Applications of halogenoalkanes*

▼ **Table 1** *Electronegativities of carbon and the halogens*

Element	Electronegativity
carbon	2.5
fluorine	4.0
chlorine	3.5
bromine	2.8
iodine	2.6

Not many halogenoalkanes occur naturally but they are the basis of many synthetic compounds. Some examples of these are PVC (used to make drainpipes), Teflon (the non-stick coating on pans), and a number of anaesthetics and solvents. Halogenoalkanes have an alkane skeleton with one or more halogen (fluorine, chlorine, bromine, or iodine) atoms in place of hydrogen atoms.

The general formula

The general formula of a halogenoalkane with a single halogen atom is $C_nH_{2n+1}X$ where X is the halogen. This is often shortened to R—X.

How to name halogenoalkanes

- The prefixes fluoro-, chloro-, bromo-, and iodo- tell us which halogen is present.
- Numbers are used, if needed, to show on which carbon the halogen is bonded:

1-chloropropane   1-iodopropane   2-bromo-2-methylpropane

- The prefixes di-, tri-, tetra-, and so on, are used to show *how many* atoms of each halogen are present.
- When a compound contains different halogens they are listed in alphabetical order, *not* in order of the number of the carbon atom to which they are bonded. For example:

is 3-chloro-2-iodopentane not 2-iodo-3-chloropentane. (C is before I in the alphabet.)

Bond polarity

Halogenoalkanes have a C—X bond. This bond is polar, $C^{\delta+}$—$X^{\delta-}$, because halogens are more electronegative than carbon. The electronegativities of carbon and the halogens are shown in Table 1. Notice that as you go down the group, the bonds get less polar.

Physical properties of halogenoalkanes

Solubility

- The polar $C^{\delta+}$—$X^{\delta-}$ bonds are not polar enough to make the halogenoalkanes soluble in water.
- The main intermolecular forces of attraction are dipole–dipole attractions and van der Waal forces.

- Halogenoalkanes mix with hydrocarbons so they can be used as dry-cleaning fluids and to remove oily stains. (Oil is a mixture of hydrocarbons.)

Boiling point

The boiling point depends on the number of carbon atoms and halogen atoms.

- Boiling point *increases* with *increased* chain length.
- Boiling point *increases* going *down* the halogen group.

Both these effects are caused by increased van der Waals forces because the larger the molecules, the greater the number of electrons (and therefore the larger the van der Waals forces).

As in other homologous series, increased branching of the carbon chain will tend to lower the melting point.

Halogenoalkanes have higher boiling points than alkanes with similar chain lengths because they have higher relative molecular masses and they are more polar.

How the halogenoalkanes react – the reactivity of the C—X bond

When halogenoalkanes react it is almost always the C—X bond that breaks. There are two factors that determine how readily the C—X bond reacts. These are:

- the $C^{\delta+}$—$X^{\delta-}$ bond polarity
- the C—X bond enthalpy.

Bond polarity

The halogens are more electronegative than carbon so the bond polarity will be $C^{\delta+}$—$X^{\delta-}$. This means that the carbon bonded to the halogen has a partial positive charge – it is electron deficient. This means that it can be attacked by reagents that are electron rich or have electron-rich areas. These are called **nucleophiles**. A nucleophile is an electron pair donor, see Topic 13.2.

The polarity of the C—X bond would predict that the C—F bond would be the most reactive. It is the most polar, so the $C^{\delta+}$ has the most positive charge and is therefore most easily attacked by a nucleophile. This argument would make the C—I bond least reactive because it is the least polar.

Bond enthalpies

C—X bond enthalpies are listed in Table 2. The bonds get weaker going down the group. Fluorine is the smallest atom of the halogens and the shared electrons in the C—F bond are strongly attracted to the fluorine nucleus. This makes a strong bond. Going down the group, the shared electrons in the C—X bond get further and further away from the halogen nucleus, so the bond becomes weaker.

The bond enthalpies would predict that iodo-compounds, with the weakest bonds, are the most reactive, and fluoro-compounds, with the strongest bonds, are the least reactive.

Experiments confirm that reactivity increases going down the group. This means that bond enthalpy is a more important factor than bond polarity.

▼ **Table 2** *Carbon–halogen bond enthalpies*

Bond	Bond enthalpy / kJ mol^{-1}	
C—F	467	stronger
[C—H	413]	
C—Cl	346	
C—Br	290	
C—I	228	

Summary questions

1 These questions are about the following halogenoalkanes:

    i $CH_3CH_2CH_2CH_2I$

    ii $CH_3CHBrCH_3$

    iii $CH_2ClCH_2CH_2CH_3$

    iv $CH_3CH_2CHBrCH_3$

  a Draw the displayed formula for each halogenoalkane and mark the polarity of the C—X bond.

  b Name each halogenoalkane.

  c Predict which of them would have the highest boiling point and explain your answer.

2 Why do the halogenoalkanes get less reactive going up the halogen group?

Learning objectives:

→ State the definition of a nucleophile.

→ Describe nucleophilic substitution.

→ Explain why ⁻OH, ⁻CN, and NH₃ behave as nucleophiles.

→ Describe the mechanism of nucleophilic substitution.

Specification reference: 3.3.3

Most reactions of organic compounds take place via a series of steps. You can often predict these steps by thinking about how electrons are likely to move. This can help you understand why reactions take place as they do and this can save a great deal of rote learning.

Nucleophiles

Nucleophiles are reagents that attack and form bonds with positively or partially positively charged carbon atoms.

- A nucleophile is either a negatively charged ion or has an atom with a $\delta-$ charge.
- A nucleophile has a lone (unshared) pair of electrons which it can use to form a covalent bond.
- The lone pair is situated on an electronegative atom.

So, in organic chemistry a nucleophile is a species that has a lone pair of electrons with which it can form a bond by donating its electrons to an electron deficient carbon atom. Some common nucleophiles are:

- the hydroxide ion, ⁻:OH
- ammonia, :NH₃
- the cyanide ion, ⁻:CN.

They will each replace the halogen in a halogenoalkane. These reactions are called **nucleophilic substitutions** and they all follow essentially the same reaction mechanism.

A reaction mechanism describes a route from reactants to products via a series of theoretical steps. These may involve short-lived intermediates.

Nucleophilic substitution

The general equation for nucleophilic substitution, using :Nu⁻ to represent any negatively charged nucleophile and X to represent a halogen atom, is:

$$R-\overset{\overset{\displaystyle H}{|}}{\underset{\underset{\displaystyle H}{|}}{C}}-X \; + \; :Nu^- \longrightarrow R-\overset{\overset{\displaystyle H}{|}}{\underset{\underset{\displaystyle H}{|}}{C}}-Nu \; + \; :X^-$$

Reaction mechanisms and curly arrows

Curly arrows are used to show how electron pairs move in organic reactions. These are shown here in red for clarity. You can write the above reaction as:

$$R-\overset{\overset{\displaystyle H}{|}}{\underset{\underset{\displaystyle H}{|}}{\overset{\delta+}{C}}}\overset{\delta-}{-}X \; + \; :Nu^- \longrightarrow R-\overset{\overset{\displaystyle H}{|}}{\underset{\underset{\displaystyle H}{|}}{C}}-Nu \; + \; :X^-$$

The lone pair of electrons of a nucleophile is attracted towards a partially positively charged carbon atom. A curly arrow starts at a lone pair of electrons and moves towards $C^{\delta+}$.

Study tip

It is important to remember that a curly arrow indicates the movement of an electron pair.

The lower curly arrow shows the electron pair in the C—X bond moving to the halogen atom, X, and making it a halide ion. The halide ion is called the **leaving group**.

The rate of substitution depends on the halogen. Fluoro-compounds are unreactive due to the strength of the C—F bond. Then, going down the group, the rate of reaction increases as the C—X bond strength decreases, see Topic 13.1.

Examples of nucleophilic substitution reactions

All these reactions are similar. Remember the basic pattern, shown above. Then work out the product with a particular nucleophile. This is easier than trying to remember the separate reactions.

Halogenoalkanes with aqueous sodium (or potassium) hydroxide

The nucleophile is the hydroxide ion, $^-$:OH.

This reaction occurs very slowly at room temperature. To speed up the reaction it is necessary to warm the mixture. Halogenoalkanes do not mix with water, so ethanol is used as a solvent in which the halogenoalkane and the aqueous sodium (or potassium) hydroxide both mix. This is called a hydrolysis reaction.

The overall reaction is:

$$R{-}X + OH^- \rightarrow ROH + X^-$$

so an alcohol, ROH, is formed.

For example:

$$C_2H_5Br + OH^- \rightarrow C_2H_5OH + Br^-$$
$$\text{bromoethane} \qquad\qquad \text{ethanol}$$

This is the mechanism:

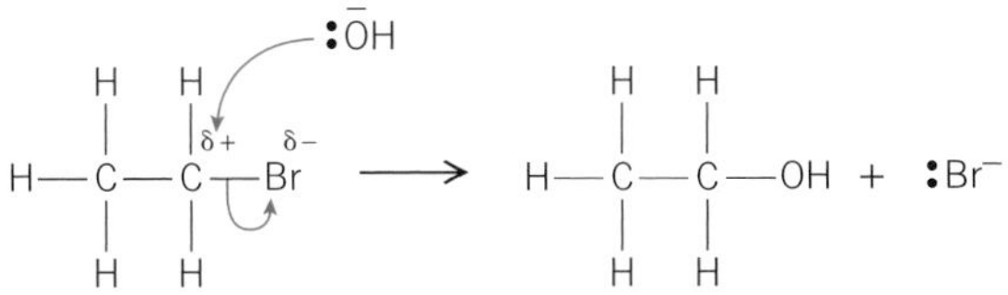

The rate of the reaction depends on the strength of the carbon–halogen bond C—F > C—Cl > C—Br > C—I (see Table 2 in Topic 13.1). Fluoroalkanes do not react at all whilst iodoalkanes react rapidly.

Halogenoalkanes with cyanide ions

When halogenoalkanes are warmed with an aqueous alcoholic solution of potassium cyanide, nitriles are formed. The nucleophile is the cyanide ion, $^-$:CN.

The reaction is:

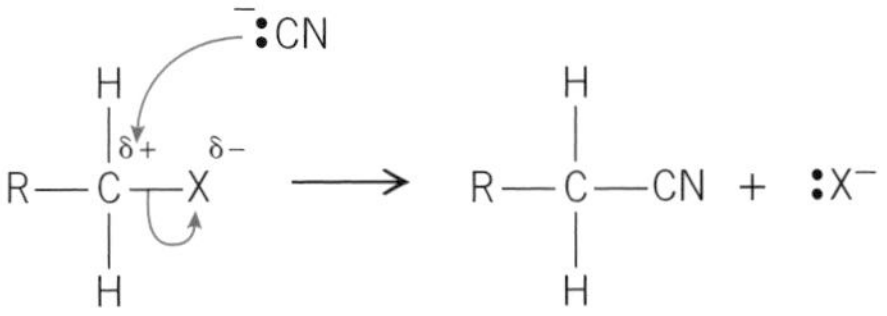

> **Hint**
>
> Nitriles have the functional group $-C{\equiv}N$. They are named using the suffix nitrile. The carbon of the $-CN$ group is counted as part of the root, so CH_3CH_2CN is propanenitrile, not ethanenitrile.

Summary questions

1 This equation represents the hydrolysis of a halogenoalkane by sodium hydroxide solution:

$$R–X + OH^- \rightarrow ROH + X^-$$

   a Why is the reaction carried out in ethanol?

   b What is the nucleophile?

   c Why is this a substitution?

   d Which is the leaving group?

   e Which would have the fastest reaction R—F, R—Cl, R—Br, or R—I?

2 **a** Starting with bromoethane, what nucleophile will produce a product with three carbon atoms?

   b Give the equation for this, using curly arrows to show the mechanism of the reaction.

   c Name the product.

The product is called a nitrile. It has one extra carbon in the chain than the starting halogenoalkane. This is often useful if you want to make a product that has one carbon more than the starting material.

Halogenoalkanes with ammonia

The nucleophile is ammonia, $:NH_3$.

The reaction of halogenoalkanes with an excess concentrated solution of ammonia in ethanol is carried out under pressure. The reaction produces an amine, RNH$_2$.

$$R—X + 2NH_3 \rightarrow RNH_2 + NH_4X$$

This is the mechanism:

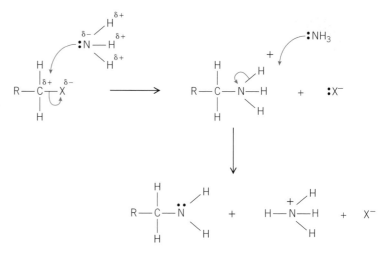

Ammonia is a nucleophile because it has a lone pair of electrons that it can donate (although it has no negative charge) and the nitrogen atom has a δ– charge. Because ammonia is a neutral nucleophile, a proton, H$^+$, must be lost to form the neutral product, called an amine. The H$^+$ ion reacts with a second ammonia molecule to form an NH$_4^+$ ion.

The uses of nucleophilic substitution

Nucleophilic substitution reactions are useful because they are a way of introducing new functional groups into organic compounds. Halogenoalkanes can be converted into alcohols, amines, and nitriles. These in turn can be converted to other functional groups.

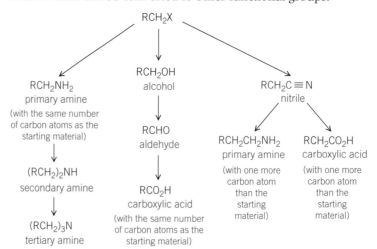

▲ **Figure 1** Uses of nucleophilic reactions

13.3 Elimination reactions in halogenoalkanes

Halogenoalkanes typically react by nucleophilic substitution. But, under different conditions they react by **elimination**. A hydrogen halide is eliminated from the molecule, leaving a double bond in its place so that an alkene is formed.

OH⁻ ion acting as a base

You saw in Topic 13.2 that the OH^- ion, from aqueous sodium or potassium hydroxide, is a nucleophile and its lone pair will attack a halogenoalkane at $C^{\delta+}$ to form an alcohol.

Under different conditions, the OH^- ion can act as a **base**, removing an H^+ ion from the halogenoalkane. In this case it is an elimination reaction rather than a substitution. In the example below, bromoethane reacts with potassium hydroxide to form ethene. A molecule of hydrogen bromide, HBr, is eliminated then the hydrogen bromide reacts with the potassium hydroxide. The reaction produces ethene, potassium bromide, and water.

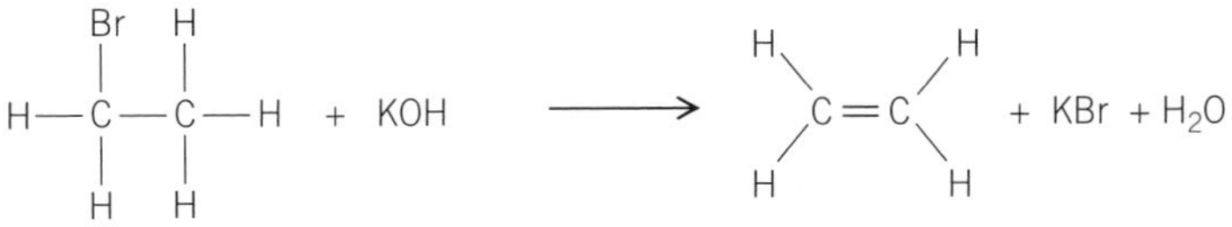

Learning objectives:
→ State the definition of an elimination reaction.
→ Describe the mechanism for elimination reactions in halogenoalkanes.
→ Describe the conditions that favour elimination rather than substitution.
→ Show when and how isomeric alkenes are formed.

Specification reference: 3.3.3

$$H-\underset{\underset{H}{|}}{\overset{\overset{Br}{|}}{C}}-\underset{\underset{H}{|}}{\overset{\overset{H}{|}}{C}}-H \;+\; KOH \longrightarrow \overset{H}{\underset{H}{}}C=C\overset{H}{\underset{H}{}} \;+\; KBr \;+\; H_2O$$

The conditions of reaction

The sodium (or potassium) hydroxide is dissolved in ethanol and mixed with the halogenoalkane. *There is no water present.* The mixture is heated. The experiment can be carried out using the apparatus shown in Figure 1.

The product is ethene. Ethene burns and also decolourises bromine solution, showing that it has a C=C bond.

The mechanism of elimination

Hydrogen bromide is eliminated as follows. The curly arrows show the movement of electron pairs:

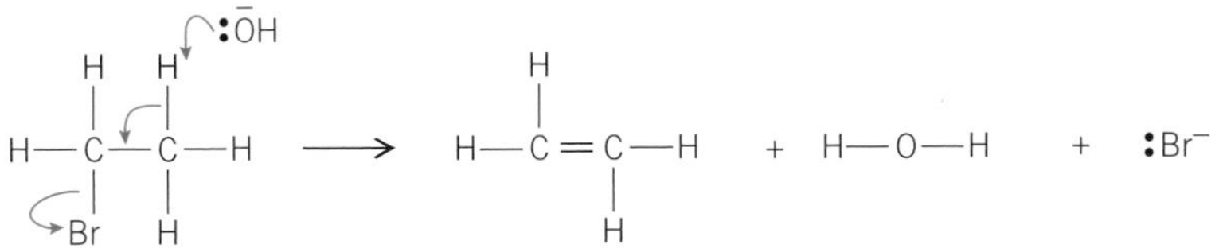

$$H-\underset{\underset{Br}{|}}{\overset{\overset{H}{|}}{C}}-\underset{\underset{H}{|}}{\overset{\overset{H}{|}}{C}}-H \longrightarrow H-\underset{\underset{H}{|}}{\overset{\overset{H}{|}}{C}}=C-H \;+\; H-O-H \;+\; :Br^-$$

- The OH^- ion uses its lone pair to form a bond with one of the hydrogen atoms on the carbon next to the C—Br bond. These hydrogen atoms are very slightly $\delta+$.
- The electron pair from the C—H bond now becomes part of a carbon–carbon double bond.
- The bromine takes the pair of electrons in the C—Br bond and leaves as a bromide ion (the leaving group).

This reaction is a useful way of making molecules with carbon–carbon double bonds.

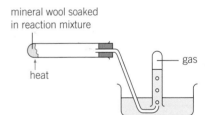

▲ **Figure 1** *Apparatus for elimination of hydrogen bromide from bromoethane*

mineral wool soaked in reaction mixture

heat

gas

Synoptic link

Decolourising bromine water is a test for an alkene. The bromine adds on across the double bond. See Topic 14.2, Reactions of alkenes.

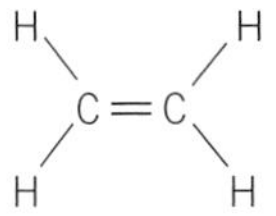

$$\overset{H}{\underset{H}{}}C=C\overset{H}{\underset{H}{}}$$

▲ **Figure 2** *A better representation of the shape of ethene*

Substitution or elimination?

Since the hydroxide ion will react with halogenoalkanes as a nucleophile *or* as a base, there is competition between substitution and elimination. In general a mixture of an alcohol and an alkene is produced. For example:

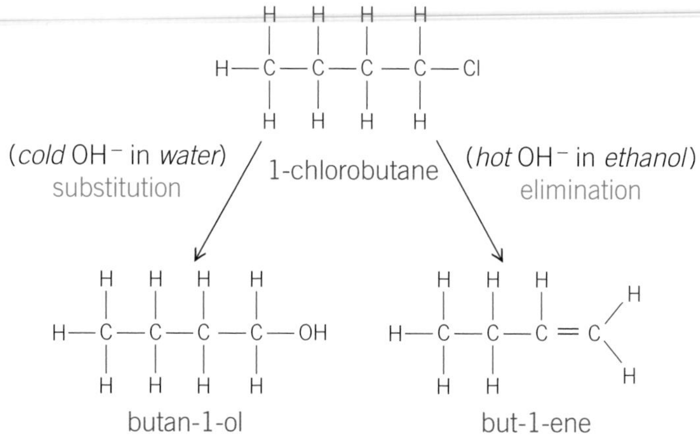

(*cold* OH⁻ in *water*)
substitution
   1-chlorobutane
(*hot* OH⁻ in *ethanol*)
elimination

butan-1-ol   but-1-ene

The reaction that predominates depends on two factors – the reaction conditions (aqueous or ethanolic solution) and the type of halogenoalkanes (primary, secondary, or tertiary).

The conditions of the reaction
- Hydroxide ions at room temperature, dissolved in water (aqueous), favour substitution.
- Hydroxide ions at high temperature, dissolved in ethanol, favour elimination.

The type of halogenoalkane
Primary halogenoalkanes tend to react by substitution and tertiary ones by elimination. Secondary will do both.

Isomeric products

In some cases a mixture of isomeric elimination products is possible.

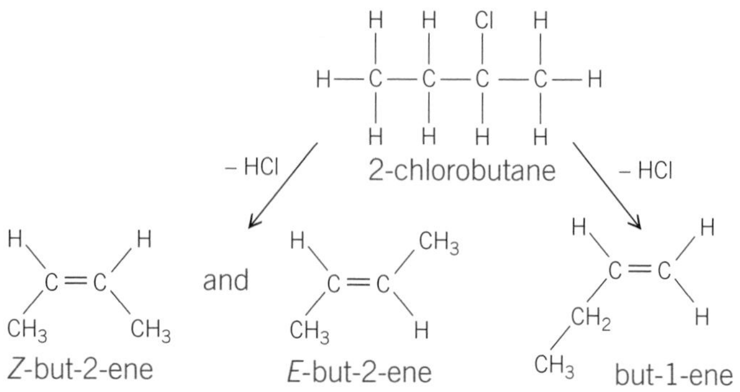

– HCl   2-chlorobutane   – HCl

Z-but-2-ene   and   *E*-but-2-ene   but-1-ene

Halogenoalkanes and the environment

Chlorofluorocarbons
Chlorofluorocarbons are halogenoalkanes containing both chlorine and fluorine atoms but no hydrogen, for example, trichlorofluoromethane, CCl_3F.

> **Hint**
>
> Halogenoalkanes are classified as primary, secondary, and tertiary according to whether the halogen atom is at the end of the hydrocarbon chain (primary), in the body of the chain (secondary) or at a branch in the chain (tertiary). The same system is used for alcohols, see Topic 15.1, Alcohols – introduction.

> **Synoptic link**
>
> The prefixes *Z* and *E* are explained in Topic 11.3, Isomerism.

- They are also called CFCs.
- They are very unreactive under normal conditions.
- The short chain ones are gases and were used, for example, as aerosol propellants, refrigerants, and blowing agents for foams like expanded polystyrene.
- Longer chain ones are used as dry cleaning and de-greasing solvents.

CFC gases eventually end up in the atmosphere where they decompose to give chlorine atoms. Chlorine atoms decompose ozone, O_3, in the stratosphere, which has caused a hole in the Earth's ozone layer. Upper atmosphere research together with laboratory research showed how ozone is broken down. Politicians were influenced by scientists and, under international agreement, CFCs are being phased out and replaced by other, safer, compounds including hydrochlorofluorocarbons, HCFCs, such as CF_3CHCl_2. However, a vast reservoir of CFCs remains in the atmosphere and it will be many years before the ozone layer recovers.

CFCs

CFCs were introduced in the 1930s by an American engineer, Thomas Midgley, for use in refrigerators. He famously demonstrated their non-toxicity and non-flammability to a scientific conference by breathing in a lungful and exhaling it to extinguish a lighted candle. It was not until long after Mldgley's death that it was realised that CFCs were involved in the depletion of the ozone layer because they release chlorine atoms in the stratosphere.

Chemists have developed less harmful replacements for CFCs. Initially these were HCFCs, which contain hydrogen, carbon, fluorine, and chlorine. One example is $CHFCl_2$. These decompose more easily than CFCs due to their C—H bonds, and the chlorine atoms are released lower in the atmosphere where they do not contribute to the destruction of the ozone layer.

The so-called second generation replacements are called HFCs (hydrofluorocarbons) such as CHF_2CF_3. These contain no chlorine and therefore do not damage the ozone layer. They are not wholly free of environmental problems though, and chemists are working on third generation compounds. Some are considering reverting to refrigerants such as ammonia, which were used before the advent of CFCs.

1   Draw a 3D representation of the formula of $CHFCl_2$. What shape is this molecule? Does it have any isomers? Explain your answer.
2   What sort of formula is CHF_2CF_3? Draw its displayed formula. What is its molecular formula?

Summary questions

1   In elimination reactions of halogenoalkanes, the OH^- group is acting as which of the following?

   A   A base

   B   An acid

   C   A nucleophile

   D   An electrophile

2   Which of the following molecules is a CFC?

   A CH_3CH_2Cl

   B $CF_2{=}CF_2$

   C CF_3CH_2Cl

   D CCl_2F_2

3   a   Name the two possible products when 2-bromopropane reacts with hydroxide ions.

   b   How could you show that one of the products is an alkene?

   c   Give the mechanism (using curly arrows) of the reaction that is an elimination.

Practice questions

1 Haloalkanes are used in the synthesis of other organic compounds.

   (a) Hot concentrated ethanolic potassium hydroxide reacts with
       2-bromo-3-methylbutane to form two alkenes that are structural isomers of each
       other. The major product is 2-methylbut-2-ene.

       (i)   Name and outline a mechanism for the conversion of 2-bromo-3-methylbutane
             into 2-methylbut-2-ene according to the equation.

             $(CH_3)_2CHCHBrCH_3 + KOH \rightarrow (CH_3)_2C{=}CHCH_3 + KBr + H_2O$

             (4 marks)

       (ii) Draw the **displayed formula** for the other isomer that is formed.

             (1 mark)

       (iii) State the type of structural isomerism shown by these two alkenes.

             (1 mark)

   (b) A small amount of another organic compound, **X**, can be detected in the reaction
       mixture formed when hot concentrated ethanolic potassium hydroxide reacts with
       2-bromo-3-methylbutane.
       Compound **X** has the molecular formula $C_5H_{12}O$ and is a secondary alcohol.

       (i)   Draw the **displayed formula** for **X**.

             (1 mark)

       (ii) Suggest **one** change to the reaction conditions that would increase the yield of **X**.

             (1 mark)

       (iii) State the type of mechanism for the conversion of 2-bromo-3-methylbutane
             into **X**.

             (1 mark)

             AQA, 2013

2 (a) Consider the following reaction.

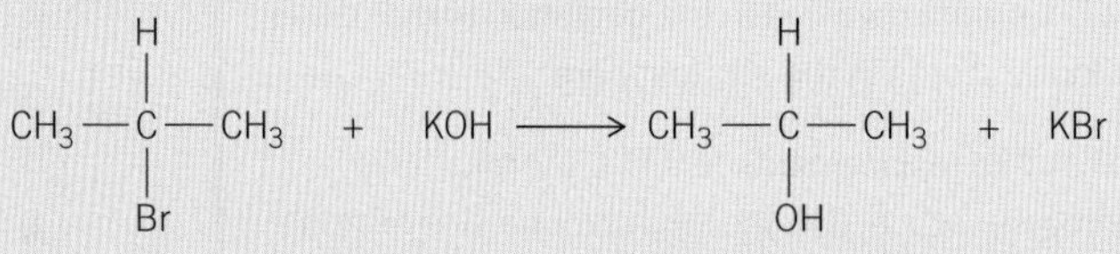

       (i)   Name and outline a mechanism for this reaction.

             (3 marks)

       (ii) Name the haloalkane in this reaction.

             (1 mark)

       (iii) Identify the characteristic of the halogenoalkane molecule that enables it to
             undergo this type of reaction.

             (1 mark)

       (iv) A student predicted that the yield of this reaction would be 90%. In an
             experiment 10.0 g of the halogenoalkane was used and 4.60 g of the organic
             product was obtained. Is the student correct? Justify your answer with
             a calculation using these data.

             (1 mark)

   (b) An alternative reaction can occur between this halogenoalkane and potassium
       hydroxide as shown by the following equation.

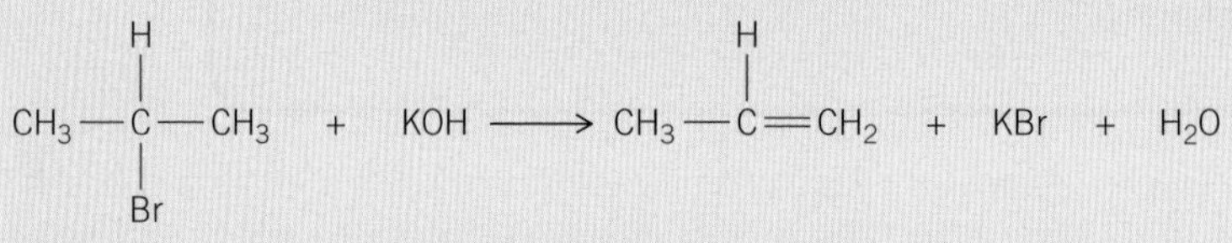

       Name and outline a mechanism for this reaction.

             (4 marks)

   (c) Give **one** condition needed to favour the reaction shown in part (b) rather than that
       shown in part (a).

             (1 mark)

             AQA, 2010

214

3 **(a)** Write a balanced symbol equation for the reaction of $CH_3CH_2CH_2Br$ with aqueous hydroxide ions.

(2 marks)

(b) Name the starting material and the product.

(2 marks)

(c) Give the formula of the leaving group in this reaction.

(1 mark)

(d) Classify the reaction as substitution, elimination or addition.

(1 mark)

(e) The hydroxide ion acts as a nucleophile in this reaction. State two features of the hydroxide ion that allow it to act as a nucleophile.

(2 marks)

(f) Draw the mechanism of the reaction using 'curly arrows' to show the movement of electron pairs.

(2 marks)

(g) How would you expect the rate of a similar reaction with $CH_3CH_2CH_2I$ to compare with that of $CH_3CH_2CH_2Br$? Explain your answer.

(2 marks)

(h) Water molecules can act as nucleophiles in a similar reaction. How do they compare with hydroxide ions as nucleophiles? Explain your answer.

(2 marks)

(i) What extra step has to occur in the reaction of a neutral nucleophile such as water compared with the reaction with a negatively charged ion such as the hydroxide ion?

(1 mark)

14 Alkenes
14.1 Alkenes

Alkenes are **unsaturated** hydrocarbons. They are made of carbon and hydrogen only and have one or more carbon–carbon double bonds. This means that alkenes have fewer than the maximum possible number of hydrogen atoms. The double bond makes them more reactive than alkanes because of the high concentration of electrons (high electron density) between the two carbon atoms. Ethene, the simplest alkene, is the starting material for a large range of products, including polymers such as polythene, PVC, polystyrene, and terylene fabric, as well as products like antifreeze and paints. Alkenes are produced in large quantities when crude oil is thermally cracked.

The general formula

The homologous series of alkenes with one double bond has the general formula C_nH_{2n}.

How to name alkenes

There cannot be a C=C bond if there is only one carbon. So, the simplest alkene is ethene, CH_2=CH_2 followed by propene, CH_3CH=CH_2.

Structure

The shape of alkenes

Ethene is a planar (flat) molecule. This makes the angles between each bond roughly 120°.

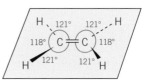

Unlike the C—C bonds in alkanes, there is no rotation about the double bond. This is because of the make-up of a double bond. Any molecules in which a hydrogen atom in ethene is replaced by another atom or group will have the same flat shape around the carbon–carbon double bond.

Why a double bond cannot rotate

As well as a normal C—C single bond, there is a p-orbital (which contains a single electron) on each carbon. These two orbitals overlap to form an orbital with a cloud of electron density above and below the single bond, see Figure 1 and Figure 2. This is called a π-orbital (pronounced pi) and its presence means the bond cannot rotate. This is sometimes called restricted rotation.

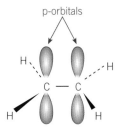
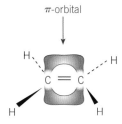

two p-orbitals produce

the π-orbital in ethene

▲ **Figure 1** *The double bond in ethene*

Isomers

Alkenes with more than three carbons can form different types of isomers and they are named according to the IUPAC system, using the suffix -ene to indicate a double bond.

As well as chain isomers like those found in alkanes, alkenes can form two types of isomer that involve the double bond:

- position isomers
- geometrical isomers.

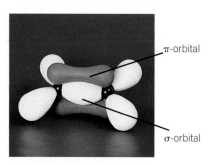

▲ **Figure 2** *Model of ethene showing orbitals*

Position isomers

These are isomers with the double bond in different positions, that is, between a pair of adjacent carbon atoms in different positions in the carbon chain.

but-2-ene

but-1-ene

The longer the carbon chain, the more possibilities there will be and therefore the greater the number of isomers.

Geometrical isomers

Geometrical isomerism is a form of stereoisomerism. The two stereoisomers have the same structural formula but the bonds are arranged differently in space. It occurs only around C=C double bonds. For example, but-2-ene, above, can exist as shown below.

Z-but-2-ene

E-but-2-ene

The isomer in which both –CH₃ groups are on the same side of the double bond is called *Z*-but-2-ene and the one in which they are on opposite sides is called *E*-but-2-ene. This type of isomerism is often called *E-Z* isomerism.

Hint

The *E-Z* system of naming is often called the CIP system after the names of its inventors – Robert Cahn, Christopher Ingold, and Vladimir Prelog.

Nomenclature – Cahn–Ingold–Prelog (CIP) notation

The number of known organic compounds is huge and increasing all the time. Finding information about these in databases, books, and journals would be almost impossible if chemists did not agree on how they should be named. This is why the Interntional Union of Pure and Applied Chemistry (IUPAC) produces rules for nomenclature. The *E-Z* notation is one example.

E-Z isomerism, until fairly recently, used to be known as *cis–trans* isomerism and the prefixes *cis-* and *trans-* were used instead of *Z-* and *E-*, respectively. So, for example, *Z*-but-2-ene was named *cis*-but-2-ene and *E*-but-2-ene was *trans*-but-2-ene. This notation is still often found in older books. However, a disadvantage of the older notation was that it did not work when there were more than two *different* substituents around a double bond. For example:

To give these two isomers different and unambiguous names the *E-Z* notation is used.

Simply, the *E-Z* notation is based on atomic numbers. Look at the atoms attached to each of the carbon atoms in the double bond. When the two atoms (of each pair) of higher atomic number (bromine and chlorine) are on the same side of the C=C, the isomer is described as *Z*, from the German word for together, *zusammen*.

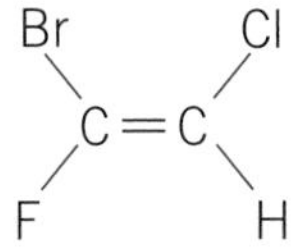

So this is *Z*-1-bromo-2-chloro-1-fluoroethene.

The other isomer has the positions of the hydrogen and chlorine atoms reversed.

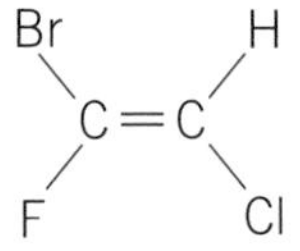

Study tip

Remember, when writing a systematic name, the groups are listed alphabetically.

So this is *E*-1-bromo-2-chloro-1-fluoroethene. See how this fits the IUPAC naming system.

The simplest interpretation of this naming system is that if the two atoms with the greatest atomic number are on the *same* side of the double bond, the name has the prefix *Z*-. If not, it has the prefix *E*-, from the German word for opposite, *entgegen*. However, the *cis–trans* notation is still commonly used when there is no possibility of confusion.

Physical properties of alkenes

The double bond does not greatly affect properties such as boiling and melting points. van der Waals forces are the only intermolecular forces that act between the alkene molecules. This means that the physical properties of alkenes are very similar to those of the alkanes. The melting and boiling points increase with the number of carbon atoms present. Alkenes are not soluble in water.

How alkenes react

The double bond makes a big difference to the reactivity of alkenes compared with alkanes. The bond enthalpy for C—C is 347 kJ mol^{-1} and that for C=C is 612 kJ mol^{-1} so you might predict that alkenes would be less reactive than alkanes. In fact alkenes are *more* reactive than alkanes.

The C=C forms an electron-rich area in the molecule, which can easily be attacked by positively charged reagents. These reagents are called **electrophiles** (electron liking). They are electron pair acceptors. An example of a good electrophile is the H$^+$ ion. As alkenes are unsaturated they can undergo addition reactions.

In conclusion, most of the reactions of alkenes are **electrophilic additions**.

> ### Study tip
>
> Learn the definition: an electrophile is an electron pair acceptor.

➕ Bond energies

Remember that a C=C bond consists of a σ-bond and a π-bond.

1   Use the bond energies in the text to calculate the strength of the π part of the bond alone.
2 Explain why the π part of the bond is weaker than the σ part.

Summary questions

1 What is the name of CH$_3$CH=CHCH$_2$CH$_2$CH$_3$?

2 Draw the structural formula for hex-1-ene.

3 There are six isomeric pentenes. Draw their displayed formulae.

4 Which of these attacks the double bond in an alkene? Choose from **a, b, c,** or **d.**

   a electrophiles

   b nucleophiles

   c alkanes

5 The double bond in an alkene can best be described by which of the following? Choose from **a, b, c,** or **d.**

   a electron-rich   c positively charged

   b electron-deficient   d acidic

14.2 Reactions of alkenes

Learning objectives:

→ Describe electrophilic addition reactions.

→ Outline the mechanism for these reactions.

Specification reference: 3.3.4

Combustion

Alkenes will burn in air:

$$\underset{\substack{\text{ethene}}}{\overset{\text{H}}{\underset{\text{H}}{>}}C=C\overset{\text{H}}{\underset{\text{H}}{<}}} \text{(g)} + 3O_2\text{(g)} \longrightarrow \underset{\text{carbon dioxide}}{2CO_2\text{(g)}} + \underset{\text{water}}{2H_2O\text{(l)}}$$

However, they are not used as fuels. This is because their reactivity makes them very useful for other purposes.

Electrophilic addition reactions

The reactions of alkenes are typically electrophilic additions. The four electrons in the carbon–carbon double bond make it a centre of high electron density. Electrophiles are attracted to it and can form a bond by using two of the four electrons in the carbon–carbon double bond (of the four electrons, the two that are in a π-bond, see Topic 14.1).

The mechanism is always essentially the same:

1 The electrophile is attracted to the double bond.
2 Electrophiles are positively charged and accept a pair of electrons from the double bond. The electrophile may be a positively charged ion or have a positively charged area.
3 A positive ion (a **carbocation**) is formed.
4 A negatively charged ion forms a bond with the carbocation.

See how the examples below fit this general mechanism.

Reaction with hydrogen halides

Hydrogen halides, HCl, HBr, and HI, add across the double bond to form a halogenoalkane. For example:

$$\underset{\substack{\text{ethene}}}{\overset{\text{H}}{\underset{\text{H}}{>}}C=C\overset{\text{H}}{\underset{\text{H}}{<}}} + \underset{\substack{\text{hydrogen}\\\text{bromide}}}{\text{HBr}} \longrightarrow \underset{\substack{\text{bromoethane}}}{H-\overset{\overset{\displaystyle H}{|}}{\underset{\underset{\displaystyle H}{|}}{C}}-\overset{\overset{\displaystyle H}{|}}{\underset{\underset{\displaystyle H}{|}}{C}}-Br}$$

- Bromine is more electronegative than hydrogen, so the hydrogen bromide molecule is polar, $H^{\delta+}-Br^{\delta-}$.
- The electrophile is the $H^{\delta+}$ of the $H^{\delta+}-Br^{\delta-}$.
- The $H^{\delta+}$ of HBr is attracted to the C=C bond because of the double bond's high electron density.
- One of the pairs of electrons from the C=C forms a bond with the $H^{\delta+}$ to form a positive ion (called a carbocation), whilst at the same time the electrons in the $H^{\delta+}-Br^{\delta-}$ bond are drawn towards the $Br^{\delta-}$.

Hint

Remember cations are positively charged.

- The bond in hydrogen bromide breaks heterolytically. Both electrons from the shared pair in the bond go to the bromine atom because it is more electronegative than hydrogen leaving a Br⁻ ion.

The Br⁻ ion attaches to the positively charged carbon of the carbocation forming a bond with one of its electron pairs.

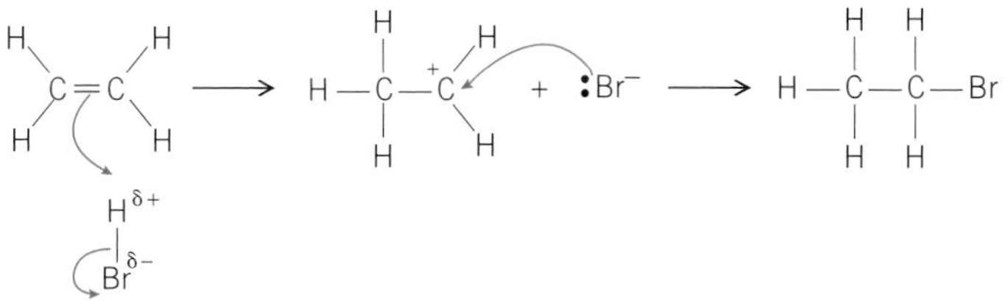

Hint

Heterolytic bond breaking means that when a covalent bond breaks, both the electrons go to one of the atoms involved in the bond and none to the other. This results in the formation of a negative ion and a positive ion.

Asymmetrical alkenes

When hydrogen bromide adds to ethene, bromoethane is the only possible product.

However, when the double bond is not exactly in the middle of the chain, there are two possible products – the bromine of the hydrogen bromide could bond to either of the carbon atoms of the double bond.

For example, propene could produce:

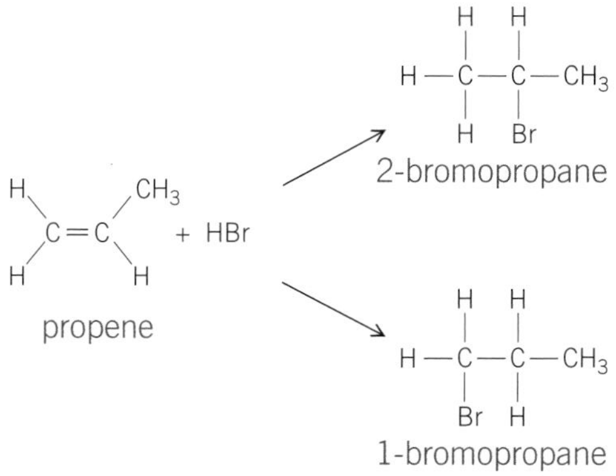

2-bromopropane

propene + HBr

1-bromopropane

Hint

There is simple way to work out the product. When hydrogen halides add on to alkenes, the hydrogen adds on to the carbon atom which already has the most hydrogens. This is called Markovnikov's rule.

In fact the product is almost entirely 2-bromopropane.

To explain this, you need to know that alkyl groups, for example, $-CH_3$ or $-C_2H_5$, have a tendency to release electrons. This is known as a **positive inductive effect** and is sometimes represented by an arrow along their bonds to show the direction of the release.

This electron-releasing effect tends to stabilise the positive charge of the intermediate carbocation. The more alkyl groups there are attached to the positively charged carbon atom, the more stable the carbocation is. So, a positively charged carbon atom which has three alkyl groups (called a tertiary carbocation) is more stable than one with two alkyl groups (a secondary carbocation) which is more stable than one with just one (a primary carbocation), see Figure 1.

The product will tend to come from the more stable carbocation.

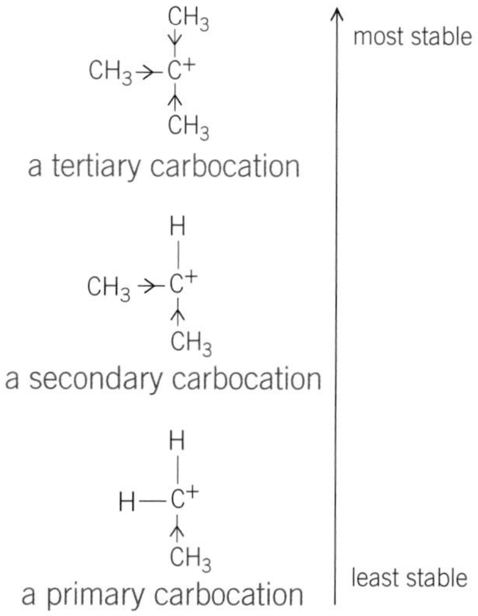

a tertiary carbocation

a secondary carbocation

a primary carbocation

▲ **Figure 1** *Stability of primary, secondary, and tertiary carbocations*

So, the two possible carbocations when propene reacts with HBr are:

a secondary carbocation a primary carbocation
(more stable, product   (less stable)
formed from this)

2-bromopropane

The secondary carbocation is more stable because it has two methyl groups releasing electrons towards the positive carbon. The majority of the product is formed from this.

Reaction of alkenes with halogens

Alkenes react rapidly with chlorine gas, or with solutions of bromine and iodine in an organic solvent, to give dihalogenoalkanes.

The halogen atoms *add* across the double bond.

In this case the halogen molecules act as electrophiles:

- At any instant, a bromine (or any other halogen) molecule is likely to have an instantaneous dipole, $Br^{\delta+}$—$Br^{\delta-}$. (An instant later, the dipole could be reversed $Br^{\delta-}$—$Br^{\delta+}$.) The $\delta+$ end of this dipole is attracted to the electron-rich double bond in the alkene – the bromine molecule has become an electrophile.

- The electrons in the double bond are attracted to the $Br^{\delta+}$. They repel the electrons in the Br—Br bond and this strengthens the dipole of the bromine molecule.

> **Hint**
>
> This dipole is also induced when a bromine molecule collides with the electron-rich double bond.

- Two of the electrons from the double bond form a bond with the $Br^{\delta+}$ and the other bromine atom becomes a Br^- ion. This leaves a carbocation, in which the carbon atom that is not bonded to the bromine has the positive charge.

- The Br^- ion now forms a bond with the carbocation.

> **Hint**
>
> The carbocation will react with any nucleophile that is present. In aqueous solution, such as bromine water, water reacts with the carbocation, forming some CH_2BrCH_2OH, 2-bromoethanol.

So the addition takes place in two steps:

1 formation of the carbocation by electrophilic addition
2 rapid reaction with a negative ion.

The test for a double bond

This addition reaction is used to test for a carbon–carbon double bond. When a few drops of bromine solution, sometimes called bromine water (which is reddish-brown) are added to an alkene, the solution is decolourised because the products are colourless.

Reaction with concentrated sulfuric acid

Concentrated sulfuric acid also adds across the double bond. The reaction occurs at room temperature and is exothermic.

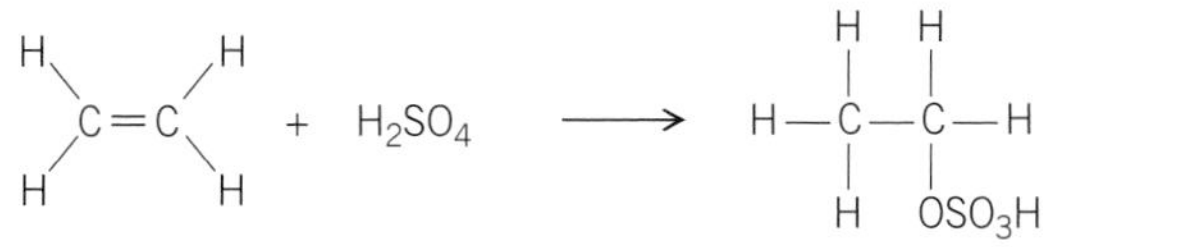

ethene   ethyl hydrogensulfate

The electrophile is a partially positively charged hydrogen atom in the sulfuric acid molecule. This can be shown as $H^{\delta+}-O^{\delta-}-SO_3H$

The carbocation which forms then reacts rapidly with the negatively charged hydrogensulfate ion.

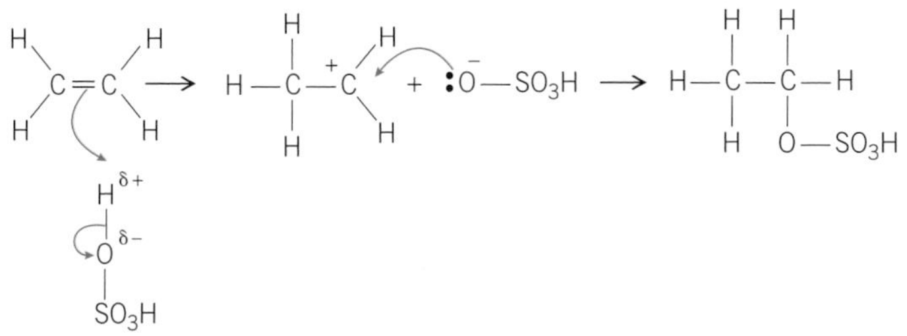

When water is added to the product an alcohol is formed and sulfuric acid reforms.

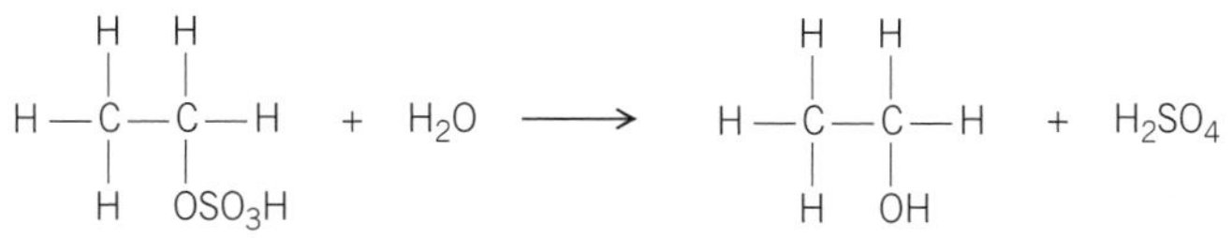

ethyl hydrogensulfate   ethanol

The overall effect is to add water H—OH across the double bond and the sulfuric acid is a catalyst for the process.

Asymmetrical alkenes

In an asymmetrical alkene, such as propene, the carbocation is exactly the same as that found in the reaction with hydrogen bromide. This means that you can predict the products by looking at the relative stability of the possible carbocations that could form.

Reaction with water

Water also adds on across the double bond in alkenes. The reaction is used industrially to make alcohols and is carried out with steam, at a suitable temperature and pressure, using an acid catalyst such as phosphoric acid, H_3PO_4.

$$CH_2{=}CH_2(g) + H_2O(g) \rightarrow CH_3CH_2OH(g)$$

The structure of sulfuric acid is:

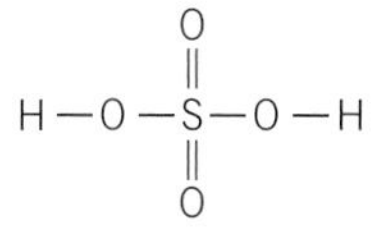

Summary questions

1 Write the equation for the complete combustion of propene.

2 Which of the following are typical reactions of alkenes?
   a Electrophilic additions
   b Electrophilic substitutions
   c Nucleophilic substitutions

3 a What are the two possible products of the reaction between propene and hydrogen bromide?
   b Which is the main product?
   c Explain why this product is more likely.

4 What is the product of the reaction between ethene and hydrogen chloride?

5 Which of the following is the test for a carbon–carbon double bond?
   a Forms a white precipitate with silver nitrate
   b Turns limewater milky
   c Decolourises bromine solution

Synoptic link

The manufacture of ethanol is discussed in Topic 15.2, Ethanol Production.

Polymers are very large molecules that are built up from small molecules, called **monomers**. They occur naturally everywhere: starch, proteins, cellulose and DNA are all polymers. The first completely synthetic polymer was Bakelite, which was patented in 1907. Since then, many synthetic polymers have been developed with a range of properties to suit them for very many applications, see Figure 1.

One way of classifying polymers is by the type of reaction by which they are made.

Addition polymers are made from a monomer or monomers with a carbon–carbon double bond (alkenes). Addition polymers are made from monomers based on ethene. The monomer has the general formula:

$$\begin{array}{ccc} H & & H \\ & C=C & \\ H & & R \end{array}$$

When the monomers polymerise, the double bond opens and the monomers bond together to form a backbone of carbon atoms as shown:

▲ **Figure 1** *Polymers around us*

This may also be represented by equations such as:

$$n\ \begin{array}{cc} H & H \\ C=C \\ H & R \end{array} \longrightarrow \begin{bmatrix} H & H \\ C-C \\ H & R \end{bmatrix}_n$$

R may be an alkyl or an aryl group.

For example, ethene polymerises to form poly(ethene)

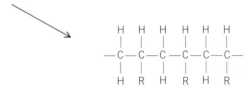

ethene

poly(ethene)

and phenylethene polymerises to poly(phenylethene):

H₂C=CH—phenyl monomers

phenylethene

↓

poly(phenylethene)

Phenylethene is sometimes called styrene, which is why poly(phenylethene) is usually called polystyrene.

Table 1 gives some examples of addition polymers based on different substituents.

▲ **Figure 2** *Both the model and the packaging are made from polystyrene*

▼ **Table 1** *Some addition polymers made from the monomer $H_2C\!\!=\!\!CHR$*

R	Monomer	Polymer	Name of polymer	Common or trade name	Typical uses
—H	$CH_2=CH_2$	$\left[\!\!-CH_2-CH_2-\!\!\right]_n$	poly(ethene)	polythene	carrier bags, washing up bowls
—CH_3	CH_3 $\|$ $CH=CH_2$	$\left[\begin{array}{c}CH_3 \\ \| \\ CH-CH_2\end{array}\right]_n$	poly(propene)	polypropylene	yoghurt containers car bumpers
—Cl	Cl $\|$ $CH=CH_2$	$\left[\begin{array}{c}Cl \\ \| \\ CH-CH_2\end{array}\right]_n$	poly (chloroethene)	PVC (polyvinyl chloride)	aprons, 'vinyl' records, drainpipes
—C≡N	CN $\|$ $CH=CH_2$	$\left[\begin{array}{c}CN \\ \| \\ CH-CH_2\end{array}\right]_n$	poly (propenenitrile)	acrylic (Acrilan, Courtelle)	clothing fabrics,
phenyl	phenyl–$CH=CH_2$	$\left[\begin{array}{c}\text{phenyl} \\ CH-CH_2\end{array}\right]_n$	poly (phenylethene)	polystyrene	packing materials, electrical insulation

Identifying the addition polymer formed from the monomer

The best way to think about this is to remember that an addition polymer is formed from monomers with carbon–carbon double bonds.

There is usually only one monomer (though it is possible to have more), and the double bond opens to form a single bond, see Table 1. This will give the repeat unit for the polymer.

▲ **Figure 3** *Bottles made from HD and LD polythene*

Many plastics are non-biodegradable

Identifying the monomer(s) used to make an addition polymer

An addition polymer must have a backbone of carbon atoms and the monomer must contain at least two carbons, so that there can be a carbon–carbon double bond. So, in the molecule below the monomer is shown in the red brackets:

Where some of the carbon atoms have substituents, the monomer must have the substituent, as well as a double bond:

Modifying the plastics

The properties of polymers materials can be considerably modified by the use of additives such as plasticisers. These are small molecules that get between the polymer chains forcing them apart and allowing them to slide across each other. For example PVC is rigid enough for use as drainpipes, but with the addition of a plasticiser it becomes flexible enough for making aprons.

Biodegradability

Polyalkenes, in spite of their name, have a backbone which is a long chain saturated alkane molecule. Alkanes have strong non-polar C—C and C—H bonds. So, they are very unreactive molecules, which is a useful property in many ways. However, this does mean that they are not attacked by biological agents – like enzymes – and so they are not biodegradable. This is an increasing problem in today's world, where waste disposal is becoming more and more difficult.

High and low density polythene

Low density poly(ethene) (polythene) is made by polymerising ethene at high pressure and high temperature via a free-radical mechanism. This produces a polymer with a certain amount of chain branching. This is a consequence of the rather random nature of free-radical reactions. The branched chains do not pack together particularly well and the product is quite flexible, stretches well and has a fairly low density. These properties make it suitable for packaging (plastic bags), sheeting and insulation for electrical cables.

High density polythene is made at temperatures and pressures little greater than room conditions and uses a Ziegler–Natta catalyst, named after its developers. This results in a polymer with much less chain branching (around one branch for every 200 carbons on the main chain). The chains can pack together well. This makes the density of the plastic greater and its melting temperature higher. Typical uses are milk crates, buckets and bottles for which low density polythene would be insufficiently rigid.

The solutions to pollution by plastics

To reduce the amount of plastic it can be reused or recycled.

Mechanical recycling

The simplest form of recycling is called mechanical recycling. The first step is to separate the different types of plastics. (Plastic containers are now collected in recycling facilities for this purpose.) The plastics are then washed and once they are sorted they may be ground up into small pellets. These can be melted and remoulded. For example, recycled soft drinks bottles made from PET (polyethylene terephthalate) are used to make fleece clothes.

Feedstock recycling

Here, the plastics are heated to a temperature that will break the polymer bonds and produce monomers. These can then be used to make new plastics.

There are problems with recycling. Poly(propene), for example, is a thermoplastic polymer. This means that it will soften when heated so it can be melted and re-used. However, this can only be done a limited number of times because at each heating some of the chains break and become shorter thus degrading the plastic's properties.

 Plasticisers

The properties of polymers can be tailored to make them suitable for a variety of application by the use of various additives. Plasticisers are small molecules that get in between the polymer chains to allow them to slide more easily past one another and make the polymer more flexible. This is how poly(chloroethene) also called polyvinyl chloride (PVC) or just vinyl can be made rigid enough for use as drainpipes and flexible enough for plastic aprons.

Summary questions

1 Which of the following monomers could form an addition polymer?

A

B F, F / C=C \ F, F

C NH₂ / CH₃—C—COOH / H

D H, H / C=C \ H, CH₃

2 **a** Draw a section of the polymer formed from the monomer

H, H / C=C \ H, Cl   showing six carbon atoms.

b What is the common name of the monomer?

c What is the systematic name of the polymer?

3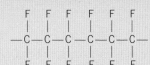

This is a section of the polymer that non-stick pans are coated with the trade name Teflon.

What is the monomer?

4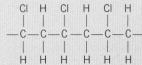

This is a section of the polymer that drainpipes are made from, trade name polyvinylchloride (PVC)

What is the monomer?

Practice questions

1 The table below gives the names and structures of three isomeric alkenes.

Name	Structure
but-1-ene	$CH_3CH_2CH{=}CH_2$
but-2-ene	$CH_3CH{=}CHCH_3$
methylpropene	$H_3C-\underset{\underset{\displaystyle CH_3}{\vert}}{C}{=}CH_2$

(a) Give the molecular formula and the empirical formula of but-2-ene.

(2 marks)

(b) Methylpropene reacts with hydrogen bromide to produce 2-bromo-2-methylpropane as the major product.

  (i) Name and outline the mechanism for this reaction.

  (ii) Draw the structure of another product of this reaction and explain why it is formed in smaller amounts.

(8 marks)

(c) Draw the structures and give the names of the two stereoisomers of but-2-ene.

(2 marks)

AQA, 2004

2 (a) Copy and complete the mechanism below by drawing appropriate curly arrows.

$H\overset{..}{O}{:}^{-}$

$H_3C-\overset{\overset{\displaystyle H}{\vert}}{\underset{\underset{\displaystyle H}{\vert}}{C}}-\overset{\overset{\displaystyle H}{\vert}}{\underset{\underset{\displaystyle H}{\vert}}{C}}-\overset{\overset{\displaystyle H}{\vert}}{\underset{\underset{\displaystyle Br}{\vert}}{C}}-CH_3$ $CH_3CH_2CH{=}CHCH_3 + H_2O + Br^-$

2-bromopentane   pent-2-ene

(3 marks)

AQA, 2005

3 (a) Propene reacts with hydrogen bromide by an electrophilic addition mechanism forming 2-bromopropane as the major product.

The equation for this reaction is shown below.

$\underset{H}{\overset{H_3C}{>}}C{=}C\underset{H}{\overset{H}{<}}$ + HBr $\longrightarrow$ $H_3C-\overset{\overset{\displaystyle Br}{\vert}}{\underset{\underset{\displaystyle H}{\vert}}{C}}-\overset{\overset{\displaystyle H}{\vert}}{\underset{\underset{\displaystyle H}{\vert}}{C}}-H$

  (i) Outline the mechanism for this reaction, showing the structure of the intermediate carbocation formed.

  (ii) Give the structure of the alternative carbocation which could be formed in the reaction between propene and hydrogen bromide.

(5 marks)

AQA, 2003

4 The reaction scheme below shows the conversion of compound **A**, 2-methylbut-1-ene, into compound **B** and then into compound **C**.

$CH_2{=}\overset{\overset{\displaystyle CH_3}{\vert}}{C}-CH_2CH_3$ $\xrightarrow[\substack{concentrated \\ H_2SO_4}]{Step\ 1}$ $CH_3-\overset{\overset{\displaystyle CH_3}{\vert}}{\underset{\underset{\displaystyle OSO_2OH}{\vert}}{C}}-CH_2CH_3$

A   **B**

Step 2 ↓

$CH_3-\overset{\overset{\displaystyle CH_3}{\vert}}{\underset{\underset{\displaystyle OH}{\vert}}{C}}-CH_2CH_3$

C

(a) The structure of **A** is shown below. Circle those carbon atoms which must lie in the same plane.

(1 mark)

$\underset{H}{\overset{H}{>}}C{=}C\underset{CH_2-CH_3}{\overset{CH_3}{<}}$

(b) Outline a mechanism for the reaction in Step 1.

(4 marks)

AQA, 2002

5 It is possible to convert but-1-ene into its structural isomer but-2-ene.

(a) State the type of structural isomerism shown by but-1-ene and but-2-ene.

(1 mark)

(b) The first stage in this conversion involves the reaction of hydrogen bromide with but-1-ene.

$CH_3CH_2CH{=}CH_2 + HBr \rightarrow CH_3CH_2CHBrCH_3$

Outline a mechanism for this reaction.

(4 marks)

(c) The second stage is to convert 2-bromobutane into but-2-ene.

$CH_3CH_2CHBrCH_3 + KOH \rightarrow CH_3CH{=}CHCH_3 + KBr + H_2O$

Outline a mechanism for this reaction.

(3 marks)

AQA, 2012

6 Alkenes are useful intermediates in the synthesis of organic compounds.

(a) (i) Complete the elimination mechanism by drawing appropriate curly arrows.

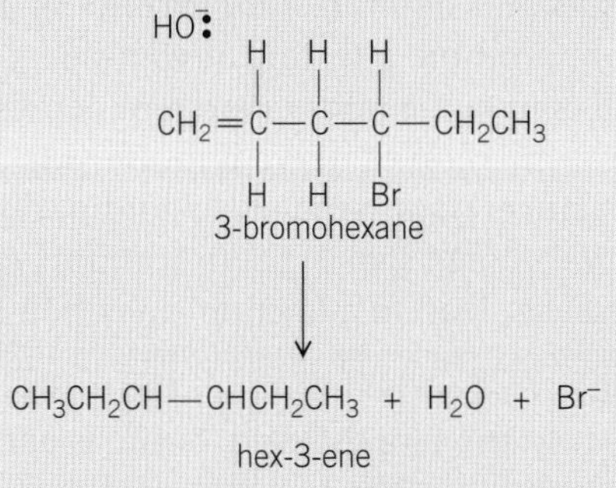
HO:
H H H
$CH_2=C—C—C—CH_2CH_3$
H H Br
3-bromohexane
$CH_3CH_2CH—CHCH_2CH_3$ + H_2O + Br^-
hex-3-ene

(3 marks)

(ii) Draw structures for the E and Z stereoisomers of hex-3-ene.

E isomer of hex-3-ene

Z isomer of hex-3-ene

(2 marks)

(iii) State the meaning of the term *stereoisomers*.

(2 marks)

(b) The equation for the first reaction in the conversion of hex-3-ene into hexan-3-ol is shown below.

$$CH_3CH_2CH{=}CHCH_2CH_3 + H_2SO_4 \rightarrow CH_3CH_2CH_2CH(OSO_2OH)CH_2CH_3$$

Outline a mechanism for this reaction.

(4 marks)

AQA, 2012

7 Propene reacts with bromine by a mechanism known as electrophilic addition.

(a) Explain what is meant by the term *electrophile* and by the term *addition*.

(2 marks)

(b) Outline the mechanism for the electrophilic addition of bromine to propene. Give the name of the product formed.

(5 marks)

AQA, 2002

8 **(a)** (i) Name the alkene $CH_3CH_2CH{=}CH_2$

(ii) Explain why $CH_3CH_2CH{=}CH_2$ does not show stereoisomerism.

(iii) Draw an isomer of $CH_3CH_2CH{=}CH_2$ which does show *E–Z* isomerism.

(iv) Draw another isomer of $CH_3CH_2CH{=}CH_2$ which does not show *E–Z* isomerism.

(4 marks)

(b) (i) Name the type of mechanism for the reaction shown by alkenes with concentrated sulfuric acid.

(ii) Write a mechanism showing the formation of the major product in the reaction of concentrated sulfuric acid with $CH_3CH_2CH{=}CH_2$.

(iii) Explain why this compound is the major product.

(6 marks)

9 The alkene (Z)-3-methylpent-2-ene reacts with hydrogen bromide as shown below.

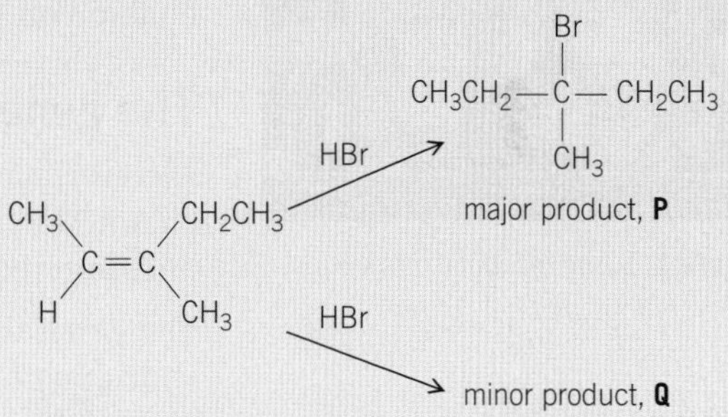
Br
$CH_3CH_2—C—CH_2CH_3$
CH_3
HBr
major product, **P**
CH_3
CH_2CH_3
C=C
H
CH_3
HBr
minor product, **Q**

(a) (i) Name the major product **P**.

(1 mark)

(ii) Name the mechanism for these reactions.

(1 mark)

(iii) Draw the displayed formula for the minor product **Q** and state the type of structural isomerism shown by **P** and **Q**.

Displayed formula for **Q**

Type of structural isomerism

(2 marks)

(iv) Draw the structure of the (E)-stereoisomer of 3-methylpent-2-ene.

(1 mark)

AQA, 2010

10 An organic compound A is shown below.

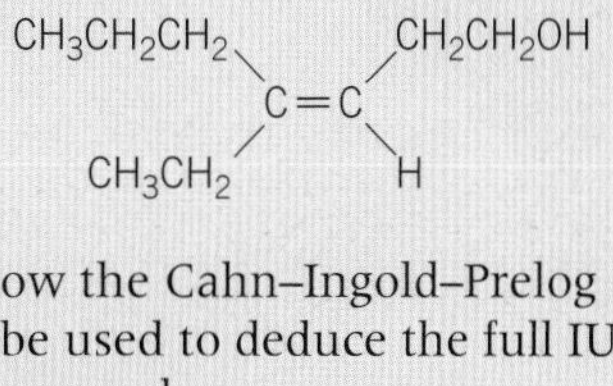
$CH_3CH_2CH_2$
CH_2CH_2OH
C=C
CH_3CH_2
H

Explain how the Cahn–Ingold–Prelog (CIP) priority rules can be used to deduce the full IUPAC name of this compound.

Learning objectives:

→ State the general formula of an alcohol.

→ Describe how alcohols are classified.

→ Describe the physical properties of alcohols.

Specification reference: 3.3.5

▲ **Figure 1** *Some alcoholic drinks and their percentage alcohol content*

Ethanol is possibly our oldest social drug as it is derived from the fermentation of sugars in fruits and so on. It is the alcohol in alcoholic drinks. It may, in moderation, promote a feeling of well-being and reduce normal inhibitions. It is in fact a nervous system depressant (i.e., it interferes with the transmission of nerve impulses). In larger amounts it leads to loss of balance, poor hand–eye coordination, impaired vision, and inability to judge speed. Large amounts can be fatal. Excessive long-term use can lead to addiction – alcoholism.

The ethanol in alcoholic drinks is absorbed through the walls of the stomach and small intestine into the bloodstream. Some is eliminated unchanged in urine and in the breath. The rest is broken down by the liver. The combined effect of these processes is that an average person can eliminate about 10 cm^3 of ethanol per hour. This is approximately the amount of ethanol in half a pint of beer, a small glass of wine (125 ml) or a shot (25 ml of spirits). So some simple arithmetic should enable you to work out how long it would take to sober up.

The general formula

Alcohols have the functional group –OH attached to a hydrocarbon chain. They are relatively reactive. The alcohol most commonly encountered in everyday life is ethanol.

The general formula of an alcohol is $C_nH_{2n+1}OH$. This is often shortened to ROH.

How to name alcohols

The name of the functional group (the –OH group) is normally given by the suffix -ol. (The prefix hydroxy- is used if some other functional groups are present.)

$$\begin{array}{ccccc} & H & H & H & H \\ & | & | & | & | \\ H- & C- & C- & C- & C-O-H \\ & | & | & | & | \\ & H & H & H & H \end{array}$$

butan-1-ol

With chains longer than ethanol, you need a number to show where the –OH group is.

$$\begin{array}{cccc} & H & H & H \\ & | & | & | \\ H- & C- & C- & C-O-H \\ & | & | & | \\ & H & H & H \end{array}$$

propan-1-ol

$$\begin{array}{ccc} & & H \\ & & | \\ & H & O & H \\ & | & | & | \\ H- & C- & C- & C-H \\ & | & | & | \\ & H & H & H \end{array}$$

propan-2-ol

If there is more than one –OH group, di-, tri-, tetra-, and so on are used to say *how many* –OH groups there are and numbers to say *where* they are.

butane-1,4-diol

propane-1,2,3-triol

Propane-1,2,3-triol is also known as glycerol, which may be obtained from the fats and oils found in living organisms.

Antifreeeze

Ethane-1,2-diol is the main ingredient in most antifreezes. These are added to the water in the cooling systems of motor vehicles so that the resulting coolant mixture does not freeze at winter temperatures.

1 Draw the structural formula of ethane-1,2-diol and indicate where hydrogen bonding with water may take place.
2 Suggest why solutions of ethane-1,2-diol have lower freezing temperatures than pure water.

The reactivity of alcohols

How an organic molecule reacts depends, among other things, on the strength and polarity of the bonds within it.

1 Draw a displayed formula for ethanol and mark on it the bond energy (mean bond enthalpy) of each bond (the amount of energy required to break it). Use a data book or database or the table of bond enthalpies in Topic 4.7 and the values C—O 336 kJ mol^{-1} and O—H 464 kJ mol^{-1}. There is a table of electronegativities in Topic 13.1, Halogenoalkanes – introduction, and the values for hydrogen and oxygen are 2.1 and 3.5 respectively.
2 Use your diagram to explain why:
  a the hydrocarbon skeleton remains intact in most reactions of ethanol
  b why the typical reactions of ethanol are nucleophilic substitutions in which the —OH group is lost.

▲ **Figure 2** *Methylated spirits contain ethanol with small amounts of methanol added to make it undrinkable. The purple dye warns of this*

Shape

In alcohols, the oxygen atom has two bonding pairs of electrons and two lone pairs. The C—O—H angle is about 105° because the 109.5° angle of a perfect tetrahedron is 'squeezed down' by the presence of the lone pairs. These two lone pairs will repel each other more than the pairs of electrons in a covalent bond.

H—C—C—O:
104.5°

Synoptic link

You learnt about the repulsion of lone pairs of electrons compared to the repulsion of bonded pairs of electrons in Topic 3.6, The shapes of molecules and ions.

Classification of alcohols

Alcohols are classified as **primary** (1°), **secondary** (2°), or **tertiary** (3°) according to how many other groups (R) are bonded to the carbon that has the –OH group.

Primary alcohols

In a primary alcohol, the carbon with the –OH group has one R group (and therefore two hydrogen atoms).

propan-1-ol is a primary alcohol

methanol, where the carbon has no R groups is counted as a primary alcohol

A primary alcohol has the –OH group at the end of a chain.

Secondary alcohols

In a secondary alcohol, the –OH group is attached to a carbon with two R groups (and therefore one hydrogen atom).

propan-2-ol is a secondary alcohol

A secondary alcohol has the –OH group in the body of the chain.

Tertiary alcohols

Tertiary alcohols have three R groups attached to the carbon that is bonded to the –OH (so this carbon has no hydrogen atoms).

2-methylpropan-2-ol is a tertiary alcohol

A tertiary alcohol has the –OH group at a branch in the chain.

Physical properties

The –OH group in alcohols means that hydrogen bonding occurs between the molecules. This is the reason that alcohols have higher melting and boiling points than alkanes of similar relative molecular mass.

The –OH group of alcohols can hydrogen bond to water molecules, but the non-polar hydrocarbon chain cannot. This means that the alcohols with short hydrocarbon chains are soluble in water because the hydrogen bonding predominates. In longer-chain alcohols the non-polar hydrocarbon chain dominates and the alcohols become insoluble in water.

15.2 Ethanol production

Industrial chemistry of alcohols

Alcohols are very important in industrial chemistry because they are used as intermediates. They are easily made and easily converted into other compounds. Methanol is made from methane (natural gas) and is increasingly being used as a starting material to make other organic chemicals.

Ethanol

Ethanol, C_2H_5OH, is by far the most important alcohol. It is used as an intermediate in the manufacture of other organic chemicals. In everyday life it is often the solvent in cosmetics, such as aftershave and perfumes. It is also used in the manufacture of drugs, detergents, inks, and coatings.

It is made industrially by reacting ethene (made from cracking crude oil) with steam, using a catalyst of phosphoric acid. It is also made from sugars by fermentation, as in the production of alcoholic drinks.

Beers have about 5% ethanol and wines about 12%. Spirits, such as gin and whisky, contain about 40% ethanol – these have been concentrated by distillation.

Making ethanol from crude oil

Ethene is produced when crude oil fractions are cracked, see Topic 12.3.

Ethene is hydrated, which means that water is added across the double bond (Figure 1).

$$CH_2{=}CH_2 + H_2O \xrightarrow[\text{catalyst}]{\text{phosphoric acid}} C_2H_5OH$$

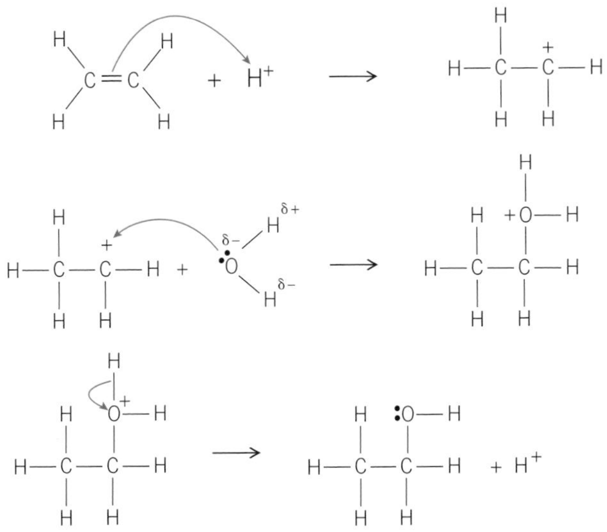

Making ethanol by fermentation

During fermentation, carbohydrates from plants are broken down into sugars and then converted into ethanol by the action of enzymes from yeast. The carbohydrates come from crops such as sugar cane and sugar beet.

The key step is the breakdown of sugar in a process called **anaerobic respiration**:

enzymes from yeast

$$C_6H_{12}O_6(aq) \rightarrow 2C_2H_5OH(aq) + 2CO_2(g)$$

glucose (a sugar)   ethanol   carbon dioxide

- The rate of this chemical reaction is affected by temperature. It is slow at low temperatures but the enzymes are made ineffective if the temperature is too high. A compromise temperature of about 35 °C, a little below our body temperature, is used.
- Air is kept out of the fermentation vessels to prevent oxidation of ethanol to ethanoic acid (the acid in vinegar).
- Once the fermenting solution contains about 15% ethanol the enzymes are unable to function and fermentation stops. Ethanol may the be distilled from this mixture by fractional distillation as its boiling temperature (78 °C) is less than that of water (100 °C).

Learning objectives:
→ Describe how ethanol is produced by fermentation.
→ Describe the economic and environmental advantages of producing ethanol by fermentation.
→ State what is meant by the term biofuel.

Specification reference: 3.3.5

▲ **Figure 1** *The mechanism for making ethanol from crude oil*

▼ **Table 1** *Different methods of producing ethanol*

	Starting material	
	Crude oil non-renewable	Carbohydrates (sugars) renewable
method	cracking and hydration	fermentation and distillation
rate of reaction	fast	slow
type of process	continuous	batch
purity	essentially pure	aqueous solution of ethanol is produced

Hint

A **biofuel** is a fuel derived or produced from renewable biological sources.

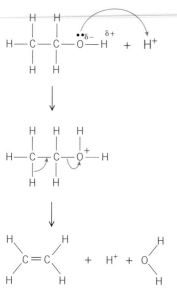

▲ **Figure 2** *The mechanism for dehydrating ethanol to make ethene*

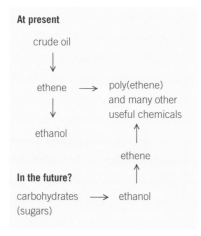

▲ **Figure 3** *Ethene and ethanol production*

A renewable source of ethene

Ethene is a vital industrial chemical; it is the starting material for poly(ethene) and many other important chemicals. Ethene can be produced by dehydrating ethanol made from sugar, giving a renewable source of ethene.

At present ethene is made from crude oil and then converted into ethanol. In the future it may become more economical to make ethene from ethanol made by fermentation, see Figure 3.

Carbon neutrality

Many conventional petrol engines will run on ethanol, or mixtures of petrol and ethanol, with little modification, and much of the petrol sold in the UK at present has 5–10% ethanol added.

Ethanol made from ethene is not a renewable fuel because it comes originally from crude oil. However, ethanol made by fermentation is renewable because the sugars come from plants such as sugar cane and beet, which can be grown annually.

Current fuels are almost all carbon-based. One concern is that they release carbon dioxide into the atmosphere. Rising carbon dioxide levels are associated with global warming and climate change. Ethanol made by fermentation is sometimes termed a **carbon-neutral** fuel. This means that the carbon dioxide released when it is burnt is balanced by the carbon dioxide absorbed by the plant from which it was originally obtained, during photosynthesis. This can be seen from Table 2. This argument concentrates on the chemistry of fuel production and use. There are inevitably other carbon costs associated with the energy needed to transport crops and the fuel, and to process the crops.

▼ **Table 2** *The carbon dioxide balance sheet for ethanol made by fermentation*

Carbon dioxide absorbed	Carbon dioxide released
Photosynthesis in the growing plant $6H_2O(l) + 6CO_2(g)$ ↓ $C_6H_{12}O_6(aq) + 6O_2(g)$ 6 molecules of CO_2 absorbed	fermentation $C_6H_{12}O_6(aq) \rightarrow 2C_2H_5OH(aq) + 2CO_2(g)$ 2 molecules of CO_2 released combustion $2C_2H_5OH(aq) + 6O_2(g) \rightarrow 4CO_2(g) + 6H_2O(l)$ 4 molecules of CO_2 released
6 molecules of CO_2 absorbed	6 molecules of CO_2 released

Summary questions

1   How can ethanol, produced by fermentation, be separated from its aqueous solution.

2   What are the advantages and disadvantages of producing ethanol from fermentation compared with its production from crude oil?

3   Explain why fermentation and distillation can only take place as a batch process rather than continuously.

15.3 The reactions of alcohols

Combustion

Alcohols burn completely to carbon dioxide and water if there is enough oxygen available. (Otherwise there is incomplete combustion and carbon monoxide or even carbon is produced.) This is the equation for the complete combustion of ethanol:

$$C_2H_5OH(l) + 3O_2(g) \rightarrow 2CO_2(g) + 3H_2O(l)$$

Ethanol is often used as a fuel, for example, in picnic stoves that burn methylated spirits. Methylated spirits is ethanol with a small percentage of poisonous methanol added to make it unfit to drink. In this way it can be sold without the tax which is levied on alcoholic drinks. A purple dye is also added to show that it should not be drunk.

Elimination reactions

Elimination reactions are ones in which a small molecule leaves the parent molecule. In the case of alcohols, this molecule is always water. The water is made from the –OH group and a hydrogen atom from the carbon next to the –OH group. So, the elimination reactions of alcohols are always dehydrations.

Dehydration

Alcohols can be dehydrated with excess hot concentrated sulfuric acid or by passing their vapours over heated aluminium oxide. An alkene is formed. For example, propan-1-ol is dehydrated to propene:

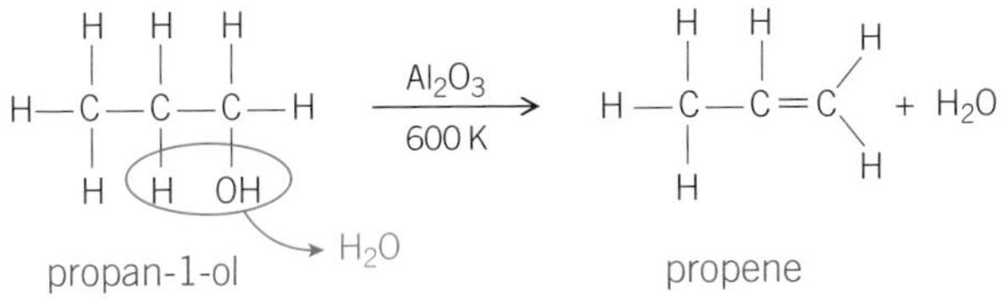

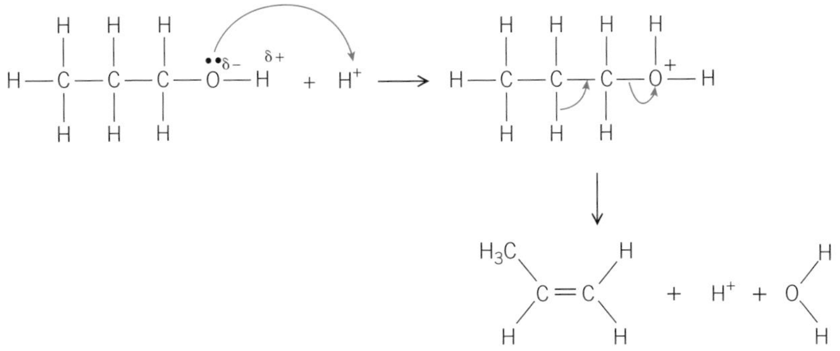

▲ **Figure 2** *The mechanism for the dehydration of propan-1-ol*

The apparatus used in the laboratory is shown in Figure 3.

Phosphoric(V) acid is an alternative dehydrating agent.

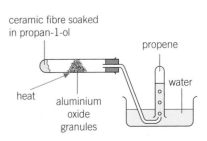

▲ **Figure 1** *Alcohol-burning stove*

Hint

In order to be dehydrated an alcohol must have a hydrogen atom on a carbon next to the —OH group.

ceramic fibre soaked in propan-1-ol

▲ **Figure 3** *Dehydration of an alcohol*

Isomeric alkenes

Dehydration of longer chain or branched alcohols may produce a mixture of alkenes, including ones with Z and E isomers, see Topic 11.3.

For example, with butan-2-ol there are three possible products:
but-1-ene, Z-but-2-ene, and E-but-2-ene.

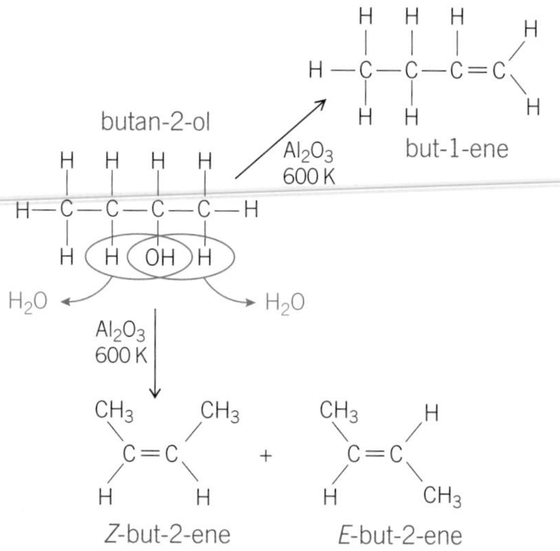

butan-2-ol

but-1-ene

Z-but-2-ene   E-but-2-ene

Name the two isomeric alkanes that are formed using the *cis/trans* notation.

cis-but-2-ene and *trans*-but-2-ene respectively

Oxidation

Combustion is usually complete oxidation. Alcohols can also be oxidised gently and in stages. Primary alcohols are oxidised to **aldehydes**, RCHO. Aldehydes can be further oxidised to carboxylic acids, RCOOH. For example:

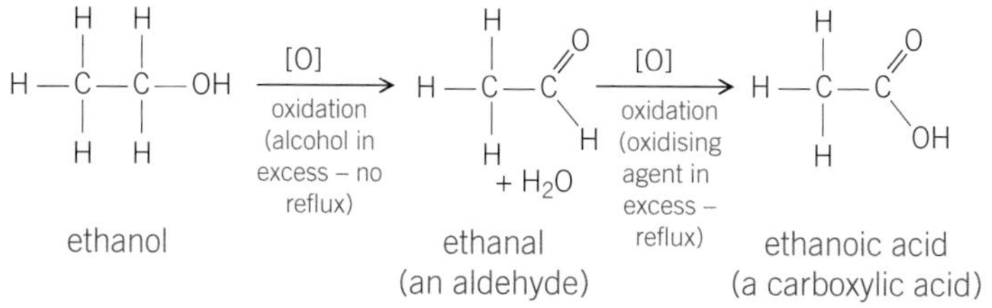

ethanol

[O]
oxidation
(alcohol in excess – no reflux)

ethanal
(an aldehyde)
+ H_2O

[O]
oxidation
(oxidising agent in excess – reflux)

ethanoic acid
(a carboxylic acid)

Secondary alcohols are oxidised to **ketones**, R_2CO. Ketones are not oxidised further.

propan-2-ol

[O]

propanone
(a ketone)
+ H_2O

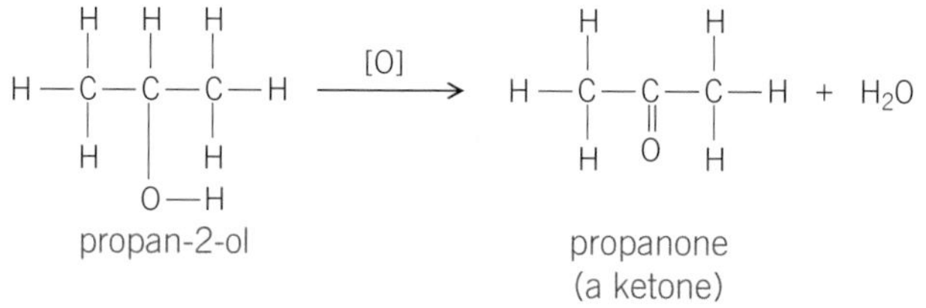

Tertiary alcohols are not easily oxidised. This is because oxidation would need a C—C bond to break, rather than a C—H bond (which is what happens when an aldehyde is oxidised). Ketones are not oxidised further for the same reason.

Many aldehydes and ketones have pleasant smells.

The experimental details

A solution of potassium dichromate, acidified with dilute sulfuric acid, is often used to oxidise alcohols to aldehydes and ketones. It is the oxidising agent. In the reaction, the orange dichromate(VI) ions are reduced to green chromium(III) ions.

To oxidise ethanol (1° alcohol) to ethanal – an aldehyde

Dilute acid and less potassium dichromate(VI) than is needed for complete oxidation to carboxylic acid are used. The mixture is heated gently in apparatus like that shown in Figure 4, but with the receiver cooled in ice to reduce evaporation of the product. Ethanal (boiling temperature 294 K, 21 °C) vaporises as soon as it is formed and distils off. This stops it from being oxidised further to ethanoic acid. Unreacted ethanol remains in the flask.

The notation [O] is often used to represent oxygen from the oxidising agent. The reaction is given by the equation:

$$CH_3CH_2OH(l) + [O] \longrightarrow CH_3CHO(g) + H_2O(l)$$
ethanol   ethanal

To oxidise ethanol (1° alcohol) to ethanoic acid – a carboxylic acid

Concentrated sulfuric acid and more than enough potassium dichromate(VI) is used for complete reaction (the dichromate(VI) is in excess). The mixture is refluxed in the apparatus shown in Figure 5. Reflux means that vapour condenses and drips back into the reaction flask.

Whilst the reaction mixture is refluxing, any ethanol or ethanal vapour will condense and drip back into the flask until, eventually, it is all oxidised to the acid. After refluxing for around 20 minutes, you can distil off the ethanoic acid (boiling temperature 391 K, 118 °C), along with any water, by rearranging the apparatus to that shown in Figure 4.

Using [O] to represent oxygen from the oxidising agent, the equation is:

$$CH_3CH_2OH(l) + 2[O] \rightarrow CH_3COOH(g) + H_2O(l)$$
ethanol   ethanoic acid

Notice that twice as much oxidising agent is used in this reaction compared with the oxidation to ethanal.

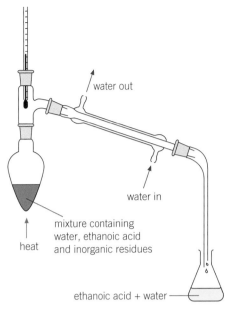

▲ **Figure 4** *Apparatus for distilling ethanoic acid from the reaction mixture*

> ### Hint
>
> Notice that even if you use the notation [O] for oxidation, the equation must still balance.

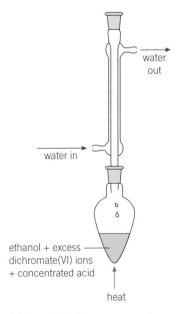

▲ **Figure 5** *Reflux apparatus for oxidation of ethanol to ethanoic acid*

➕ The breathalyser

Since 1967 it has been illegal to drive with a blood ethanol concentration of more than 80 mg of alcohol in 100 cm^3 of blood. This corresponds to 35 μg of ethanol per 100 cm^3 of breath or 107 mg of ethanol per 100 cm^3 of urine.

Before the law could be introduced, a quick method of

▲ **Figure 6** *A modern fuel cell breathalyser*

measuring alcohol levels by the roadside was needed. Taking blood or urine samples was not practical, so breath was chosen.

The oxidation of ethanol was the basis of the original roadside breath test. The suspect was asked to breathe into an inflatable bag through a tube containing orange acidified potassium dichromate crystals and a silver nitrate catalyst.

As the breath passes over the crystals, any ethanol is oxidised by the potassium dichromate and the yellow Cr(VI) is reduced to green Cr(III). The more ethanol, the more crystals change colour. If the colour change passed a pre-determined line, the suspect was deemed to be potentially over the limit and was required to have a further test in a police station by a more accurate method. In modern roadside breath testers the oxidation takes place in a fuel cell and generates an electric current that can be measured.

1   What is 1 µg? Write 35 µg in grams in standard form.
2   Explain the reason for the breath sample being breathed into a bag?
3   Why is a catalyst needed?

Oxidising a secondary alcohol to a ketone

Secondary alcohols are oxidised to ketones by acidified dichromate. You do not have to worry about further oxidation of the ketone.

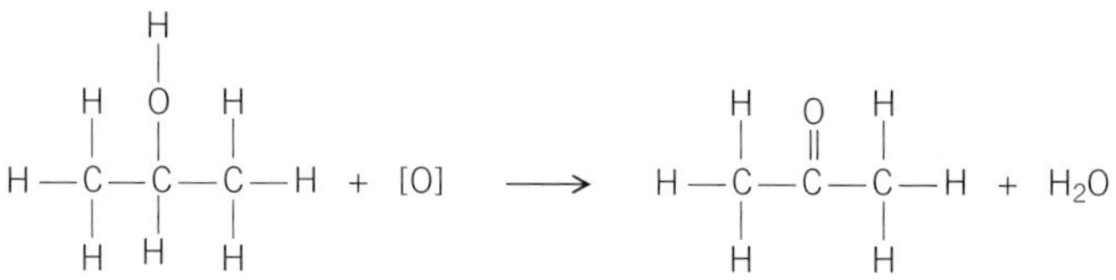

Aldehydes and ketones

Aldehydes and ketones both have the C=O group. This is called the carbonyl group.

In aldehydes it is at the end of the hydrocarbon chain:

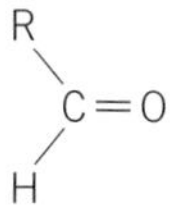

In ketones it is in the body of the hydrocarbon chain:

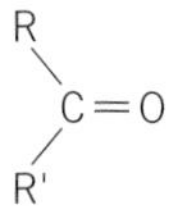

Aldehydes are usually named using the suffix –al and ketones with the suffix –one.

So CH_3CHO is ethan*al* (two carbons) and CH_3COCH_3 is propan*one* (three carbons).

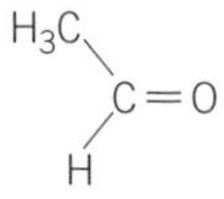

ethanal 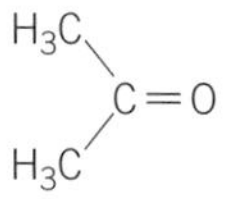   propanone

Tests for aldehydes and ketones

Aldehydes and ketones have similar physical properties but there are two tests that can tell them apart. Both these tests involve gentle oxidation.

- Aldehydes are oxidised to carboxylic acids – RCHO + [O] → RCOOH (This is the second stage of the oxidation of a primary alcohol.)
- Ketones are not changed by gentle oxidation.

The Tollens' (silver mirror) test

Tollens' reagent is a gentle oxidising agent. It is a solution of silver nitrate in aqueous ammonia. It oxidises aldehydes but has no affect on ketones. It contains colourless silver(I) complex ions, containing Ag^+, which are reduced to metallic silver, Ag, as the aldehyde is oxidised.

On warming an aldehyde with Tollens' reagent, a deposit of metallic silver is formed on the inside of the test tube – the silver mirror, see Figure 7. This reaction was once used commercially for making mirrors.

▲ **Figure 7** *The silver mirror test*

The Fehling's test

The Fehling's reagent and is a gentle oxidising agent. It contains blue copper(II) complex ions which will oxidise aldehydes but not ketones. During the oxidation, the blue solution gradually changes to a brick red precipitate of copper(I) oxide:

$$Cu^{2+} + e^- \rightarrow Cu^+.$$

On warming an aldehyde with blue Fehling's solution a brick red precipitate gradually forms.

Summary questions

1 State what happens in each case when the following alcohols are oxidised as much as possible, by acidified potassium dichromate.

   a a primary alcohol

   b a secondary alcohol

2 Why is a tertiary alcohol not oxidised by the method outlined in question **1**?

3 What is the difference between distilling and refluxing?

4 Suggest how you would distinguish between a primary alcohol and a secondary alcohol, using Tollens' reagent or Fehling's solution.

5 Write the equation for the elimination of water from ethanol and name the product.

6 What are the possible products of dehydrating pentan-2-ol?

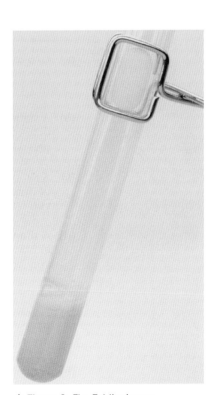

▲ **Figure 8** *The Fehling's test*

Practice questions

1 Glucose, $C_6H_{12}O_6$, can be converted into ethanol. Ethanol can be used as a fuel or can be converted into ethene by acid-catalysed dehydration. Most of the ethene used by industry is formed by the thermal cracking of alkanes.

   (a) State **four** essential conditions for the conversion of glucose into ethanol. Name the process and give an equation for the reaction which takes place. Write an equation for the complete combustion of ethanol.

(7 marks)

   (b) Explain what is meant by the term *dehydration*. Identify a catalyst which could be used in the acid-catalysed dehydration of ethanol. Write an equation for the reaction which takes place.

(3 marks)
AQA, 2006

2 Consider the following pairs of structural isomers.

Molecular formula	Structure	Structure
$C_4H_{10}O$	Isomer **A** $\begin{array}{c} CH_3 \\ \mid \\ H_3C - C - CH_3 \\ \mid \\ OH \end{array}$	Isomer **B** $CH_3CH_2CH_2CH_2OH$
	Isomer **C** $\begin{array}{c} CH_3CH_2 - C = O \\ \mid \\ H \end{array}$	Isomer **D** $\begin{array}{c} H_3C - C - CH_3 \\ \parallel \\ O \end{array}$
C_6H_{12}	Isomer **E** $\begin{array}{c} CH_2 \\ H_2C \quad CH_2 \\ H_2C \quad CH_2 \\ CH_2 \end{array}$	Isomer **F** $CH_3CH_2CH = CHCH_2CH_3$

   (a) (i) Explain what is meant by the term *structural isomers*.
      (ii) Complete the table to show the molecular formula of isomers **C** and **D**.
      (iii) Give the empirical formula of isomers **E** and **F**.

(4 marks)

   (b) A simple chemical test can be used to distinguish between separate samples of isomer **A** and isomer **B**. Suggest a suitable test reagent and state what you would observe in each case.

(3 marks)

   (c) A simple chemical test can be used to distinguish between separate samples of isomer **C** and isomer **D**. Suggest a suitable test reagent and state what you would observe in each case.

(3 marks)

   (d) A simple chemical test can be used to distinguish between separate samples of isomer **E** and isomer **F**. Suggest a suitable test reagent and state what you would observe in each case.

(3 marks)
AQA, 2006

3 (a) Pentanal, $CH_3CH_2CH_2CH_2CHO$, can be oxidised to a carboxylic acid.
      (i) Write an equation for this reaction. Use [O] to represent the oxidising agent.
      (ii) Name the carboxylic acid formed in this reaction.

(4 marks)

   (b) Pentanal can be formed by the oxidation of an alcohol.
      (i) Identify this alcohol.
      (ii) State the class to which this alcohol belongs.

(2 marks)
AQA, 2006

4   Some alcohols can be oxidised to form aldehydes, which can then be oxidised further to form carboxylic acids.
    Some alcohols can be oxidised to form ketones, which resist further oxidation. Other alcohols are resistant to oxidation.
    (a) Draw the structures of the **two** straight-chain isomeric alcohols with molecular formula, $C_4H_{10}O$
        (2 marks)
    (b) Draw the structures of the oxidation products obtained when the two alcohols from part (a) are oxidised separately by acidified potassium dichromate(VI). Write equations for any reactions which occur, using [O] to represent the oxidising agent.
        (6 marks)
    (c) Draw the structure and give the name of the alcohol with molecular formula $C_4H_{10}O$ which is resistant to oxidation by acidified potassium dichromate(VI).
        (2 marks)
        AQA, 2005

5   Consider the following reaction schemes involving two alcohols, **A** and **B**, which are position isomers of each other.

$$CH_3CH_2CH_2CH_2OH \rightarrow CH_3CH_2CH_2CHO \rightarrow CH_3CH_2CH_2COOH$$
$$\quad\quad\;\textbf{A} \quad\quad\quad\quad\quad\quad \text{butanal} \quad\quad\quad\quad \text{butanoic acid}$$

$$CH_3CH_2CH(OH)CH_3 \rightarrow CH_3CH_2COCH_3$$
$$\quad\quad\;\textbf{B} \quad\quad\quad\quad\quad\quad\quad \textbf{C}$$

   (a) State what is meant by the term *position isomers*.
       (2 marks)
   (b) Name compound **A** and name the class of compounds to which **C** belongs.
       (2 marks)
   (c) Each of the reactions shown in the schemes above is of the same type and uses the same combination of reagents.
       (i)   State the type of reaction.
       (ii) Identify a suitable combination of reagents.
       (iii) State how you would ensure that compound **A** is converted into butanoic acid rather than into butanal.
       (iv) Draw the structure of an isomer of compound **A** which does not react with this combination of reagents.
       (v)   Draw the structure of the carboxylic acid formed by the reaction of methanol with this combination of reagents.
       (6 marks)
   (d) (i) State a reagent which could be used to distinguish between butanal and compound **C**.
       (ii) Draw the structure of another aldehyde which is an isomer of butanal.
       (2 marks)
       AQA, 2005

6   Glucose can be used as a source of ethanol. Ethanol can be burned as a fuel or can be converted into ethene.

$$C_6H_{12}O_6 \rightarrow CH_3CH_2OH \rightarrow H_2C{=}CH_2$$
$$\text{glucose} \quad\quad \text{ethanol} \quad\quad \text{ethene}$$

   (a) Name the types of reaction illustrated by the two reactions above.
       (2 marks)
   (b) (i) State what must be added to an aqueous solution of glucose so that ethanol is formed.
       (ii) Identify a suitable catalyst for the conversion of ethanol into ethene.
       (2 marks)
   (c) (i) State the class of alcohols to which ethanol belongs.
       (ii) Give **one** advantage of using ethanol as a fuel compared with using a petroleum fraction.
       (2 marks)
   (d) Most of the ethene used by industry is produced when ethane is heated to 900 °C in the absence of air. Write an equation for this reaction.
       (1 mark)
       AQA, 2005

When you are identifying an organic compound you need to know the functional groups present.

Chemical reactions

Some tests are very straightforward.

- Is the compound acidic (suggests carboxylic acid)?
- Is the compound solid (suggests long carbon chain or ionic bonding), liquid (suggests medium length carbon chain or polar or hydrogen bonding), or gas (suggests short carbon chain, little or no polarity)?
- Does the compound dissolve in water (suggests polar groups) or not (suggests no polar groups)?

Some specific chemical tests are listed in Table 1.

▼ **Table 1** *Chemical tests for functional groups*

Functional group	Test	Result
alkene —C=C—	shake with bromine water	orange colour disappears
halogenoalkane R—X	1. add NaOH(aq) and warm 2. acidify with HNO_3 3. add $AgNO_3$(aq)	precipitate of AgX
alcohol R—OH	add acidified $K_2Cr_2O_7$	orange colour turns green with primary or secondary alcohols (also with aldehydes)
aldehydes R—CHO	warm with Fehling's solution or warm with Tollens' solution	blue colour turns to red precipitate silver mirror forms
carboxylic acids R—COOH	add $NaHCO_3$(aq)	CO_2 given off

Summary questions

1. How could you tell if R—X was a chloroalkane, a bromoalkane, or an iodoalkane?

2. In the test for a halogenoalkane:
   a. Explain why it is necessary to acidify with dilute acid before adding silver nitrate.
   b. Why would acidifying with hydrochloric acid not be suitable?

3. A compound decolourises bromine solution and fizzes when sodium hydrogencarbonate solution is added:
   a. What two functional groups does it have?
   b. Its relative molecular mass is 72. What is its structural formula?
   c. Give equations for the two reactions.

You saw in Topic 1.4 how mass spectrometry is used to measure the relative *atomic* masses of atoms. It is also the main method for finding the relative *molecular* mass of organic compounds. The compound enters the mass spectrometer in solution. It is ionised and the positive ions are accelerated through the instrument as a beam of ionised molecules. These then fly through the instrument towards a detector.

Their times of flight are measured. These depend on the mass to charge ratio m/z of the ion.

The output is then presented as a graph of relative abundance (vertical axis) against mass/charge ratio (horizontal axis). However, since the charge on the ions is normally +1, the horizontal axis is effectively relative mass. This graph is called a mass spectrum.

Learning objectives:

→ State what is meant by the term molecular ion.

→ Describe what the mass of a molecular ion shows.

→ Explain what a high resolution mass spectrum can tell us.

Specification reference: 3.3.6

+ Fragmentation

There are many techniques for mass spectrometry. In some of these the ions of the sample break up of fragment as they pass through the instrument.

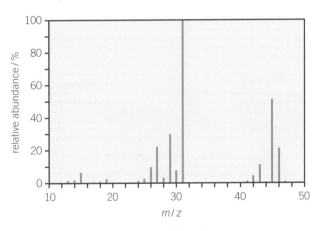

▲ **Figure 1** *The mass spectrum of ethanol*

A mass spectrum of ethanol is shown in Figure 1. Notice that it contains many lines and not just one as we might expect. When ethanol is ionised it forms the ion $C_2H_5OH^+$ ($CH_3CH_2OH^+$). This is called the **molecular ion**. Many of these ions will then break up because some of their bonds break as they are ionised, so there are other ions of smaller molecular mass. This process is called fragmentation. Each of these fragment ions produces a line in the mass spectrum. These can provide information that will help to deduce the structure of the compound. They also act as a 'fingerprint' to help identify it.

Bearing in mind the fact that the ethanol molecule, CH_3CH_2OH is breaking up, suggest formulae for the fragments represented by the peaks at m/z 46, 45, 31, and 29.

46: $CH_3CH_2OH^+$; 45: $CH_3CH_2O^+$ or $CH_3CH_2OH^+$ or CH_3CHOH^+; 31: CH_2OH^+; 29: $CH_3CH_2^+$

Mass spectrometry and sport

One of the many applications of mass spectrometry is testing athletes for the presence of drugs in urine samples. It is also used in forensic work.

Hint

• In any spectrum of an organic compound there will be a tiny peak one mass unit to the right of the molecular ion. This is caused by ions containing the ^{13}C isotope.

GCMS

Gas Chromatography Mass Spectrometry (GCMS) is one of the most powerful analytical techniques used currently. It is used in forensic work and also to detect drugs used by athletes and doping of racehorses. It is a combination of two techniques.

Gas chromatography is a technique for separating mixtures which uses a stream of gas to carry a mixture of vapours through a tube packed with a powdered

solid. The different components of the mixture emerge from the tube (called a column) at different times. As the components emerge from the column, their amounts are measured and they are fed straight into a mass spectrometer which produces the mass spectrum of each and allows them to be identified. So the amount and identity of each component in a complex mixture can be found.

Hint

Remember that the mass spectrometer detects isotopes separately.

High resolution mass spectrometry

Mass spectra often show masses to the nearest whole number only. However, many mass spectrometers can measure masses to three or even four decimal places. This method allows us to work out the molecular formula of the parent ion. It makes use of the fact that isotopes of atoms do not have exactly whole number atomic masses (except for carbon-12 which is exactly twelve by definition), for example, $^{16}O = 15.99491$ and $^1H = 1.007829$.

A parent ion of mass 200, to the nearest whole number, could have the molecular formulae of: $C_{10}H_{16}O_4$, or $C_{11}H_4O_4$, or $C_{11}H_{20}O_3$

Adding up the accurate atomic masses gives the following molecular masses:

$$C_{10}H_{16}O_4 = 200.1049$$

$$C_{11}H_4O_4 = 200.0110$$

$$C_{11}H_{20}O_3 = 200.1413$$

These can easily be distinguished by high resolution mass spectrometry. A computer database can be used to identify the molecular formula from the accurate relative molecular mass.

Water sampling

Water boards sample the water from the rivers in their areas to monitor pollutants. The pollutants are separated by chromatography and fed into a mass spectrometer. Each pollutant can be identified from its spectrum; a computer matches its spectrum with known compounds in a library of spectra.

Summary questions

1 a How are ions formed from molecules in a mass spectrometer?

b What sign of charge do the ions have as a result of this?

2 A compound was found to have a molecular ion with a mass to charge ratio of 136.125. Which of the following molecular formulae could it have? $C_9H_{12}O$ or $C_{10}H_{16}$

You will need to work out the accurate M_r of each of these molecules.

Infrared (IR) spectroscopy is often used by organic chemists to help them identify compounds.

How infrared spectroscopy works

A pair of atoms joined by a chemical bond is always vibrating. The system behaves rather like two balls (the atoms) joined by a spring (the bond). Stronger bonds vibrate faster (at higher frequency) and heavier atoms make the bond vibrate more slowly (at lower frequency). Every bond has its own unique natural frequency that is in the infrared region of the electromagnetic spectrum.

When you shine a beam of infrared radiation (heat energy) through a sample, the bonds in the sample can absorb energy from the radiation and vibrate more. However, any particular bond can only absorb radiation that has the same frequency as the natural frequency of the bond. Therefore, the radiation that emerges from the sample will be missing the frequencies that correspond to the bonds in the sample, see Figure 1.

The infrared spectrometer

This is what happens in an infrared spectrometer:

1   A beam of infrared radiation containing a spread of frequencies is passed through a sample.
2   The radiation that emerges is missing the frequencies that correspond to the types of bonds found in the sample.
3   The instrument plots a graph of the intensity of the radiation emerging from the sample, called the transmittance, against the frequency of radiation.
4   The frequency is expressed as a wavenumber, measured in cm^{-1}.

The infrared spectrum

A typical graph, called an infrared spectrum is shown in Figure 2. The dips in the graph (confusingly, they are usually called peaks) represent particular bonds. Figure 3 and Table 1 show the wavenumbers of some bonds commonly found in organic chemistry.

These can help us to identify the functional groups present in a compound. For example:

*   the O—H bond produces a broad peak at about between 3230 and 3550 cm^{-1} and this is found in alcohols, ROH; and a very broad O—H peak between 2500 and 3000 cm^{-1} in carboxylic acids, RCOOH.
*   the C=O bond produces a peak between 1680 and 1750 cm^{-1}. This bond is found in aldehydes, RCHO, ketones, R_2CO, and carboxylic acids, RCOOH.

▲ **Figure 1** *Schematic diagram of an infrared spectrometer*

Hint

Wavenumber is proportional to frequency.

▼ **Table 1** *Characteristic infra-red absorptions in organic molecules*

Bond	Location	Wavenumber/ cm^{-1}
C—O	alcohols, esters	1000–1300
C=O	aldehydes, ketones, carboxylic acids, esters	1680–1750
O—H	hydrogen bonded in carboxylic acids	2500–3000 (broad)
N—H	primary amines	3100–3500
O—H	hydrogen bonded in alcohols, phenols	3230–3550

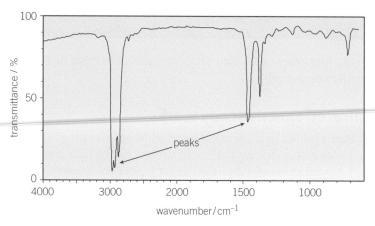

▲ **Figure 2** *A typical infrared spectrum. Note that wavenumber gets smaller going from left to right*

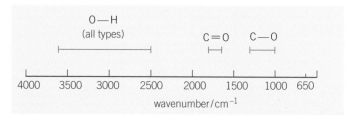

▲ **Figure 3** *The ranges of wavenumbers at which some bonds absorb infrared radiation*

Data about the frequencies that correspond to different bonds can be found on the data sheet at the back of the book.

Figures 4, 5, and 6 show the infrared spectra of ethanal, ethanol, and ethanoic acid with the key peaks marked.

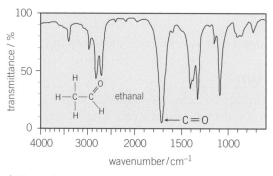

▲ **Figure 4** *Infrared spectrum of ethanal*

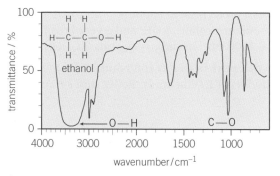

▲ **Figure 5** *Infrared spectrum of ethanol*

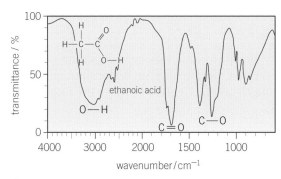

▲ **Figure 6** *Infrared spectrum of ethanoic acid*

Greenhouse gases

The greenhouse effect, which contributes to global warming, is caused by gases in the atmosphere that absorb the infrared radiation given off from the surface of the Earth and would otherwise be lost into space. The table gives some data about some of these gases. The infrared radiation is absorbed by bonds in these gases in the same way as in an infrared spectrometer. Carbon dioxide has two $C = O$ bonds which absorb in the infrared region of the spectrum.

Gas	Relative greenhouse effect per molecule	Concentration in the atmosphere / parts per million (ppm)
carbon dioxide, CO_2	1	350
methane, CH_4	30	1.7
nitrous oxide (dinitrogen monoxide, NO)	160	0.31
ozone, O_3	2000	0.06
trichlorofluoromethane (a CFC)	21 000	0.000 26
dichlorodifluoromethane (a CFC)	25 000	0.000 24

Water vapour is a powerful greenhouse gas, absorbing IR via its O—H bonds. It is not included in the table because its concentration in the atmosphere is very variable.

1 Write the displayed formulae of trichlorofluoromethane and of dichlorodifluoromethane (showing all the atoms and the bonds).

2 What bonds are present in these compounds? Suggest why their relative effects are so similar.

3 Suggest why the concentration of water vapour is so variable.

4 One way of comparing the overall greenhouse contribution of a gas would be to multiply its concentration by its relative effect. Use this method to compare the contribution of carbon dioxide and methane.

The fingerprint region

The area of an infrared spectrum below about 1500 cm⁻¹ usually has many peaks caused by complex vibrations of the whole molecule. This shape is unique for any particular substance. It can be used to identify the chemical, just as people can be identified by their fingerprints. It is therefore called the **fingerprint region**.

Chemists can use a computer to match the fingerprint region of a sample with those on a database of compounds. An exact match confirms the identification of the sample.

Figures 7 and 8 show the IR spectra of two very similar compounds, propan-1-ol and propan-2-ol.

They are as expected, very similar overall. However, superimposing the spectra, Figure 9, shows that their fingerprint regions are quite distinct. This is shown more clearly in Figure 10, where the fingerprint region has been enlarged.

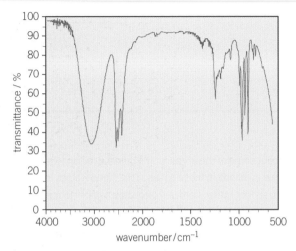

▲ **Figure 7** *Infrared spectrum of propan-1-ol*

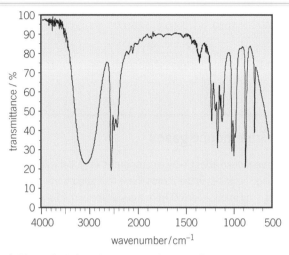

▲ **Figure 8** *Infrared spectrum of propan-2-ol*

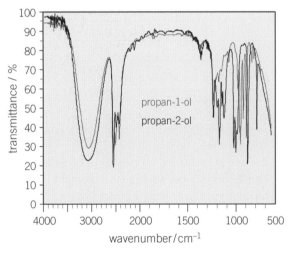

▲ **Figure 9** *Infrared spectra of propan-1-ol superimposed on propan-2-ol*

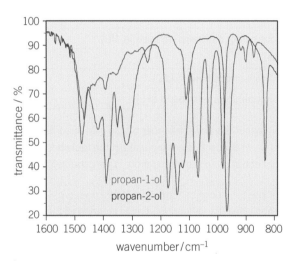

▲ **Figure 10** *The fingerprint region of the infrared spectra of propan-1-ol superimposed on propan-2-ol enlarged*

✚ Identifying impurities

Infra-red spectra can also be used to show up the presence of impurities. These may be revealed by peaks that should not be there in the pure compound. Figures 11 and 12 show the spectrum of a sample of pure caffeine and that of caffeine extracted from tea. The broad peak at around 3000 cm^{-1} in the impure sample (Figure 12) is an O—H stretch caused by water in the sample that has not been completely dried. Notice that there are no O—H bonds in caffeine (Figure 13).

In practice, analytical chemists will often use a combination of spectroscopic techniques to identify unknown compounds.

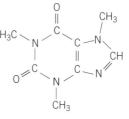

▲ **Figure 13** *The structural formula of caffeine. It has no O—H bonds.*

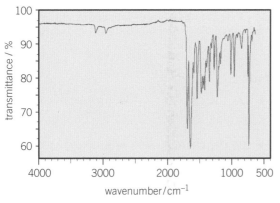

▲ **Figure 11** *The infrared spectrum of pure caffeine*

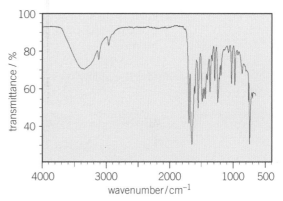

▲ **Figure 12** *The infrared spectrum of impure caffeine*

Summary questions

1 An organic compound has a peak in the IR spectrum at about 1725 cm^{-1}. Which of the following compounds could it be?

    a   b   c

2 Explain your answer to question **1**.

3 An organic compound has a peak in the IR spectrum at about 3300 cm^{-1}. Which of the compounds in question **1** could it be?

4 Explain your answer to question **3**.

5 An organic compound has a peak in the IR spectrum at 1725 and 3300 cm^{-1}. Which of the compounds in question **1** could it be?

6 Explain your answer to question **5**.

Practice questions

1 (a) The infra-red spectrum of compound A, $C_3H_6O_2$, is shown below.

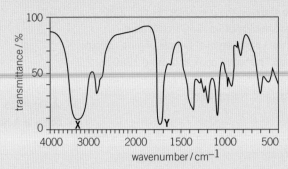

Identify the functional groups which cause the absorptions labelled **X** and **Y**.
Using this information draw the structures of the three possible structural isomers
for A. *(5 marks)*

AQA, 2006

2 Consider the following scheme of reactions.

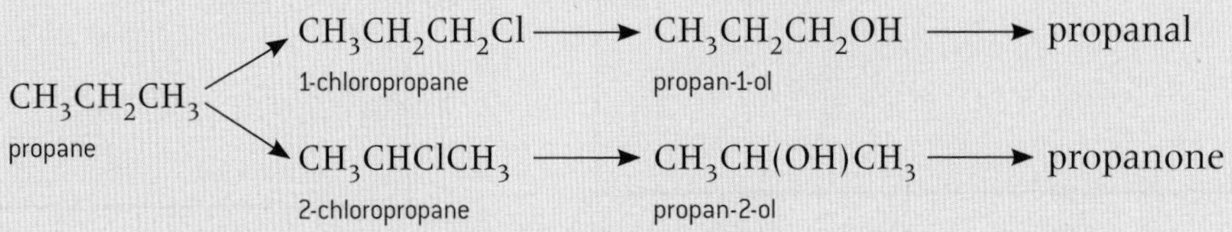

$CH_3CH_2CH_3$ (propane)
→ $CH_3CH_2CH_2Cl$ (1-chloropropane) → $CH_3CH_2CH_2OH$ (propan-1-ol) → propanal
→ $CH_3CHClCH_3$ (2-chloropropane) → $CH_3CH(OH)CH_3$ (propan-2-ol) → propanone

(a) State the type of structural isomerism shown by propanal and propanone. *(1 mark)*

(b) A chemical test can be used to distinguish between separate samples of propanal and
propanone.
Identify a suitable reagent for the test.
State what you would observe with propanal and with propanone. *(3 marks)*

(c) State the structural feature of propanal and propanone which can be identified
from their infrared spectra by absorptions at approximately 1720 cm^{-1}.
You may find it helpful to refer to **Table 1** on the Data Sheet. *(1 mark)*

(d) The reaction of chlorine with propane is similar to the reaction of chlorine with
methane.
(i) Name the type of mechanism in the reaction of chlorine with methane.

(1 mark)

(ii) Write an equation for each of the following steps in the mechanism for the
reaction of chlorine with propane to form 1-chloropropane ($CH_3CH_2CH_2Cl$).
Initiation step
First propagation step
Second propagation step
A termination step to form a molecule with the empirical formula C_3H_7

(4 marks)

(e) High resolution mass spectrometry of a sample of propane indicated that it was
contaminated with traces of carbon dioxide.
Use the data in the table to show how precise M_r values can be used to prove that the
sample contains both of these gases.

Atom	Precise relative atomic mass
^{12}C	12.00000
^{1}H	1.00794
^{16}O	15.99491

(2 marks)

AQA, 2010

3 The table below shows the structures of three isomers with the molecular formula $C_5H_{10}O$.

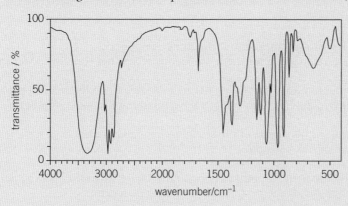

Isomer 1	(E)-pent-3-en-2-ol
Isomer 2	pentanal
Isomer 3	

(a) Complete the table by naming Isomer **3**.

(1 mark)

(b) State the type of structural isomerism shown by these three isomers.

(1 mark)

(c) The compound (Z)-pent-3-en-2-ol is a stereoisomer of (E)-pent-3-en-2-ol.
   (i) Draw the structure of (Z)-pent-3-en-2-ol.

(1 mark)

   (ii) Identify the feature of the double bond in (E)-pent-3-en-2-ol and that in (Z)-pent-3-en-2-ol that causes these two compounds to be stereoisomers.

(1 mark)

(d) A chemical test can be used to distinguish between separate samples of Isomer **2** and Isomer **3**.
   Identify a suitable reagent for the test.
   State what you would observe with Isomer **2** and with Isomer **3**.

(3 marks)

(e) The following is the infrared spectrum of one of the isomers **1**, **2**, or **3**.

<!-- infrared spectrum graph: transmittance / % (y-axis, 0 to 100) vs wavenumber/cm⁻¹ (x-axis, 4000 to 500) -->

   (i) Deduce which of the isomers (**1**, **2**, or **3**) would give this infrared spectrum.

(1 mark)

   (ii) Identify two features of the infrared spectrum that support your deduction.
      In each case, identify the functional group responsible.

(2 marks)

AQA, 2011

Section 3 practice questions

1 Trifluoromethane, CHF_3, can be used to make the refrigerant chlorotrifluoromethane, $CClF_3$.
(a) Chlorotrifluoromethane is formed when trifluoromethane reacts with chlorine.
$$CHF_3 + Cl_2 \rightarrow CClF_3 + HCl$$
The reaction is a free-radical substitution reaction similar to the reaction of methane with chlorine.
(i) Write an equation for each of the following steps in the mechanism for the reaction of CHF_3 with Cl_2 *(4 marks)*
Initiation step
First propagation step
Second propagation step
Termination step to form hexafluoroethane
(ii) Give **one** essential condition for this reaction. *(1 mark)*
AQA, 2014

(b) A small amount of $CClF_3$ with a mass of $2.09 \times 10^{-4}\,kg$ escaped from a refrigerator into a room with a volume of $200\,m^3$. Calculate the number of $CClF_3$ molecules in a volume of $500\,cm^3$. Assume that the $CClF_3$ molecules are evenly distributed throughout the air in the room. Give your answer to the appropriate number of significant figures.
The Avogadro constant $= 6.02 \times 10^{23}\,mol^{-1}$. *(3 marks)*

2 Some oil-fired heaters use paraffin as a fuel.
One of the compounds in paraffin is the straight-chain alkane, dodecane, $C_{12}H_{26}$.
(a) Give the name of the substance from which paraffin is obtained.
State the name of the process used to obtain paraffin from this substance.
(2 marks)

(b) The combustion of dodecane produces several products.
Write an equation for the **incomplete** combustion of dodecane to produce gaseous products only.
(1 mark)

(c) Oxides of nitrogen are also produced during the combustion of paraffin in air.
(i) Explain how these oxides of nitrogen are formed.
(2 marks)

(ii) Write an equation to show how nitrogen monoxide in the air is converted into nitrogen dioxide.
(1 mark)

(iii) Nitric acid, HNO_3, contributes to acidity in rainwater.
Deduce an equation to show how nitrogen dioxide reacts with oxygen and water to form nitric acid.
(1 mark)

(d) Dodecane, $C_{12}H_{26}$, can be cracked to form other compounds.
(i) Give the general formula for the homologous series that contains dodecane.
(1 mark)

(ii) Write an equation for the cracking of one molecule of dodecane into equal amounts of two different molecules each containing the same number of carbon atoms.
State the empirical formula of the straight-chain alkane that is formed. Name the catalyst used in this reaction.
(3 marks)

(iii) Explain why the melting point of dodecane is higher than the melting point of the straight-chain alkane produced by cracking dodecane.
(2 marks)

(e) Give the IUPAC name for the following compound and state the type of structural isomerism shown by this compound and dodecane. *(2 marks)*

(f) Dodecane can be converted into halododecanes.
Deduce the formula of a substance that could be reacted with dodecane to produce 1-chlorododecane and hydrogen chloride only. *(1 mark)*
AQA, 2014

3 The following table shows the boiling points of some straight-chain alkanes.

	CH_4	C_2H_6	C_3H_8	C_4H_{10}	C_5H_{12}
Boiling point / °C	−162	−88	−42	−1	36

(a) State a process used to separate an alkane from a mixture of these alkanes. *(1 mark)*
(b) Both C_3H_8 and C_4H_{10} can be liquefied and used as fuels for camping stoves.
Suggest, with a reason, which of these two fuels is liquefied more easily. *(1 mark)*
(c) Write an equation for the complete combustion of C_4H_{10} *(1 mark)*
(d) Explain why the complete combustion of C_4H_{10} may contribute to environmental problems. *(1 mark)*
(e) Balance the following equation that shows how butane is used to make the compound called maleic anhydride.

_____ $CH_3CH_2CH_2CH_3$ + _____ O_2 → _____ $C_2H_2(CO)_2O$ + _____ H_2O *(1 mark)*

(f) Ethanethiol, C_2H_5SH, a compound with an unpleasant smell, is added to gas to enable leaks from gas pipes to be more easily detected.
 (i) Write an equation for the combustion of ethanethiol to form carbon dioxide, water and sulfur dioxide. *(1 mark)*
 (ii) Identify a compound that is used to react with the sulfur dioxide in the products of combustion before they enter the atmosphere.
 Give **one** reason why this compound reacts with sulfur dioxide. *(2 marks)*
 (iii) Ethanethiol and ethanol molecules have similar shapes.
 Explain why ethanol has the higher boiling point. *(2 marks)*

(g) The following compound **X** is an isomer of one of the alkanes in the table on page 8.

 (i) Give the IUPAC name of **X**. *(1 mark)*
 (ii) **X** has a boiling point of 9.5 °C.
 Explain why the boiling point of **X** is lower than that of its straight-chain isomer. *(2 marks)*
 (iii) The following compound **Y** is produced when **X** reacts with chlorine.

 Deduce how many **other** position isomers of **Y** can be formed. Write the number of **other** position isomers in this box. *(1 mark)*

(h) Cracking of one molecule of an alkane **Z** produces one molecule of ethane, one molecule of propene and two molecules of ethene.
 (i) Deduce the molecular formula of **Z**. *(1 mark)*
 (ii) State the type of cracking that produces a high proportion of ethene and propene.
 Give the **two** conditions for this cracking process. *(2 marks)*
AQA, 2013

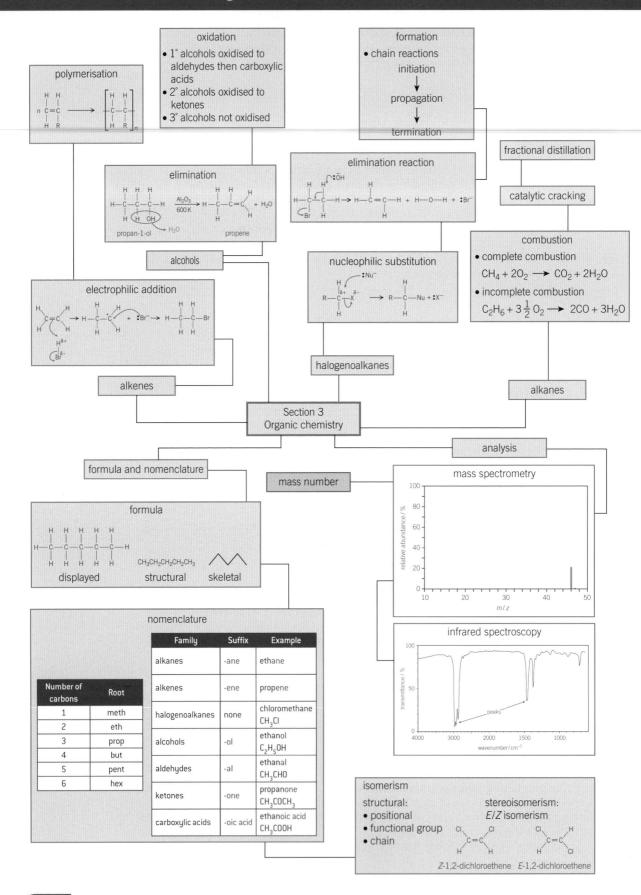

polymerisation

oxidation
- 1° alcohols oxidised to aldehydes then carboxylic acids
- 2° alcohols oxidised to ketones
- 3° alcohols not oxidised

formation
- chain reactions
  initiation
  ↓
  propagation
  ↓
  termination

fractional distillation

catalytic cracking

elimination

propan-1-ol → propene + H_2O

elimination reaction

combustion
- complete combustion
  $CH_4 + 2O_2 \longrightarrow CO_2 + 2H_2O$
- incomplete combustion
  $C_2H_6 + 3\frac{1}{2}O_2 \longrightarrow 2CO + 3H_2O$

alcohols

nucleophilic substitution

electrophilic addition

alkenes

halogenoalkanes

alkanes

**Section 3
Organic chemistry**

analysis

formula and nomenclature

mass number

mass spectrometry

formula

displayed   structural   skeletal

$CH_3CH_2CH_2CH_2CH_3$

infrared spectroscopy

peaks

nomenclature

Family	Suffix	Example
alkanes	-ane	ethane
alkenes	-ene	propene
halogenoalkanes	none	chloromethane CH_3Cl
alcohols	-ol	ethanol C_2H_5OH
aldehydes	-al	ethanal CH_3CHO
ketones	-one	propanone CH_3COCH_3
carboxylic acids	-oic acid	ethanoic acid CH_3COOH

Number of carbons	Root
1	meth
2	eth
3	prop
4	but
5	pent
6	hex

isomerism

structural:
- positional
- functional group
- chain

stereoisomerism:
E/Z isomerism

Z-1,2-dichloroethene E-1,2-dichloroethene

Practical skills

In this section you have met the following ideas:

- Performing fractional distillation
- Carrying out hydrolysis of halogenoalkanes to find their relative rates of reaction
- Testing organic compounds for unsaturation using bromine water
- Making a polymer, such as poly(phenylethene) from its monomer.
- Producing ethanol by fermentation and purifying it by distillation
- Preparing aldehydes and carboxylic acids by the oxidation of a primary alcohol
- Preparing cyclohexene from cyclohexanol by acid-catalysed elimination
- Carry out tests to identify alcohols, aldehydes, alkenes and carboxylic acids

Maths skills

In this section you have met the following maths skills:

- Balancing symbol equations
- Representing 2D and 3D forms using two-dimensional diagrams
- Identifying and drawing different isomers of a substance by its formula

Extension

1 The production of ethene from ethanol by dehydration is considered a more sustainable than by cracking naptha. Write:

   a a summary of its use for industrial applications

   b current and future challenges. These could be, for example, related to sustainability or new applications.

2 Research how advances in computer technology and robotics have changed the way in which research of organic compounds is conducted, particularly the development of new medicines.

Section 4
Practical skills

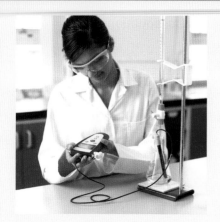

Practical work is firmly at the heart of any chemistry course. It helps you to understand new ideas and difficult concepts, and helps you to appreciate the importance of experiments in testing and developing scientific theories. In addition, practical work develops the skills that scientists use in their everyday work. Such skills involve planning, researching, making and processing measurements, analysing, and evaluating experimental results, as well as the ability to use a variety of apparatus and chemicals safely.

During this course you will be carrying out a number of practicals using a range of apparatus and techniques. Table 1 lists some of the practical activities you will do in your AS/1st year A level course, and questions may be set on these in the written exam. In carrying out these activities, you should become proficient in all the practical skills assessed directly and indirectly in your AS/1st year A-level course. For each activity, the table references the relevant topic or topics in this book. (In addition there will be a teacher assessed pass/fail endorsement of your practical skills.)

▼ **Table 1** *Some practical activities you will cover as part of your course*

	Practical	Topic
1	Making up a volumetric solution and carrying out a titration	2.5, Balanced equations an related calculations
2	Measurement of an enthalpy change	4.3, Measuring enthalpy changes 4.4, Hess's law
3	Investigation of how the rate of a reaction changes with temperature	5.2, The Maxwell-Boltzmann distribution
4	Carrying out test-tube reactions to identify cations and anions in aqueous solutions	9.1, The alkaline earth elements
5	Distillation of a product from a reaction	15.2, Ethanol production
6	Tests for alcohol, aldehyde, alkene, and carboxylic acid	15.3, The reaction of alcohols

Practice questions

The following questions will give you some practice – you may not have done the actual experiment mentioned but you should think about similar practical work that you have done.

Practical 1

1   Describe how to make $250\,cm^3$ of a $0.10\,mol\,dm^{-3}$ solution of blue, hydrated copper sulfate, $CuSO_4.5H_2O$.
    A_r Cu = 63.5, S = 32.1, H = 1.0, O = 16.0

2 a Describe carefully the steps needed to find the concentration of a sodium hydroxide solution, using $0.10\,\text{mol dm}^{-3}$ hydrochloric acid and phenolphthalein indicator, which is pink in alkali and colourless in acid. Choose the apparatus used for each step explaining any measures you take to ensure safety and accuracy Select from the following apparatus

> conical flask   funnel   goggle   burette
>
> $25.0\,\text{cm}^3$ pipette with filler   dropping pipette   white tile

b $25.00\,\text{cm}^3$ of sodium hydroxide solution were neutralised by $30.00\,\text{cm}^3$ of acid.
  i Write the equation for the reaction.
  ii Was the sodium hydroxide solution more or less concentrated than the acid? Explain your answer.
  iii Find the concentration of the sodium hydroxide solution.

c Why is phenolphthalein a better indicator than universal indicator for titrations?

Practical 2

Determining an enthalpy change that cannot be measured directly

Anhydrous (white) copper sulfate reacts with water to form hydrated (blue) copper sulfate:

$$CuSO_4(s) + 5H_2O(l) \rightarrow CuSO_4.5H_2O(s) \quad \Delta H_3 = ?$$

The enthalpy change cannot be measured directly, but it can be determined using the Hess's Law cycle below and measuring the enthalpy changes ΔH_1 and ΔH_2, the enthalpies of solution of the two copper sulfates.

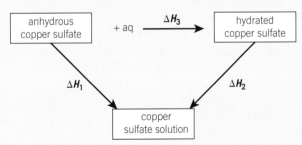

A student weighed out $4.00\,\text{g}$ of anhydrous copper sulfate on a top pan balance that read to $0.01\,\text{g}$. She measured out $50.0\,\text{cm}^3$ water into an expanded polystyrene beaker using a measuring cylinder that measured to $0.5\,\text{cm}^3$. The water had been standing in the laboratory for over 1 hour. She recorded the temperature of the water. She quickly stirred the mixture until all the solid had dissolved and recorded the highest temperature attained using a thermometer that read to $0.1\,^{\circ}\text{C}$.

She then repeated the procedure using 6.25 g of hydrated copper sulfate and 47.5 cm³ water (this allows for the water present in the hydrated copper sulfate). This time the temperature dropped.

Results

Anhydrous copper sulfate
Initial temperature of water = 18.7 °C;
Highest temperature attained = 25.9 °C

Hydrated copper sulfate
Initial temperature of water = 18.7 °C;
Lowest temperature attained = 17.5 °C

Calculation

1   Calculate the number of moles of a) anhydrous copper sulfate b) hydrated copper sulfate that were used. Use $A_r(Cu) = 63.5$, $S = 32.1$, $O = 16.0$, $H = 1.0$
2   How many moles of water does this amount of hydrated copper sulfate contain?
3   Calculate the temperature change in each experiment.
4   Calculate the heat produced by the anhydrous copper sulfate using $q = m \times c \times \Delta T$. Remember that the mass of the solution includes the mass of the copper sulfate added. Use a value of $4.18\,J\,g^{-1}\,°C^{-1}$ for c, the specific heat of a dilute aqueous solution.
5   Calculate the heat absorbed by the hydrated copper sulfate using $q = m \times c \times \Delta t$
6   Calculate the enthalpy change of solution in each case in $kJ\,mol^{-1}$ giving the correct sign for each.
    What is ΔH_3?

Questions

1   Explain why the heat of hydration of anhydrous copper sulfate cannot be measured directly.
2   What is the percentage error in **a** weighing the anhydrous copper sulfate **b** weighing the hydrated copper sulfate **c** measuring the 50.0 cm³ water **d** measuring the temperature rise **e** measuring the temperature drop?
3   Why had the water been left to stand for one hour before the experiment?
4   Why were the experiments carried out in expanded polystyrene beakers?
5   Outline how a cooling curve could be used to allow for heat loss – see Topic 4.3, Measuring enthalpy changes.
6   Bearing in mind the number of significant figures in each of the measurements, how many significant figures can be given in the value for ΔH_3?

Practical 3
When equal volumes of $0.1\,mol\,dm^{-3}$ sodium thiosulfate solution and $0.5\,mol\,dm^{-3}$ hydrochloric acid are mixed, the solution gradually

becomes more cloudy as a suspension of sulfur particles is produced in the following reaction:

$$Na_2S_2O_3(aq) + 2HCl(aq) \rightarrow 2NaCl(aq) + SO_2(g) + H_2O(l) + S(s)$$

These are the results of an experiment to find out how temperature affects the rate of this reaction. $10\,cm^3$ of each solution was mixed in a boiling tube and the time for the mixture to become opaque was noted.

Results

1 Write instructions for this experiment making sure you say how to measure the solutions, the temperature and time taken for the reaction.
2 Plot a graph of these results with temperature on the horizontal axis.
   a At which temperature does the result seem not to fit the pattern?
   b Suggest possible causes of this anomalous result.
3 This experiment was carried out in a well-ventilated laboratory. Why was this a sensible instruction?

Temperature / °C	Time / s
15	140
20	74
25	70
30	45
40	25
50	12
60	7

Practical 4

An unlabeled bottle of white powder is found in a school chemical store. Close by is a label "magnesium bromide". What tests would you carry out to confirm that the label belongs to the bottle?

Practical 5
Distillation of ethanol

The diagram shows a distillation apparatus set up to produce pure ethanol (boiling temperature 78 °C) from a mixture resulting from the fermentation of a solution containing sugar (sucrose) and yeast.

Look at the diagram and answer the following questions.

1 There are four faults in the experimental set up. Point out each one and explain why it is a problem and what should be done to correct it.
2 What is the purpose of the ground glass beads in the boiling flask?
3 Over which of these temperature ranges should the product be collected?
   a 76–80 °C,   b 70–85 °C,
   c all the vapour should be collected

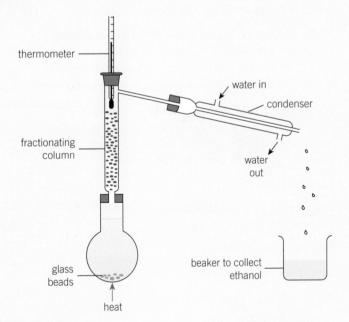

Practical 6

You have to identify three liquids and two gases. You know you have an alcohol, an alkene, an aldehyde, a carboxylic acid, and air.

These are the tests you decide to do:

1 Add a few drops of bromine water to each gas to identify the gases.
2 Add a sample of each liquid to water and test the pH.
3 Add Benedict's solution to the remaining two liquids and warm in a water bath.

Explain how this would work.

Section 5
Mathematical skills

Units in calculations

Examiners expect you to use the correct units in calculations

Units

You still describe the speed of a car in miles per hour. The units miles per hour could be written miles/hour or miles hour^{-1} where the superscript '$^{-1}$' is just a way of expressing per something. In science you use the metric system of units, and speed has the unit metres per second, written m s^{-1}. In each case, you can think of per as meaning divided by.

Units can be surprisingly useful

A mile is a unit of distance and an hour a unit of time, so the unit miles per hour reminds you that speed is distance divided by time.

In the same way, if you know the units of density are grams per cubic centimetre, usually written g cm^{-3}, where cm^{-3} means per cubic centimetre, you can remember that density is mass divided by volume.

Multiplying and dividing units

When you are doing calculations, units cancel and multiply just like numbers. This can be a guide to whether you have used the right method.

For example:

The density of a liquid is 0.8 g cm^{-3}. What is the volume of a mass of 1.6 g of it?

Density = mass / volume

So volume = mass / density

Putting in the values and the units:

volume = 1.6 g / 0.8 g cm^{-3}

Cancelling the gs

volume = 2.0 / cm^{-3}

volume = 2.0 cm^3

If you had started with the wrong equation, such as

volume = density/mass or

volume = mass × density, you would not have ended up with the correct units for volume.

Units to learn

It is a good idea to learn the units of some basic quantities by heart.

	Unit	Comment
volume	dm^3	1 dm^3 is 1 litre, L, which is 1000 cm^3
concentration	mol dm^{-3}	
pressure	Pascals, Pa = N m^{-2}	N m^{-2} are newtons per square metre
enthalpy	kJ mol^{-1}	kJ is kilojoule. Occasionally J mol^{-1} is used.
entropy	J K^{-1} mol^{-1}	joules per kelvin per mole

Standard form

This is a way of writing very large and very small numbers in a way that makes calculations and comparisons easier.

The number is written as number multiplied by ten raised to a power. The decimal point is put to the right of the first digit of the number.

For example:

22 000 is written 2.2×10^4.

0.000 002 2 is written 2.2×10^{-6}.

How to work out the power to which ten must be raised

Count the number of places you must move the decimal point in order to have one digit before the decimal point.

For example:

$0.000\,51 = 5.1 \times 10^{-4}$

$51\,000 = 5.1 \times 10^4$

Moving the decimal point to the right gives a negative index (numbers less than 1), and to the left a positive index (numbers greater than one).

(The number 1 itself is 10^0, so the numbers 1–9 are followed by $\times 10^0$ when written in standard form. Can you see why?)

Multiplying and dividing

When *multiplying* numbers expressed in this way, *add* the powers (called indices) and when *dividing*, *subtract* them.

Worked examples

Calculate

a $2 \times 10^5 \times 4 \times 10^6$

b $\dfrac{8 \times 10^3}{4 \times 10^2}$

c $\dfrac{5 \times 10^8}{2 \times 10^{-6}}$

Answer

a $2 \times 10^5 \times 4 \times 10^6$

Multiply $2 \times 4 = 8$. Add the indices to give 10^{11}

Answer $= 8 \times 10^{11}$

b $\dfrac{8 \times 10^3}{4 \times 10^2}$

Divide 8 by 4 = 2. Subtract the indices to give 10^1

Answer $= 2 \times 10^1 = 20$

Answer $= 18 \times 10^{-2} = 1.8 \times 10^{-1}$

c $\dfrac{5 \times 10^8}{2 \times 10^{-6}}$

Divide 5 by 2 = 2.5. Subtract the indices $(8 - (-6))$ to give 10^{14}

Answer $= 2.5 \times 10^{14}$

A handy hint for non-mathematicians

Non-mathematicians sometimes lose confidence when using small numbers such as 0.002. If you are not sure whether to multiply or divide, then do a similar calculation with numbers that you are happy with, because the rule will be the same.

Example:

How many moles of water in 0.000 1 g? A mole of water has a mass of 18 g.

Do you divide 18 by 0.000 1 or 0.000 1 by 18?

If you have any doubts about how to do this, then in your head change 0.000 1 g into a more familiar number such as 100 g.

How many moles of water in 100 g? A mole of water has a mass of 18 g.

Now you can see that you must divide 100 by 18. So in the same way you must divide 0.000 1 by 18 in the original problem.

$$\frac{0.000\,1}{18} = 5.6 \times 10^{-6}$$

Prefixes and suffixes

In chemistry you will often encounter very large numbers (such as the number of atoms in a mole) or very small numbers (such as the size of an atom).

Prefixes and suffixes are often used with units to help express these numbers. You will come across the following which multiply the number by a factor of 10^n. The red ones are the ones you are most likely to use.

Prefix	Conversion Factor	Symbol
pico	10^{-12}	p
nano	10^{-9}	n
micro	10^{-6}	μ
milli	10^{-3}	m
centi	10^{-2}	c
deci	10^{-1}	d
kilo	10^{3}	k
mega	10^{6}	M

So $5400\,g = 5.4 \times 10^3\,g = 5.4\,kg$

Converting to base units

If you want to convert a number expressed with a prefix to one expressed in the base unit, multiply by the conversion factor. If you have a very small or very large number (and have to handle several zeros) the easiest way is to first convert the number to standard form.

Worked example

Convert a) 2 cm and b) 100 000 000 mm to metres

a $2 \text{ cm} = 2 \times 10^{-2} \text{ m} = 0.02 \text{ m}$

b $100\,000\,000 \text{ mm} = 1 \times 10^{8} \text{ mm}$
$= 1 \times 10^{8} \times 10^{-3} \text{ m} = 1 \times 10^{5} \text{ m}$
conversion factor

Base units

The SI system is founded on base units. The ones you will meet in chemistry are:

Unit	Symbol	Used for
metre	m	length
kilogram	kg	mass
second	s	time
ampere (amp)	A	electric current
kelvin	K	temperature
mole	mol	amount of substance

Handling data

Sorting out significant figures

Many of the numbers used in chemistry are measurements – for example, the volume of a liquid, the mass of a solid, the temperature of a reaction vessel – and no measurement can be exact. When you make a measurement, you can indicate how uncertain it is by the way you write it. For example a length of 5.0 cm means that you have used a measuring device capable of reading to 0.1 cm, a value of 5.00 cm means that you have measured to the nearest 0.01 cm and so on. So the numbers 5, 5.0 and 5.00 are different, you say they have different numbers of *significant figures*.

What exactly is a significant figure?

In a number that has been found or worked out from a measurement, the significant figures are all the digits known for certain, *plus the first uncertain one* (which may be a zero). The last digit is the uncertain one and is at the limit of the apparatus used for measuring it (Figure 1).

▲ **Figure 1** *A number with four significant figures*

For example if you say a substance has a mass of 4.56 grams it means that you are certain about the 4 and the 5 but not the 6 as you are approaching the limit of accuracy of our measuring device (you will have seen the last figure on a top pan balance fluctuate). The number 4.56 has three significant figures (s.f.).

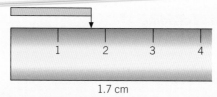

1.7 cm

This ruler gives an answer to two significant figures

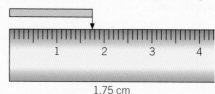

1.75 cm

This ruler gives an answer to three significant figures

▲ **Figure 2** *Rulers with different precision*

When a number contains zeros, the rules for working out the number of significant figures are given below.

- Zeros between digits are significant.
- Zeros to the left of the first non-zero digit are not significant (even when there is a decimal point in the number).
- When a number with a decimal point ends in zeros to the right of the decimal point, these zeros are significant.
- When a number with no decimal point ends in several zeros, these zeros may or may not be significant. The number of significant figures should ideally be stated. For example 20 000 (to 3 s.f.) means that the number has been measured to the nearest 100 but 20 000 (to 4 s.f.) means that the number has been measured to the nearest 10.

The following examples should help you to work out the number of significant figures in your data.

Worked examples

What is the number of significant figures in each of the following?

a 11.23

Answer

4 s.f. all non-zero digits are significant.

b 1100

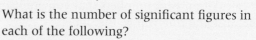

Answer

2 s.f. (but it could be 2, 3, or 4 significant figures). The number has no decimal point so the zeros may or may not be significant. With numbers with zeros at the end it is best to state the number of significant figures.

c 1100.0

Answer

5 s.f. the decimal point implies a different accuracy of measurement to example (b).

d 0.025

Answer

2 s.f. zeros to the left of the decimal point only fix the position of the decimal point. They are not significant.

Question

1 How many significant figures?
    a   40 000
    b   1.030
    c   0.22
    d   22.00

Using significant figures in answers

When doing a calculation, it is important that you don't just copy down the display of your calculator, because this may have a far greater number of significant figures that the data in the question justifies. Your answer cannot be more certain than the least certain of the information that you used to calculate it. So your answer should contain the same number of significant figures as the measurement that has the smallest number of them.

Worked example

81.0 g (3 s.f.) of iron has a volume of 10.16 cm³ (4 s.f.). What is its density?

Answer

$$\text{Density} = \frac{\text{mass}}{\text{volume}} = \frac{81.0 \text{ g}}{10.16 \text{ cm}^3}$$

= 7.972 440 94 g cm⁻³ (this number has 9 s.f.) Since our least certain measurement was to 3 s.f., our answer should have 3 s.f., ie 7.97 g cm⁻³

If our answer had been 7.976 440 94, you would have rounded it ip to 7.98 because the fourth significant figure (6) is five or greater.

The other point to be careful about is *when* to round up. This is best left to the very end of the calculation. Don't round up as you go along as this could make a difference to your final answer.

Decimal places and significant figures

The apparatus you use in the laboratory usually reads to a given number of decimal places (for example hundredths or thousandths of a gram). For example, the top pan balances in most schools and colleges usually weigh to 0.01 g which is to two decimal places.

The number of significant figures of a measurement obtained by using the balance depends on the mass you are finding. A mass of 10.38 g has 4 s.f. but a mass of 0.08 has only 1 s.f. Check this with the rules above.

Hint

Calculator displays usually show numbers in standard form in a particular way. For example 2.6×10^{-4} may appear as 2.6 − 04, a shorthand form which is not acceptable as a way of writing and answer. It is an error that examiners often complain about.

Algebra

Equations

You can write an equation if you can show a connection between sets of measurements (variables).

For example, at a *fixed* volume, if you double the temperature (in Kelvin) of a gas, the pressure doubles too.

Mathematically speaking, the pressure p is directly proportional to the temperature T.

$P \propto T$

The symbol $\propto$ means is proportional to.

This is shown in the data in Table 1

▼ **Table 1**

Temperature/K	Pressure/Pa
100	1000
150	1500
200	2000
250	2500

This also means that the pressure, P, is equal to some constant, k, multiplied by the temperature:

P = kT

In this case, the constant is 10 and if you multiply the temperature in K by 10, you get the pressure in Pa.

Pressure and volume of a gas also vary. At constant temperature, as the pressure of a gas goes up, its volume goes down. More precisely, if you double the pressure, you halve the volume. Mathematically speaking, volume V, is *inversely* proportional to pressure P.

$$V \propto \frac{1}{P}$$

So $$V = k \times \frac{1}{P}$$

Or simply $$V = \frac{k}{P}$$

This is shown by the data in Table 2.

In this case the constant is 24. If you multiply $\frac{1}{P}$ by 24, you get the volume.

▼ Table 2

Pressure/Pa	Volume/l	1/P =1/Pa
1	24	1.00
2	12	0.50
3	8	0.33
4	6	0.25

Mathematical symbols	
Symbol	Meaning
$\rightleftharpoons$	equilibrium
<	less than
≪	much less than
>	greater than
≫	much greater than
~	approximately equal to
$\propto$	proportional to

Question

2a If two variables, x and y, are directly proportional to each other, what happens to one if you quadruple the other?

b Write an expression that means x is inversely proportional to y.

c What happens to the volume of a gas if you triple the pressure at constant temperature?

Handling equations
Changing the subject of an equation

If you can confidently do the next exercise, go straight to the section Substituting into equations

and try that. Otherwise work through Rearranging equations.

Question

3 The equation that connects the pressure P, volume V and temperature T of a mole of gas is
PV = RT
Where P, V, and T are variables and R is a constant called the gas constant.
Rearrange the equation to find:

    a P in terms of V, R and T

    b V in terms of P, R and T

    c T in terms of P, V and R

    d R in terms of P, V and T

Rearranging equations

Start with a simple relation because the rules are the same however complicated the equation.

$$a = \frac{b}{c}$$

where $b = 10$ and $c = 5$,

It is easy to see that substituting these values into the expression

$$a = \frac{10}{5} = 2.$$

But what do you do if you need to find b or c from this equation?

you need to rearrange the equation so that b (or c) appears on its own on the left hand side of the equation like this

$$b = ?$$

$a = \frac{b}{c}$ means $a = b \div c$

Step 1: Multiply both sides of the equation by c, because b is *divided* by c, so to get b on it own you must *multiply* by c.

Remember that to keep an equation valid whatever you do to one side you must do to the other – think of it as a see-saw, with the = sign as the pivot.

So now $c \times a = \frac{b \times c}{c}$

usually written $ca = \frac{bc}{c}$

Now cancel the c's on the right since b is being both multiplied and divided by c

Which leaves $c \times a = b$

Or $b = c \times a$ usually written $b = ca$

You can now rearrange this equation in the same way to find c.

Step 2: $b = c \times a$. Divide both sides by a

$$\frac{b}{a} = \frac{cax}{a}$$

Now cancel the a's on the right

So $\frac{b}{a} = c$

Notice that because c started on the bottom, a two-step process was necessary. You found as expression for b first and then found one for c.

Question

4 Find the variable in brackets in terms of the others.

a $p = \frac{q}{r}$ (q)

b $n = mt$ (m)

c $g = \frac{fe}{h}$ (h)

d $\frac{pr}{e} = s$ (r)

Substituting into equations

When you are asked to substitute numerical values into an equation, it is essential that you carry out the mathematical operations in the right order.

A useful aid to remembering the order is the word **BIDMAS**:

Brackets

Indices

Division

Multiplication

Addition

Subtraction

Graphs

The graph in Figures 3 and 4 show the rate of reaction between hydrochloric acid and magnesium to produce hydrogen gas.

The gradient of a straight line section of a graph is found by dividing the length of the line A (vertical) by the length of line B (horizontal) (see Figure 11.2).

In the case of experiment 1 above this tells us the rate at which hydrogen is produced, between 0 ands 10 seconds. It will have units:

Rate = 20 cm³/10 s = 2 cm³/s

or 2 cm³ s⁻¹

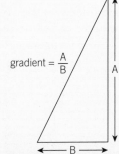

gradient = $\frac{A}{B}$

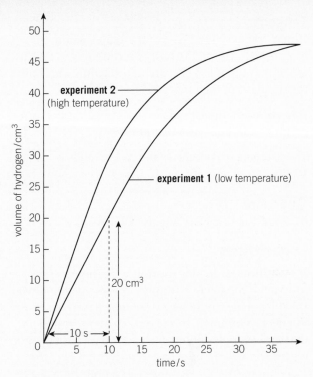

▲ **Figure 3** *Reaction rate graph*

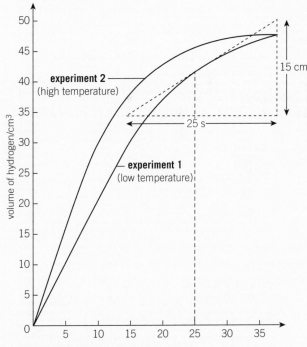

▲ **Figure 4** *Reaction rate graph*

The steeper the line is, the greater the rate.

Question

5 Find the rate for experiment 2 over the first 10 seconds.

Tangents

When you get to the part of the graph where the line starts to curve, the best you can do is to take a tangent at the point you are investigating and find the gradient of the tangent.

A tangent is a line drawn so that it just touches the graph line at one particular point.

The rate in experiment 1, after 25 seconds is
$$\frac{15\,\text{cm}^3}{25\,\text{s}} = 0.6\ \text{cm}^3\,\text{s}^{-1}$$

Questions

6 Find the rate after 30 s in experiment 2.

Logarithms

A logarithm, or log for short, is a mathematical function – the log of a number represents the power to which a base number (often ten) has to be raised to give the number. This is easy to do for numbers that are multiples of ten such as 100 or 10 000. 100 is 10^2, so the log to the base ten (written $\log_{10}$ or just log) of 100 is 2 and $\log_{10}$ of 10,000 is 4. Logs can also be negative numbers. The log of $\frac{1}{1000}$ is -3 as $\frac{1}{1000}$ is 10^{-3}. With other numbers you must use a calculator to find the log. Log_{10} 72.33 is 1.859 3.

Make sure that you are confident using your calculator to find logs.

Question

7 What is the $\log_{10}$ of:
   a 1000
   b $\frac{1}{100}$
   c 0.0001
   d 48.2
   e 0.037

You will need to use a calculator for the last two you can go back from the log to the original number by using the antilog, or inverse log function, of the calculator. Log (10^{27}) is 27 (which doesn't need a calculator to work out) and log (21379.6) is 4.33.

Make sure that you are confident to use your calculator to find antilogs (inverse logs).

Question

8 What is the antilog (inverse log) of:
   a 3
   b −2
   c 14
   d 8.2
   e 0.37

You will need to use a calculator for the last two. The log function turns very large or very small numbers into more manageable numbers without losing the original number (which you can recover using the antilog function). So $\log_{10} 6 \times 10^{23} = 23.778$ and $\log_{10} 1.6 \times 10^{-19} = -18.79$.

This can be very useful in plotting graphs as it can allow numbers with a wide range in magnitudes to be fitted onto a reasonable size of graph. For example, the successive ionisation energies of sodium range from 496 to 159 079 whereas their logs range from just 2.695 to 5.201.

Another important use of logs in chemistry is the pH scale, which measures acidity, and depends on the concentration of hydrogen ions (H^+) in a solution. This can vary from around $5\ \text{mol}\,\text{dm}^{-3}$ to around $5 \times 10^{-15}\ \text{mol}\,\text{dm}^{-3}$ – an enormous range. If you use a log scale, this becomes 0.698 9 to −14.301 – a much more manageable range.

When multiplying numbers you add their logs and when dividing you subtract them. This is easy to see with numbers that are multiples of ten.

$100 \times 10\,000 = 1\,000\,000$, that is, $10^2 \times 10^4 = 10^6$ and $\frac{10^4}{10^2} = 10^2$, but the same rules apply for more awkward numbers.

Geometry and trigonometry

Simple molecules adopt a variety of shapes. The most important of these are shown below with the relevant angles. When drawing representations of three dimensional shapes, the convention is to show bonds coming out of the paper as wedges which get thicker as they come towards you. Bonds going into the paper are usually drawn as dotted lines or reverse wedges.

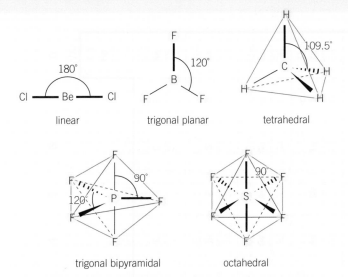

180°

Cl — Be — Cl

linear

F

120°

B

F   F

trigonal planar

H

109.5°

C

H   H

H

tetrahedral

F

90°

F

120°

P   F

F

F

trigonal bipyramidal

F

90°

F   F

S

F   F

F

octahedral

▲ **Figure 5** *Three-dimensional drawings of molecular shapes*

The three-dimensional shapes are based on
geometrical solid figures as shown

Glossary

A

Activation energy The minimum energy that a particle needs in order to react; the energy (enthalpy) difference between the reactants and the transition state.

Aldehyde An organic compound with the general formula RCHO.

Alkaline earth metals The metals in Group 2 of the periodic table.

Alkane A hydrocarbon with C—C and C—H single bonds only, with the general formula C_nH_{2n+2}.

Allotropes Pure elements which can exist in different physical forms in which their atoms are arranged differently. For example, diamond, graphite and buckminsterfullerene are allotropes of carbon.

Anaerobic respiration The process by which energy is released and new compounds formed in living things in the absence of oxygen.

Atom economy This describes the efficiency of a chemical reaction by comparing the total number of atoms in the product with the total number of atoms in the starting materials. It is defined by:

$$\% \text{ Atom economy} = \frac{\text{mass of desired product}}{\text{total mass of reactants}} \times 100$$

Atomic orbital A region of space around an atomic nucleus where there is a high probability of finding an electron.

Avogadro constant The total number of particles in a mole of substance. Also called the **Avogadro number**. It is numerically equal to 6.022×10^{23}.

B

Bond dissociation enthalpy The enthalpy change required to break a covalent bond with all species in the gaseous state.

C

Calorimeter An instrument for measuring the heat changes that accompany chemical reactions.

Catalyst A substance that alters the rate of a chemical reaction but is not used up in the reaction.

Catalytic cracking The breaking, with the aid of a catalyst, of long-chain alkane molecules (obtained from crude oil) into shorter chain hydrocarbons (some of which are alkenes).

Carbocation An organic ion in which one of the carbon atoms has a positive charge.

Carbon-neutral A process, or series of processes, in which as much carbon dioxide is absorbed from the air as is given out.

Chemical feedstock The starting materials in an industrial chemical process.

Co-ordinate bonding Covalent bonding in which both the electrons in the bond come from one of the atoms in the bond. (Also called dative covalent bonding.)

Covalent bonding Describes a chemical bond in a pair of electrons are shared between two atoms.

D

Dative covalent bonding Covalent bonding in which both the electrons in the bond come from one of the atoms in the bond. (Also called co-ordinate bonding.)

Delocalised Describes electrons that are spread over several atoms and help to bond them together.

Dipole–dipole force An intermolecular force that results from the attraction between molecules with permanent dipoles.

Displacement reaction A chemical reaction in which one atom or group of atoms replaces another in a compound, for example, $Zn + CuO \rightarrow ZnO + Cu$.

Displayed formula The formula of a compound drawn out so that each atom and each bond is shown.

Disproportionation Describes a redox reaction in which the oxidation number of some atoms of a particular element increases and that of other atoms of the same element decreases.

Dynamic equilibrium A situation in which the composition of a constant concentration reaction mixture does not change because both forward and backward reactions are proceeding at the same rate.

E

Electron density The probability of electrons being found in a particular volume of space.

Electron pair repulsion theory A theory which explains the shapes of simple molecules by assuming that pairs of electrons around a central atom repel each other and thus take up positions as far away as possible from each other in space.

Electronegativity The power of an atom to attract the electrons in a covalent bond.

Electrophile An electron-deficient atom, ion or molecule that takes part in an organic reaction by attacking areas of high electron density in another reactant.

Electrophilic addition A reaction in which a carbon–carbon double bond is saturated, by the carbon–carbon double bond attacking an electrophile.

Electrostatic forces The forces of attraction and repulsion between electrically charged particles.

Elimination A reaction in which an atom or group of atoms is removed from a reactant.

Empirical formula The simplest whole number ratio of atoms of each element in a compound.

Endothermic Describes a reaction in which heat is taken in as the reactants change to products; the temperature thus drops.

Enthalpy change A measure of heat energy given out or taken in when a chemical or physical change occurs at constant pressure.

Enthalpy diagrams Diagrams in which the enthalpies (energies) of the reactants and products of a chemical reaction are plotted on a vertical scale to show their relative levels.

Equilibrium mixture The mixture of reactants and products formed when a reversible reaction is allowed to proceed in a closed container until no further change occurs. The forward and backward reactions are still proceeding but at the same rate.

Exothermic Describes a reaction in which heat is given out as the reactants change to products; the temperature thus rises.

F

Fingerprint region The area of an infra-red spectrum below about $1500\,cm^{-1}$. It is caused by complex vibrations of the whole molecule and is characteristic of a particular molecule.

Fraction A mixture of hydrocarbons collected over a particular range of boiling points during the fractional distillation of crude oil.

Free radical A chemical species with an unpaired electron – usually highly reactive.

Functional group An atom or group of atoms in an organic molecule which is responsible for the characteristic reactions of that molecule.

G

Group A vertical column of elements in the periodic table. The elements have similar properties because they have the same outer electron arrangement.

H

Half equation An equation for a redox reaction which considers just one of the species involved and shows explicitly the electrons transferred to or from it.

Homologous series A set of organic compounds with the same functional group. The compounds differ in the length of their hydrocarbon chains.

Hydrogen bonding A type of intermolecular force in which a hydrogen atom ($H^{\delta+}$) interacts with a more electronegative atom with a $\delta-$ charge.

I

Incomplete combustion A combustion reaction in which there is insufficient oxygen for all the carbon in the fuel to burn to carbon dioxide. Carbon monoxide and/or carbon (soot) are formed.

Ionic bonding Describes a chemical bond in which an electron or electrons are transferred from one atom to another, resulting in the formation of oppositely charged ions with electrostatic forces of attraction between them.

Ionisation energy The energy required to remove a mole of electrons from a mole of isolated gaseous atoms or ions.

Isomer One of two (or more) compounds with the same molecular formula but different arrangement of atoms in space.

K

Ketone An organic compound with the general formula R_2CO in which there is a $C{=}O$ double bond.

L

Lattice A regular three-dimensional arrangement of atoms, ions or molecules.

Leaving group In an organic substitution reaction, the leaving group is an atom or group of atoms that is ejected from the starting material, normally taking with it an electron pair and forming a negative ion.

Lone pair A pair of electrons in the outer shell of an atom that is not involved in bonding.

M

Maxwell–Boltzmann distribution The distribution of energies (and therefore speeds) of the molecules in a gas or liquid.

Mean bond enthalpy The average value of the bond dissociation enthalpy for a given type of bond taken from a range of different compounds.

Metallic bonding Describes a chemical bond in which outer electrons are delocalised within the lattice of metal ions.

Mole A quantity of a substance that contains the Avogadro number (6.022×10^{23}) of particles (e.g., atoms, molecules or ions).

Molecular formula A formula that tells us the actual numbers of atoms of each different element that make up a molecule of a compound.

Molecular ion In mass spectrometry this is a molecule of the sample which has been ionised but which has not broken up during its flight through the instrument.

Monomer A small molecule that combines with many other monomers to form a polymer.

N

Nucleons Protons and neutrons – the sub-atomic particles found in the nuclei of atoms.

Nucleophile An ion or group of atoms with a negative charge or a partially negatively-charged area that takes part in an organic reaction by attacking an electron-deficient area in another reactant.

Nucleophilic substitution An organic reaction in which a molecule with a partially positively charged carbon atom is attacked by a reagent with a negative charge or partially negatively charged area (a nucleophile). It results in the replacement of one of the groups or atoms on the original molecule by the nucleophile.

Nucleus The tiny, positively charged centre of at atom composed of protons and neutrons.

O

Oxidation A reaction in which an atom or group of atoms loses electrons.

Oxidation state The number of electrons lost or gained by an atom in a compound compared to the uncombined atom. It forms the basis of a way of keeping track of redox (electron transfer) reactions. Also called oxidation number.

Oxidising agent A reagent that oxidises (removes electrons from) another species.

P

Percentage yield In a chemical reaction this is the actual amount of product produced divided by the theoretical amount (predicted from the chemical equation) expressed as a percentage.

Period A horizontal row of elements in the periodic table. There are trends in the properties of the elements as we cross a period.

Periodicity The regular recurrence of the properties of elements when they are arranged in atomic number order as in the periodic table.

Polar Describes a molecule in which the charge is not symmetrically distributed so that one area is slightly positively charged and another slightly negatively charged.

Positive inductive effect Describes the tendency of some atoms or groups of atoms to release electrons via a covalent bond.

Proton number The number of protons in the nucleus of an atom; the same as the atomic number.

R

Redox reaction Short for reduction–oxidation reaction, it describes reactions in which electrons are transferred from one species to another.

Reducing agent A reagent that reduces (adds electrons to) another species.

Reduction A reaction in which an atom or group of atoms gain electrons.

Relative atomic mass, A_r

$$A_r = \frac{\text{average mass of an atom}}{\frac{1}{12}\text{th mass of 1 atom of } ^{12}\text{C}}$$

Relative formula mass M_r

$$M_r = \frac{\text{average mass of an entity}}{\frac{1}{12}\text{th mass of 1 atom of } ^{12}\text{C}}$$

Relative molecular mass M_r

$$M_r = \frac{\text{average mass of a molecule}}{\frac{1}{12}\text{th mass of 1 atom of } ^{12}\text{C}}$$

S

Saturated hydrocarbon A compound containing only hydrogen and carbon with only C—C and C—H single bonds, i.e. one to which no more hydrogen can be added.

Specific heat capacity c The amount of heat needed to raise the temperature of $1\,g$ of substance by $1\,K$.

Spectator ions Ions that are unchanged during a chemical reaction, that is, they take no part in the reaction.

Standard molar enthalpy change of combustion $\Delta_c H^{\ominus}$ The enthalpy change when 1 mole of a substance is completely burned in oxygen with all reactants and products in their standard states ($298\,K$ and $100\,kPa$).

Standard molar enthalpy change of formation
$\Delta_f H^{\ominus}$ The enthalpy change when 1 mole of substance is formed from its elements with all reactants and products in their standard states (298 K and 100 kPa).

Stereoisomer Isomers with the same molecular formula and the same structure, but a different position of atoms in space.

Stoichiometry Describes the simple whole number ratios in which chemical species react.

Strong nuclear force The force that holds protons and neutrons together within the nucleus of the atom.

Structural formula A way of writing the formula of an organic compound in which bonds are not shown but each carbon atom is written separately with the atoms or groups of atoms attached to it.

Structural isomer Isomers with the same molecular formula but a different structure.

T

Thermochemical cycle A sequence of chemical reactions (with their enthalpy changes) that convert a reactant into a product. The total enthalpy change of the sequence of reactions will be the same as that for the conversion of the reactant to the product directly (or by any other route).

V

van der Waals force A type of intermolecular force of attraction that is caused by instantaneous dipoles and acts between all atoms and molecules.

Answers to summary questions

1.1

1 a i proton, neutron   ii proton, neutron
    iii proton, electron   iv neutron   v electron

  b Because they have opposite charges of the same size and the atom is neutral.

1.2

1 a 1 proton, 1 neutron, 1 electron

  b 1 proton, 2 neutrons, 1 electron

2 X and Z

3

Element		W	X	Y	Z
a	Number of protons	15	7	8	7
b	Mass number	31	14	16	15
c	Number of neutrons	16	7	8	8

Carbon dating

1 17 190 years (three half-lives)
2 Not necessarily – it tells us when the tree from which the wood of bowl was made died. The bowl may have been made later than this and would therefore not be so old.

1.3

1

  a   b   c

2 a 2,2   b 2,8,3   c 2,8,8

3 A^{2+}, C^-, E^+
  A: Mg, B: He, C: Cl, D: Ne, E: Li

1.4

1 Because they have lost one (or more) electrons (which have a negative charge)

2 They are attracted by a negatively charged plate.

3 The ions pass through a series of holes or slits.

4 Ions with the smallest m/z.

5 72.63

6 63.6

Relative abundance

Probability of chlorine molecule being
^{35}Cl—^{35}Cl (m/z = 70) is $\frac{3}{4} \times \frac{3}{4} = \frac{9}{16}$

Probability of chlorine molecule being
^{37}Cl—^{37}Cl (m/z = 74) is $\frac{1}{4} \times \frac{1}{4} = \frac{1}{16}$

Probability of chlorine molecule being
^{35}Cl—^{37}Cl (m/z = 72) is $\frac{3}{4} \times \frac{1}{4} = \frac{3}{16}$

Probability of chlorine molecule being
^{37}Cl—^{35}Cl (m/z = 72) is $\frac{1}{4} \times \frac{3}{4} = \frac{3}{16}$

Probability of ions of m/z = 72 can be added together:
$\frac{3}{16} + \frac{3}{16} = \frac{6}{16}$

1.5

1 a $1s^2 2s^2 2p^6 3s^2 3p^3$ b [Ne] $3s^2 3p^3$

2 a i $1s^2 2s^2 2p^6 3s^2 3p^6$ ii $1s^2 2s^2 2p^6$

  b [Ar], [Ne]

1.6

1 Second electron is removed from a positively charged ion whilst the first is removed from a neutral atom. More energy is needed to overcome the additional attractive force and so the second ionisation energy is higher.

2

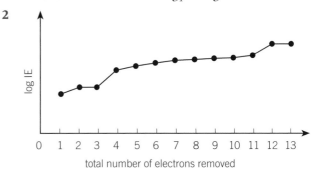

3 a Group 4

  b The large jump in ionisation energy comes after the removal of the fourth electron showing that there are four electrons in the outer shell.

2.1

1 a 16.0   b 106.0   c 58.3   d 132.1

2 Many answers possible such that the relative atomic masses add up to 16, for example, hydrogen atoms; or one carbon atom and 4 hydrogen atoms.

3 a 2   b 0.05   c 0.1

4 4g O_2 5 11g CO_2

2.2

1 a 1 mol dm^{-3} b 0.125 mol dm^{-3}

   c 10 mol dm^{-3}

2 a 0.002   b 0.025   c 0.05

3 a 58.5   b 0.004   c 0.016

2.3

1 a approximately 8.75×10^9

   b The same as part **a.**

2 a 50 360 cm^3 b 113 000 Pa

3 1.94

4 The same as in question **3**. The same number of moles of any gas has the same volume under the same conditions of temperature and pressure.

2.4

1 a H$_2$SO$_4$ sulfuric acid

   b Ca(OH)$_2$ calcium hydroxide

   c MgCl$_2$ magnesium chloride

2 a 0.16 mol Mg, 0.16 mol O   b MgO

3 a CH$_2$ b CHCl   c CH

4 C$_2$H$_6$O$_2$ 5 C$_3$H$_6$O

6 a CH   b C$_6$H$_6$

Finding the empirical formula of copper oxide

1 The flame burns off excess hydrogen to prevent it entering the laboratory.

2 The green colour is characteristic of copper ions.

3 Water droplets form due to the reaction of oxygen from the copper oxide reacting with the hydrogen. They condense while the tube is still cool.

4 This prevents oxygen from the air entering the tube and reacting with the copper while it is still hot and converting some of it back to copper oxide.

Erroneous results

1 $\dfrac{0.635}{63.5} = 0.01$ moles copper

   $0.735 - 0.635 = 0.100$ g oxygen

   $\dfrac{0.100}{16.0} = 0.006\,25$ moles oxygen

   This gives the formula Cu$_{1.6}$O

2 The student had not allowed the reaction to go to completion, or they had allowed air back into the reduction tube while the copper was still hot.

Another oxide of copper

1 mass of copper = 1.27 g

   mass of oxygen = 6.16 g

2 moles of copper = 0.02

   moles of oxygen = 0.01

3 Cu$_2$O

2.5

1 a 2Mg + O$_2$ → 2MgO

   b Ca(OH)$_2$ + 2HCl → CaCl$_2$ + 2H$_2$O

   c Na$_2$O + 2HNO$_3$ → 2NaNO$_3$ + H$_2$O

2 0.25 mol dm^{-3}

3 a Yes, there are 0.107 mol Mg. This would be enough to react with 0.0.214 mol HCl, but there is only 0.100 mol HCl.

   b 1238 cm^3

4 a i H$_2$SO$_4$(aq) + 2NaOH(aq) →

$$Na_2SO_4(aq) + 2H_2O(l)$$

      ii 2H$^+$ + SO$_4^{2-}$ + 2Na$^+$ + 2OH$^-$ →

$$2Na^+ + SO_4^{2-} + 2H_2O$$

   b Na$^+$ and SO$_4^{2-}$

2.6

1 CaCOOO → CaO + COO   56.0%

2 79.8%

3 100%   All the reactants are incorporated into the desired product.

4 a 1 mol   b 5.6 g   c 64.3%

3.1

1 **b** and **c**, they are both metal/non-metal compounds

2 Because they have strong electrostatic attraction between the ions that extends through the whole structure.

3 When they are molten or in aqueous solution.

4 a

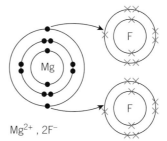

Mg^{2+}, 2F$^-$

   b

2Na^{2+}, O^{2-}

5 MgF_2 Na_2O **6** neon

Noble gas compounds

In xenon, outer electrons are further from the nucleus and more sub-shells means the outer electrons experience more shielding. Less energy required to remove an outer electron.

3.2

1 A pair of electrons shared between two non-metal atoms (usually) that holds the atoms together.

2 **b** and **d**, they are both non-metal/non-metal compounds

3

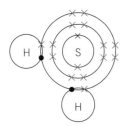

4

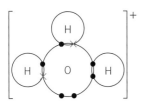

3.3

1 For example: metals conduct heat and electricity well, non-metals do not; metals are shiny, malleable and ductile, non-metals are not.

2 $1s^2\ 2s^2\ 2p^6\ 3s^2\ 3p^6\ 4s^2$

3 The two 4s electrons.   **4** 2

5 **a** Sodium would have a lower melting temperature because there are fewer electrons in the delocalised system and the charge on the ions is smaller.

 b Magnesium would be stronger as there are more electrons in the delocalised system and the charge on the ions is greater.

3.4

1 Fluorine is a smaller atom and when it forms a covalent bond, the shared electrons are closer to the nucleus.

2 $H^{\delta+}-Cl^{\delta-}$

3 **a** **i** and **ii** In both cases, the two atoms in the molecule are the same and therefore the electrons in the bond are equally shared.

4 **a** H—N < H—O < H—F

 b The order of the polarity is the same as the order of electronegativity of the second atom.

3.5

1 He, Ne, Ar, Kr. The van der Waals forces increase as the number of electrons in the atom increases.

2 H_2. It cannot have a permanent dipole because both atoms are the same.

3 Hexane (C_6H_{12}) is a larger molecule than butane (C_4H_{10}) and so has more electrons. This means that there are larger van der Waals forces between the molecules.

4 Because covalent bonds are localised between the two atoms that they bond and there is little attraction between the individual molecules.

5
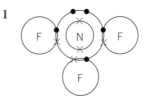

6 HBr

7 **a** There is no sufficiently electronegative atom.

 b There is no hydrogen atom.

8   **a** 2   **b** 2

 c A hydrogen bond requires both a lone pair of electrons on an atom of N, O, or F and a hydrogen atom. In water there are two lone pairs and two hydrogen atoms, allowing the formation of two hydrogen bonds. In ammonia, although there are three hydrogen atoms, there is only one lone pair of electrons on the N. This means that only one hydrogen bond can form.

3.6

1

eight electrons gives four pairs of electrons. This is a triangular pyramid.

2 BF_3 has three electron pairs in its outer shell and is trigonal planar. NF_3 has four electron pairs in its outer shell and its shape is based on that of a tetrahedron (the bond angle is 'squeezed down' by a couple of degrees because one of the electron pairs is a lone pair).

3

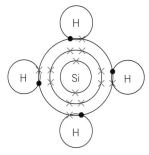

The shape is perfectly tetrahedral.

4 109.5°

5 Angular. It is similar to a water molecule.

3.7

1 a In a molecular crystal, there is strong covalent bonding between the atoms within the molecules but weaker intermolecular forces between the molecules. In a macromolecular crystal, all the atoms within the crystal are covalently bonded.

  b Macromolecular crystals have higher melting and boiling temperatures.

2 Both elements consist of molecular crystals. The van der Waals forces between the S_8 molecules are greater than those between P_4 because the sulfur molecules have more electrons.

3 The layers of carbon atoms are held together by weak van der Waals forces which allow the layers to slide over one another. They may also allow other molecules such as oxygen to penetrate between the layers.

4 Electricity is conducted via the delocalised electrons that spread along the layers of carbon atoms. Graphite conducts well along the layers but poorly at right angles to them. Metals conduct well in all directions.

5 Both have giant structures in which covalent bonding occurs between many atoms.

6 a A, C, D **b** B **c** A **d** B, D
  e C **f** D

4.1

1 445 kJ **2** endothermic

3 It is the reverse of the reaction in question **1**.

4 1.6 g

The energy values of fuels

1 $CH_3OH + 2O_2 \rightarrow CO_2 + 2H_2O$

2 No carbon dioxide formed.

3 Carbon dioxide is a greenhouse gas.

4.2

1 a \change **b** enthalpy (heat energy)

  c That the enthalpy change is measured at 298 K.

  d That heat is given out. **e** exothermic

  f enthalpy

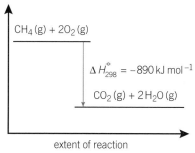

extent of reaction

4.3

1 −672 kJ mol^{-1}

2 a −46.2 kJ mol^{-1}

  b It will be smaller (less negative).

  c This is caused by heat loss.

3 a The enthalpy change.

  b The mass of water (or other substance in which the heat is collected).

  c The specific heat capacity of the water (or other substance).

  d The temperature change.

4.4

1 a −70 kJ mol^{-1} **b** −217 kJ mol^{-1}
  c −97 kJ mol^{-1} **d** −191 kJ mol^{-1}
  e +301 kJ mol^{-1}

4.5

1 a via $\Delta_f H^\ominus$ −85 kJ mol^{-1}
  b via $\Delta_c H^\ominus$ −86 kJ mol^{-1}

(The difference is due to rounding errors in the data.)

4.6

1 a −70 kJ mol^{-1} **b** 217 kJ mol^{-1}
  c −97 kJ mol^{-1} **d** −191 kJ mol^{-1}
  e +301 kJ mol^{-1}

4.7

1 a

$$H-\underset{\underset{H}{|}}{\overset{\overset{H}{|}}{C}}-\underset{\underset{H}{|}}{\overset{\overset{H}{|}}{C}}-H \ + \ Br-Br \longrightarrow H-\underset{\underset{H}{|}}{\overset{\overset{H}{|}}{C}}-\underset{\underset{H}{|}}{\overset{\overset{H}{|}}{C}}-Br \ + \ H-Br$$

2 a $1 \times$ C—C, $6 \times$ C—H, $1 \times$ Br—Br

  b 3018 kJ mol^{-1}

3 a $1 \times$ C—C, $5 \times$ C—H, $1 \times$ C—Br, $1 \times$ H—Br

  b 3063 kJ mol^{-1}

4 45 kJ mol^{-1}

5 a -45 kJ mol^{-1} **b** exothermic

5.1

1 temperature, concentration of reactants, surface area of solid reactants, pressure of gaseous reactants, catalyst

2 a reactants   **b** products

  c transition state or activated complex

  d activation energy

3 a exothermic

  b The products have less energy (enthalpy) than the reactants.

5.2

1 a Fraction of particles with energy E.

  b energy E

  c The number of particles with enough energy to react.

  d moves to the right   **e** no change

5.3

1 a A = enthalpy, B = extent of reaction, C = transition state with catalyst, D = transition state without catalyst, R = reactants, P = products

  b The activation energies without and with catalyst respectively

  c Exothermic

6.1

1 a true   **b** false   **c** true   **d** false

2 They are the same.

6.2

1 a Yes. This is a gas phase reaction with different numbers of particles on each side of the arrow.

  b No. This is not a gas phase reaction.

  c No. This is a gas phase reaction with the same number of particles on each side of the arrow.

2 a move to the left   **b** no change

  c Equilibrium would be set up more quickly but the final position would be unchanged.

  d High pressure. It would force the equilibrium in the direction of fewest particles.

6.3

1 For increased surface area.

2 The raw material for fermentation is sugar which is a crop which can be grown regularly. Ethene is made from crude oil which is non-renewable.

3 No, because synthesis gas is made from methane or propane which are derived from crude oil and/or natural gas.

Ammonia, NH$_3$

Making ammonia

1 a The reaction is exothermic, so low temperature would move the equilibrium to the right so that heat is given out.

  b High pressure forces the equilibrium to the side with fewer molecules, i.e. the ammonia side.

The Haber process

2 Approximately 40%

3 A lower temperature would reduce the reaction rate unacceptably. A higher temperature would require more expensive plant both in construction to withstand the higher pressure and in the costs of running compressors etc.

6.4

1 a $K_c = \dfrac{[C]_{eqm}}{[A]_{eqm}\,[B]_{eqm}}$

  b $K_c = \dfrac{[C]_{eqm}}{[A]_{eqm}{}^2\,[B]_{eqm}}$

  c $K_c = \dfrac{[C]_{eqm}{}^2}{[A]_{eqm}{}^2\,[B]_{eqm}{}^2}$

2 a mol^{-1} dm^3

  b mol^{-2} dm^6

  c mol^{-2} dm^6

3 a $K_c = 0.503$ (no units)

  b they cancel out

  c further to the right

6.5

1 a 1.01 mol

  b 1.067 mol

  c 2.067 mol

6.6

1 a increase

  b no change

  c no change

2 More water (and ethyl ethanoate) will be produced to oppose the change so that the equilibrium is maintained and K_c remains constant.

3 a increases

  b does not change

  c increases

  d does not change

  e does not change

7.1

1 a Bromine   b Calcium

  c Calcium   d Bromine

  e $Ca \rightarrow Ca^{2+} + 2e^-$ $Br_2 + 2e^- \rightarrow 2Br^-$

  f Bromine   g Calcium

7.2

1 a Pb +2, Cl −1   b C +4, Cl −1

  c Na +1, N +5, O −2

2 −2 before and after

3 before 0, after −2

4 before +2, after +3

5 a +5   b +5   c −3

7.3

1 a +2   0   +3   −1

  $Fe^{2+} + \frac{1}{2}Cl_2 \rightarrow Fe^{3+} + Cl^-$

  b Iron, because its oxidation number has increased.

  c Chlorine, because its oxidation number has decreased.

  d $Fe^{2+} \rightarrow Fe^{3+} + e^-$ $\frac{1}{2}Cl_2 + e^- \rightarrow Cl^-$

2 a i $3Cl_2 + 6NaOH \rightarrow NaClO_3 + 5NaCl + 3H_2O$

  ii $Sn + 4HNO_3 \rightarrow SnO_2 + 4NO_2 + 2H_2O$

b i $2\frac{1}{2}Cl_2 + 5e^- \rightarrow 5Cl^-$

  ii $\frac{1}{2}Cl_2 + 6OH^- \rightarrow ClO_3^- + 5e^- + 3H_2O$

  ii $Sn + 2H_2O \rightarrow SnO_2 + 4H^+ + 4e^-$

  $4HNO_3 + 4H^+ + 4e^- \rightarrow 4NO_2 + 4H_2O$

8.1

1 a i any two from Br, K, Fe

  ii Br and Cl or K and Cs

  iii Br, Cl

2 b i Fe   ii K or Cs

3 a Xe   b Ge   c Sr   d Ge or Xe   e W

8.2

1 a left   b right

2 a 0 (8 or 18)   b 4

3 a A: Na, B: Si

8.3

1 decrease   2 increase   3 increase

4 Because they have the highest nuclear charge.

The discovery of argon

1 5

2 A_r of argon is 50% greater than that of nitrogen and 1 mole of any gas occupies approximately the same volume.

3 Pass the air over a heated metal, e.g., copper.

4 42

5 It did not fit into the Periodic Table as it was understood then.

8.4

1 a $1s^2 2s^2$ b $1s^2 2s^2 2p^1$

2 a 2s   b 2p

3 Level 2p is of higher energy than 2s.

9.1

1 a +2   b They all lose their two outer electrons when they form compounds.

2 The outer electrons become further from the nucleus and are thus more easily lost.

3 Electrons are transferred from calcium to chlorine.

4 $\overset{0}{Ca(s)} + \overset{+1\,-2}{2H_2O(l)} \rightarrow \overset{+2\,-2\,+1}{Ca(OH)_2(aq)} + \overset{0}{H_2(g)}$

5 a more vigorous

b Less vigorous. The Group 2 metals become more reactive as we descend the group due, in part, to the fact that the outer electrons become further from the nucleus and are thus more easily lost.

6 a most soluble

b Least soluble. These are the trends found in the rest of the group.

Lime kilns

1 14 tonnes

2 Carbon dioxide produced in the reaction and by burning the fuel.

3 To increase its surface area.

4 11.9 tonnes

5 a 1 **b** 1

6 a 56 g **b** 76 g

7 Lime is more efficient.

8 cost, safety, etc.

The extraction of titanium

1 $2Fe_2O_3(s) + 3C(s) \rightarrow 3CO_2(g) + 4Fe(s)$

2 It is liquid – suggests $TiCl_4$ exists as covalently bonded molecules.

3 $TiO_2(s) + 2C(s) + 2Cl(g) \rightarrow Ti\,Cl_4(g) + 2CO_2(g)$

4 Carbon has been oxidised (0 to +2). Chlorine has been reduced (0 to –1)

10.1

1 a solid, very dark colour

b largest atom

c least electronegative

2 These properties can be predicted by extrapolating the trends observed with the halogens from F to I.

3 a approximately 600 K

b It has the most electrons and therefore the strongest van der Waals forces.

10.2

1 a mixture ii only

b Chlorine is a better oxidising agent than iodine and bromine. Therefore it can displace iodine from iodide salt, but bromine cannot displace chlorine from chloride salts.

c $Cl_2(aq) + 2NaI(aq) \rightarrow I_2(s) + 2NaCl(aq)$

Extraction of iodine from kelp

1 It is an oxidation – the oxidation state of the iodine atom goes from −1 to 0.

2 It is added to prevent thyroid problems which can be caused by lack of iodine in the diet.

10.3

1 a It has the least reducing power.

b The F^- ion is the smallest halide ion which means that it is hardest for it to lose an electron (which comes from an outer shell close to the nucleus).

c $NaF(s) + H_2SO_4(l) \rightarrow NaHSO_4(s) + HF(g)$

d No, the oxidation state of the fluorine remains as –1.

2 a Formation of a cream precipitate.

b $AgNO_3(aq) + NaBr(aq) \rightarrow NaNO_3(aq) + AgBr(s)$

c The precipitate would dissolve.

d To remove ions such as carbonate and hydroxide which would also produce a precipitate.

e Chloride ions (from HCl) and sulfate ions (from H_2SO_4) would also form precipitates with silver ions.

f Silver fluoride is soluble in water and does not form a precipitate.

10.4

1 a $Br_2(g) + H_2O(l) \rightarrow HBrO(aq) + HBr(aq)$

b $Br_2(g) + 2NaOH(aq) \rightarrow NaBrO\,(aq) + NaBr(aq) + H_2O(l)$

2 To kill micro-organisms and make the water safe to drink.

3 a hydrochloric acid and oxygen

b $\overset{0}{2Cl_2(g)} + \overset{+1\,-2}{2H_2O(l)} \rightarrow \overset{+1\,-1}{4HCl(aq)} + \overset{0}{O_2(g)}$

c oxygen has been oxidised

d chlorine has been reduced

e chlorine

f oxygen

11.1

1 a 0.4 mol **b** 1.0 mol **c** C_2H_5 **d** C_4H_{10}

e $CH_3CH_2CH_2CH_3$

f

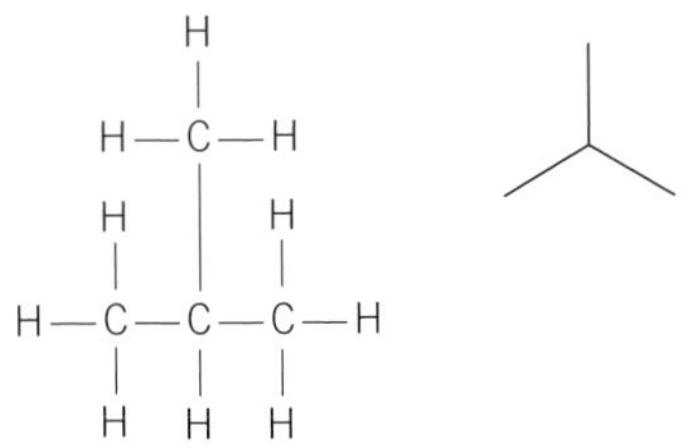

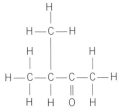

From inorganic to organic

1 A reaction in which the reactant and product have the same molecular formula but a different arrangement of atoms.

2 100% – there is just one starting material and one product.

11.2

1 **a** 1-chloropropane **b** pentane
   c pent-2-ene **d** 2-methylpentane

2 **a**

b

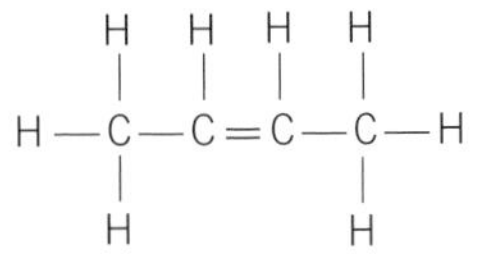

c

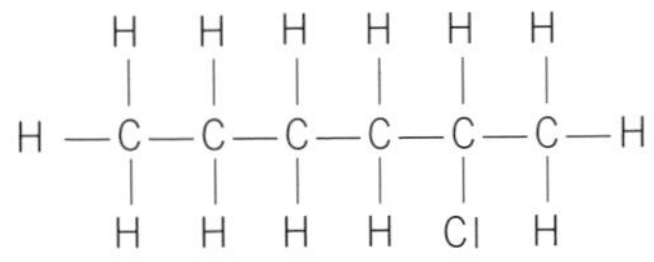

d

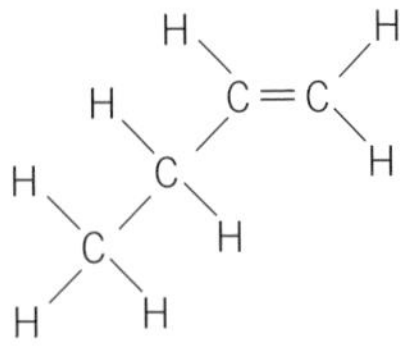

What's in a name?

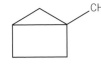
eave-methylhousane

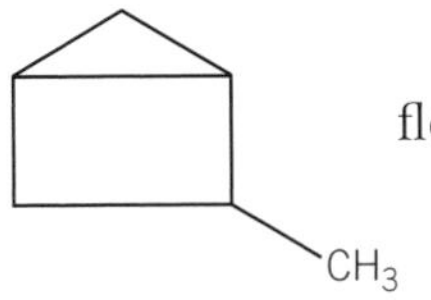

floor-methylhousane

11.3

1 **a** B **b** A **c** C

2 **a**

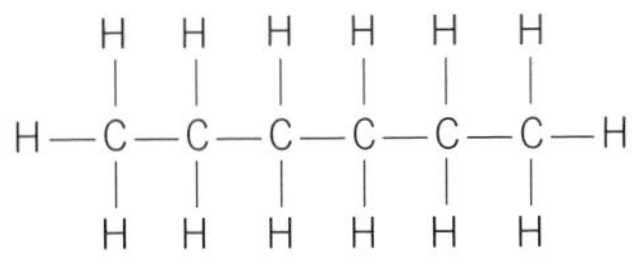

CH₃CH₂CH₂CH₂CH₂CH₃

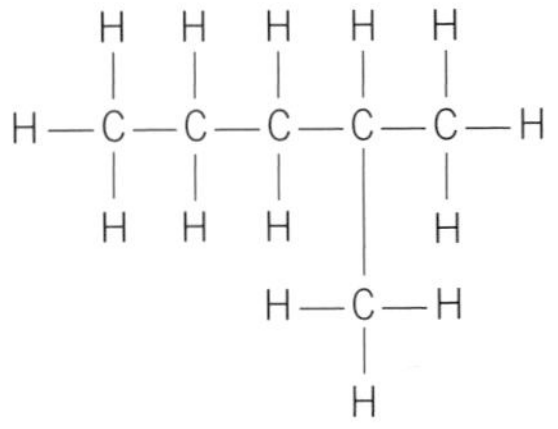

$CH_3CH_2CH_2CH_2CH_2CH_3$

$CH_3CH_2CH_2CH(CH_3)CH_3$

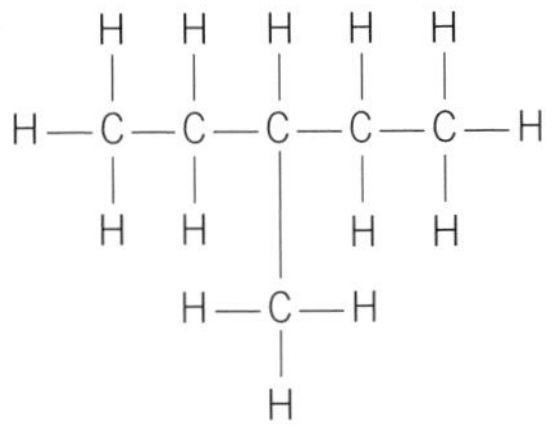

$CH_3CH_2CH(CH_3)CH_2CH_3$

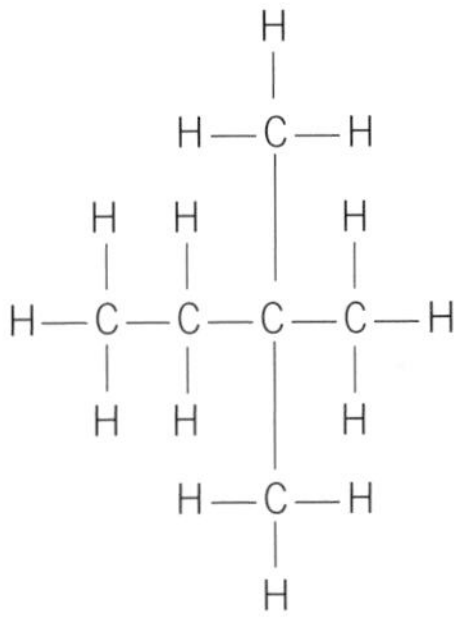

$CH_3CH_2C(CH_3)_2CH_3$

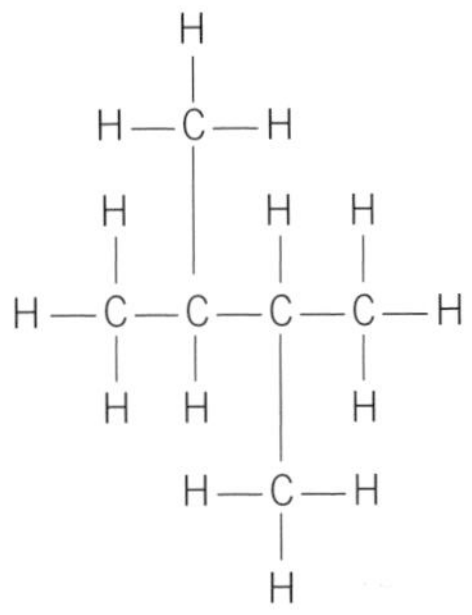

$CH_3CH(CH_3)CH(CH_3)CH_3$

  b hexane 2-methylpentane 3-methylpentane
        2,2-dimethylbutane 2,3-dimethylbutane

3 D

4 **a** *E*-pent-2-ene
   b *Z*-pent-2-ene

12.1

1 Methylbutane

2

$CH_3CH_2CH_2CH_2CH(CH_3)CH_3$

3 Heptane

4 Heptane will have a higher melting temperature because its straight chains will pack together more closely.

12.2

1

$CH_3CH_2CH_2CH_2CH_2CH_3$

2 Petrol

3 Fractional distillation separates a mixture into several components with different ranges of boiling temperatures whereas distillation simply separates all the volatile components of a mixture from the non-volatile ones.

4 e.g., methane, ethane, propane, butane

12.3

1 decane → octane + ethene

2 Many of the products are gases rather than liquids.

3 Octane itself has a short enough chain length to be in demand.

4 By using a catalyst

5 Short chain products are in greater demand than long chain ones. Alkenes are more useful than alkanes as starting materials for further chemical reactions.

12.4

1 a propane + oxygen → carbon dioxide + water

$C_3H_8(g) + 5O_2(g) \rightarrow 3CO_2(g) + 4H_2O(l)$

   b butane + oxygen → carbon monoxide + water

$C_4H_{10}(g) + 4\frac{1}{2}O_2 \rightarrow 4CO(g) + 5H_2O(l)$

2 a They produce carbon dioxide (a greenhouse gas) when they burn. They may produce poisonous carbon monoxide when burnt in a restricted supply of oxygen. They are in general non-renewable resources. They may produce nitrogen oxides and sulfur oxides when burnt. Cancer-causing carbon particulates may be produced and unburnt hydrocarbons (which contribute to photochemical smog) may be released into the atmosphere.

   b Burn as little fuel as possible and/or offset the CO_2 produced by planting trees, for example. Ensure that burners are serviced and adjusted to burn the fuel completely. Remove sulfur from the fuel before burning or remove SO_2 from the combustion products (by reacting with calcium oxide, for example).

3 Possibilities include wind power, wave power and nuclear power.

12.5

1 a Termination   b Propagation
   c Propagation   d Initiation

2 a $Cl\bullet + O_3 \rightarrow ClO\bullet + O_2$ and
      $ClO\bullet + O_3 \rightarrow 2O_2 + Cl\bullet$

   b $Cl\bullet + Cl\bullet$ $ClO\bullet + ClO\bullet$ $Cl\bullet + ClO\bullet$

13.1

1 a

A

B

C

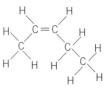

D

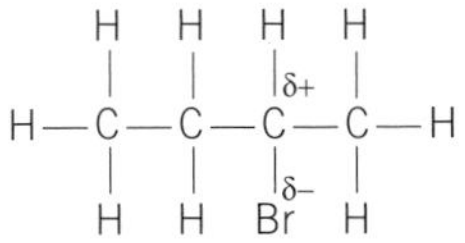

b A 1-iodobutane

   B 2-bromopropane

   C 1-chlorobutane

   D 2-bromobutane

c A, because it has the highest M_r and therefore most electrons and highest van der Waals forces.

2 Because the C—X bond becomes stronger

13.2

1 a Because haloalkanes do not dissolve in aqueous solutions

b OH^-

c Because the OH group replaces the halogen atom

d X^-

e R–I

2 a CN^-

b

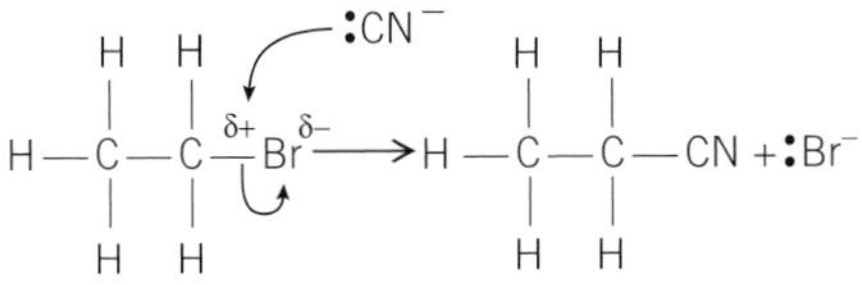

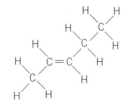

c Propanenitrile

13.3

1 A

2 D

3 a Propan-2-ol, propene

b Show that it decolourises a solution of bromine.

c

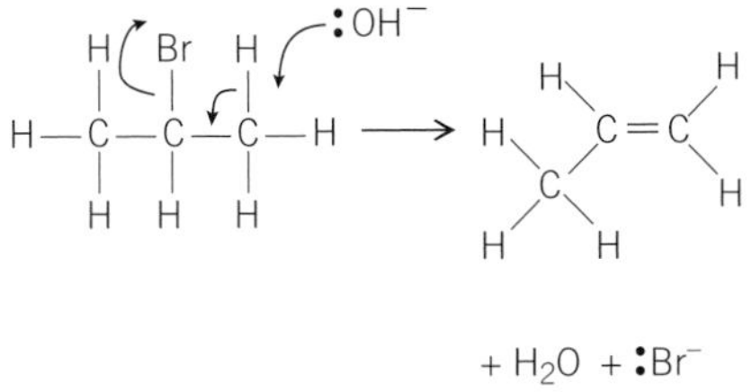

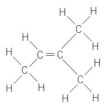

$+ H_2O + \textbf{:}Br^-$

CFCs

1

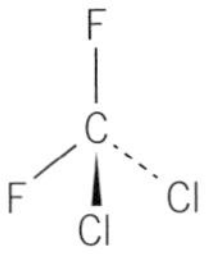

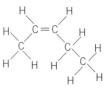

Tetrahedral with all atoms positioned around the carbon atom equivalent. No isomers.

2 structural

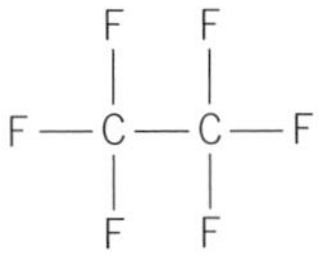

C_2F_5

14.1

1 Hex-2-ene

2 $CH_3CH_2CH_2CH_2CH{=}CH_2$

3

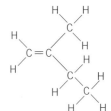

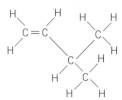

4 a electrophiles

5 a electron-rich

Bond energies

1 $612 - 347 = 265$ kJ mol^{-1}

2 The electron density in the σ-orbital is concentrated between the nuclei and holds them together better than the electron density of the π-orbital which is above and below the plane of the molecule.

14.2

1 $CH_2\!\!=\!\!CHCH_3 + 4\frac{1}{2}O_2 \rightarrow 3CO_2 + 3H_2O$

2 a electrophilic additions

3 a 1-bromopropane and 2-bromopropane

   b 2-bromopropane

   c It is formed from the more stable of the two intermediate carbocations.

4 chloroethane

5 c Bromine solution is decolourised.

14.3

1 A, B, and D

2 a

   b vinyl chloride

   c poly(chloroethene)

3 $CF_2\!\!=\!\!CF_2$

4 $CH_2\!\!=\!\!CHCl$

15.1

1

butan-2-ol

2 primary: methanol
secondary: butan-2-ol
tertiary: 2-methylpentan-2-ol

3 Because the oxygen atom has two lone pairs which repel more than bonding pairs.

Antifreeze

1

2 The ethane-1,2-diol molecules can form hydrogen bonds with water but cannot fit into the regular three-dimensional structure of ice, see Topic 3.5.
So these solutions remain liquid at lower temperatures than pure water.

The reactivity of alcohols

1

2 The C—C and C—H bonds are strong and relatively non-polar. The weakest bond is C—O which is also polarised $^{\delta+}$C—O$^{\delta-}$ so the C$^{\delta+}$ can be attached by nucleophiles.

15.2

1 By distillation. The ethanol has a lower boiling temperature than water and would distil off first.

2 Advantages: It is renewable. The process takes place at a low temperature.

   Disadvantages: The process is slow. It is a batch process rather than a continuous one. An aqueous solution of ethanol is produced rather than pure ethanol.

3 Fermentation takes place at a relatively low temperature and distillation at a much higher one. So the two cannot occur at the same time.

15.3

1 a A carboxylic acid is formed.

   b A ketone is formed.

2 This would require a C—C bond to break.

3 In distillation, the vapour is removed from the original flask and condensed in a different one. In refluxing, the vapour is condensed and returned to the original flask.

4 Gently oxidise the alcohols. In the case of the primary alcohol, an aldehyde will be formed that will give a positive silver mirror or Benedict's test. In the case of the secondary alcohol, a ketone will be formed that will not give a positive silver mirror or Benedict's test.

5 $CH_3CH_2OH \rightarrow CH_2\!\!=\!\!CH_2 + H_2O$ ethene

6 Pent-1-ene and pent-2-ene (*E*- and *Z*-isomers)

The breathalyser

1 1×10^{-6} g, 3.4×10^{-5} g

2 It measures to volume of the breath sample.

3 To make sure the reaction is complete within a few seconds.

16.1

1 Test the solubility of the precipitate in ammonia – see Topic 10.3.

2 **a** To remove any CO_3^{2-} or OH^- ions present which would also form a precipitate.

b This would form a precipitate of AgCl.

3 **a** alkane and carboxylic acid

b $CH_2{=}CHCO_2H$

c $CH_2{=}CHCO_2H + Br_2 \rightarrow CH_2Br{-}CHBrCO_2H$

$CH_2{=}CHCO_2H + NaHCO_3 \rightarrow CH_2CHCO_2Na + H_2O + CO_2$

16.2

1 **a** A solution of the molecules passes through a positively charged hollow needle.

b positive

2 $C_{10}H_{16}$

16.3

1 **a** or **b**

2 This IR peak is caused by $C{=}O$ which is present in both **a** and **b** but not **c**.

3 **b** or **c**

4 This IR peak is caused by $O{-}H$ which is present in **b** and **c** but not **a**.

5 **b**

6 This compound has both $C{=}O$ and $O{-}H$.

Greenhouse gases

1

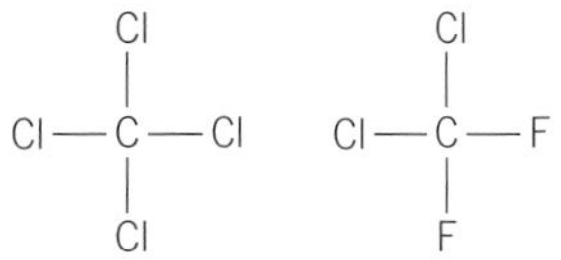

2 $C{-}Cl$ and $C{-}F$.

3 Temperature will vary as well as other climatic conditions.

4 CO_2: $1 \times 350 = 350$

CH_4: $30 \times 1.7 = 51$

Index

acids
  carboxylic acids 182, 236–238, 242
  equilibrium constants 114–117, 120
  Group 2 element reactions 156–157
  halide reactions 164–167
  ionic equations 36–37
  neutralisation of 72–73
activated complexes 97
activation energy 96–100
addition reactions 219–227
adsorption, catalysts 102
alcohols
  breathalysers 237–238
  classification 232
  combustion 235–236
  dehydration 235–236
  elimination reactions 235–236
  equilibria 112–117, 120
  formulae 230–231
  industry 112–113, 233–234
  infrared spectroscopy 245–248
  naming 182, 230–231
  oxidation 236–239
  physical properties 232
  reactions 235–239, 242
  reactivity 231
  structure 231
aldehydes 182, 236–239, 242
alkaline earth metals 150, 154–159
alkalis 167
alkanes 181, 182, 190–215
  boiling points 192, 207
  bonding 190–192, 206–207
  branched chains 190–191
  chain length 193
  chain reactions 201–203
  combustion 198–200
  cracking 196–197
  crude oil fractional distillation 193–195
  environmental problems 198–200, 212–213
  formulae 190–191, 206
  fuels 193–200
  halogenoalkanes 182, 201–203, 206–215, 242
  isomers 191, 211–213
  melting points 192

  naming 181, 182, 190–191, 206
  physical properties 192, 206–207
  polar bonds 206–207
  polarity 192, 206–207
  reactions 182, 192, 201–203, 206–215
  reactivity 192, 207
  solubility 192, 206–207
  structure 191
alkenes 181, 182, 196, 216–229, 242
  addition reactions 219–227
  CIP notation 218
  combustion 220–223
  electrophiles 219–223
  formulae 216
  halogen reactions 220–223
  hydrogen halide reactions 220–222
  isomers 217–218, 235–236
  naming 181, 182, 216, 218
  physical properties 219
  polymerisation 224–227
  reactions 219–224, 242
  reactivity 219
  structure 216–217
  sulfuric acid reactions 223
  water reactions 223
allotropes of carbon 66
ammonia 62
  equilibrium reactions 111–112, 120–121
  halogenoalkanes 210
  oxidation states 128
ammonium ions 61
amount of substances 22–43
  balanced equations 35–41
  concentrations 25–26
  formula determination 30–41
  gases 27–29
  moles 23–29
  reactions 35–41
  solutions 25–26, 38
  volumes 25–29
analytical methods
  combustion analysis 33–34
  organic analysis 242–253
aqueous solution 70, 131–132, 209
asymmetrical alkenes 220–222, 223

atmosphere pollution 198–200, 246
atom economy of reactions 39–40, 176
atomic number 6, 155, 160
atomic orbitals 14–16
atomic radii 147–148, 154, 160–161
atomic structure 2, 4–21
  electrons 4–9
  periodicity 147–148, 154, 160–161
atoms
  atomic structure 4–5
  electrons 4–5, 8–9
  number of 24
Avogadro constant/number 10, 23

balanced equations 35–41, 130–133
biofuels 233–234
biogradability 226
blocks of the Periodic Table 143–144, 145, 154, 160
boiling points
  alkanes 192, 207
  alkenes 219
  bonding 56–57, 64–65, 69
  gases 65
  hydrides 57
  hydrogen bonding 57
  intermolecular forces 56, 57
  liquids 65
  metals 51
  periodicity 145–146, 155, 160–165
  states of matter 64
  van der Waals forces 56
  water 57
Boltzmann distribution 98–99
bond angles 60–63, 190
bond dissociation enthalpy 91
bond energy 219
bond enthalpies 64–65, 91–93, 207
bonding 2, 44–71
  alcohols 231
  boiling points 56–57, 64–65, 69
  carbon compounds 176–180
  covalent compounds 44, 47–49, 52–53, 65–70
  crystals 65–67

dipole–dipole forces 54–55, 65, 69

electrical conductivity 48, 50–51, 68, 70

electron pair repulsion 60–63

electrostatic forces 45, 48, 55–56, 65

enthalpy change 64–65

halogenoalkanes 206–207

halogens 160

hydrogen bonding 48, 54, 56–58, 65, 69

infrared spectroscopy 245–246

intermolecular forces 54–58, 69

ionic compounds 44–46, 60–63, 65, 67–70

liquids 64–65, 70

lone pairs 49, 61–63

melting points 51, 64–65, 69–70

metals 44, 50–51, 65, 67–69

molecular shape 54–55, 60–63, 68–70

noble gas compounds 44

organic chemistry 176–189, 211, 216–226

physical properties 64–70

polarity 52–55, 93, 206–207

simple molecular structures 60–61, 68–70

solids 64–66, 70

states of matter 64–67, 70

van der Waals forces 54, 55–56, 65, 66–67, 69

bond strength 46, 48–49, 51, 57–58

Boyle's law 27

branched chains 176, 190–191

bromide test 201, 211

bromine extraction 163

buckminsterfullerenes/ buckyballs 67

calorimeters 78–79

carbocation 220–222

carbon compounds 176–187

  bonding 176–184

  formulae 177–180

  fuels 193–200

  reaction mechanisms 180, 208–209

carbon dating 7

carbon dioxide 200, 234, 246, 247

carbon isotopes 6–7

carbon monoxide 198

carbon neutrality 200, 234

carboxylic acids 182, 236–238, 242

catalysts 96, 100–103, 110, 120–121

catalytic converters 101–103, 199–200

catalytic cracking 197

chain isomerism 183–184, 186

chain reactions 201–203

charge, balanced equations 36

Charles' law 27

chemical equilibria 107

chemical formulae see formulae

chlorine isotopes 11–12

chlorine reactions 162, 167, 201–203

chlorofluorocarbons (CFCs) 100–101, 203, 212

CIP notation 218

cis–trans-isomerism 187, 218

classification

  alcohols 232

  halogenoalkanes 232

collision theory 96–97

combustion

  alcohols 235–236

  alkanes 198–200

  alkenes 220–223

  analysis 33–34

compound identity 245–246, 248–249

concentration

  equilibria 108–109, 114–120

  moles 25–26, 38

  solutions 25–26, 38

  titrations 38

constant volume law 27–29

co-ordinate bonds 47, 49

copper oxides 31–32, 128

covalent bonding 44, 47–49, 52–53, 65–70

  bond strength 48–49

  co-ordinate bonds 47, 49

  crystals 65–66

  dot-and-cross diagrams 66–67

  electrical conductivity 48, 68, 70

  electron density 52–53

  electronegativity 52–53

  electrostatic forces 48

ions 49

macromolecular structures 66–69

molecular bonding 47–49, 52–53, 65–67

periodicity 145

polarity 52–53, 93, 206–207

cracking 196–197, 216

crude oil 193–197, 216, 233

crystals 64, 65–67

curly arrows 180, 208–209, 211

cyanide ions 209–210

dative covalent bonding 47, 49

d-block 143–144

dehydration of alcohols 235–236

delocalised electrons 50

density

  Group 2 elements 155

  ice 57

diamond structure 66–67, 83

diatomic molecules 160–167

diffusion 64

dipole–dipole forces 54–56, 65, 69

dipole moments 54–55

displacement reactions 80–81, 162–163

displayed formulae 178, 180

disproportionation 132, 167

distillation 258

DNA 58

dot-and-cross diagrams 44–45, 61–63, 66–67

double bonds 48–49, 178, 181, 211, 216–226

dynamic equilibrium 106–123

electrical conductivity 48, 50–51, 68, 70

electronegativity 52–53, 160–161

electronic structures 16, 143–151, 154–155, 160–161

electron pair repulsion theory 60–63

electrons

  affinity 17

  arrangement 8–9, 143–151, 154–155, 160–161

  atomic orbitals 14–16

  atomic structures 4–9, 14–19

  bonding 44–71

  density 52–53

  diagrams 9, 60

energy levels 14–19
ionisation energy 17–19
lone pairs 49, 61–63
orbitals 14–16, 47, 150–151
redox reactions 124–126,
130–133
shells 9, 14–16, 60
spin 15
electrophiles 219–223
electrophilic additions
219–223
electrostatic forces 4, 45, 48,
55–56, 65
elements
enthalpy level diagrams 87
mass spectrometers 11–13
oxidation states 127–129
Periodic Table 143–149
elimination reactions 213–215,
235–236
empirical formulae 30–32,
177, 180
endothermic reactions 72–95, 97,
109–110
energetics 2, 72–95, 98–99
energy
activation energy 96–100
heating 64–65
levels 14–19, 76
states of matter 64–67
values of fuels 73–74
enthalpy change ΔH 73–95,
257–258
catalysts 100
of combustion 74, 77, 85–86
displacement reactions 80–81
of formation 77, 83–84, 86
of fusion 64–65
heat loss allowances 80–81
Hess's law 82–84, 91
mean bond enthalpies 91–93
measuring 77–81, 257–258
of melting 64–65
neutralisation of acids 72–73,
79–81
reactions 64–65, 73–95
of vaporisation 65
enthalpy H 73–95
bond enthalpies 64–65, 91–93,
209
level diagrams 76, 87–90, 97,
100
environmental problems 198–200,
212–213, 246–247

equations
balanced equations 35–41,
130–133
half equations 124–126,
132–133
redox 130–133
equilibria 2, 106–123
alcohols 112–117, 120
catalyst effects 110, 120–121
concentration effects 108–109,
114–120
conditions of 107, 108–110,
119–123
endothermic/exothermic
reactions 109–110
gases 111–112, 120–123
industrial reactions/
processes 111–113
Le Chatelier's principle 108–113,
119–123
pressure effects 109, 121–123
temperature effects 109–110,
119, 122–123
equilibrium constant K_c
114–123
equilibrium law 120
equilibrium mixtures 106–110
ethanoic acid 114–117, 120,
237, 247
ethanol 230–231, 247
distillation 256
equilibria 112–117, 120
industrial reactions 112–113
mass spectrometers 243
oxidation 236–238
production 233–234
ethene 234
exothermic reactions 72–95, 97,
109–110
E/Z isomerism 187, 218

f-block 145–146
Fehling's tests 239
fermentation 233–234
fingerprint region 245, 247–249
first ionisation energy 148–151,
154–155
flame calorimeter 78
flue gas desulfurisation 199
formulae
alcohols 230–231
alkanes 190–191, 206
alkenes 216
carbon compounds 177–180

determination 30–41
empirical 30–32
ionic compounds 36–37
molecular 30, 32–34
organic chemistry 177–191,
206, 216, 230–231
fracking 195
fractional distillation 193–195
fragmentation 243
free radicals 180, 201–203
fuels 73–74, 193–200
functional groups 182–184,
186, 242
fundamental particles 4–5
fusion 64–65

gas chromatography mass
spectrometry (GCMS) 244
gases
boiling points 65
enthalpy change 76
equilibria 111–112, 120–123
greenhouse gases 200, 246
ideal gas equation 27–29
moles 28–29
relative molecular mass 28–29
states of matter 64
volumes 27–29
Gay-Lussace law (constant volume
law) 27
geometrical isomerism 217
giant structures 66–69, 70,
145–146
global warming 200, 247
graphene/graphite 66–67, 83
greenhouse gases/effect 200, 246
Group 2 elements 150, 154–159
Group 7 elements 160–167
groups of Periodic Table 144,
148–151, 154–171

Haber process 111, 120–121
half equations 124–126, 132–133
halides 164–169, 220–222
halogenoalkanes
boiling points 192, 207
bonding 206–207
classification 232
elimination reactions 211–213
formation 201–203, 220–221
formulae 206
hydroxide reactions 209,
211–212
naming 182

nucleophilic substitution 209–210, 212

physical properties 206–207

potassium hydroxides 209, 211–212

reactivity 207

sodium hydroxide 209, 211–212

solubility 206–207

test-tube reactions 242

halogens

alkane reactions 182, 201–203, 206–215, 242

alkene reactions 220–223

reactions 164–169

heat

conductivity in metals 50

energy changes 64–65

enthalpy changes 72–95

loss allowances 80–81

see also temperature

Hess's law 82–84, 91

heterolytic bond breaking 221

high density polythene 226

high resolution mass spectrometry 244

homogeneous systems 116

homologous series of hydrocarbons 185

hydrides 57

hydrocarbons

boiling points 56

chains 176, 183–184, 186

formulae 177–189

isomers/isomerism 176, 183–184, 186–189

naming 181–185

rings 176, 182

saturation 181, 182, 190, 216

hydrogen

bonding 48, 54, 56–58, 65, 69

Periodic Table 144

hydrogen bromide 211–212

hydrogen chloride 48, 54–55

hydrogen halide reactions 220–222

hydroxides 157, 209, 211–212

ice, density/structure 57

ideal gas equation 27–29

impurity identification 245, 238–249

incomplete combustion 198–199

induced dipoles 55–56

industry

alcohols 112–113, 233–234

cracking 196–197

equilibrium reactions 111–113

infrared (IR) spectroscopy 245–249

initiation of chain reactions 201

inorganic chemistry 140–171, 176

instantaneous dipoles 55–56

intermolecular forces 54–58, 65–66, 69, 219

iodine 66, 163

ionic compounds

bonding 44–46, 60–63, 65, 67–70

bond strength 46

dot-and-cross diagrams 44–45, 61–63

electrical conductivity 68, 70

electrostatic forces 45, 65

formulae and equations 36–37

giant structures 68–69

properties 46

shapes 60–63

ionic crystals 65

ionic radii 164

ionisation energy (IE) 17–19, 148–155

ions

covalent bonding 49

shapes 60–63

IUPAC rules 181, 217

isomers/isomerism 176, 183–191, 211–213, 217–218, 235–236

isotopes 6–7, 11–12

Kelvin 27–29, 78

ketones 182, 236, 238–239

kinetic energy 77

kinetics 2, 96–105

catalysts 96, 100–103

collision theory 96–97

Maxwell–Boltzmann distribution 98–99

lattice structures 45, 65

Le Chatelier's principle 108–113, 119–123

liquids

boiling points 65

bonding 64–65, 70

electrical conductivity 70

enthalpy change 75–76

heating 65

locant 183

lone pairs of electrons 49, 61–63

low density polythene 226

low resolution mass spectrometry 11–12

macromolecular structures, bonding 66–69, 70

magnesium 156

mass

balanced equations 36–37

moles 23–26, 30

mass number 6–7

mass spectrometry 10–13, 243–244

Maxwell–Boltzmann distribution 98–99

mean bond enthalpy 91–93

melting points

alkanes 192

alkenes 219

bonding 51, 64–65, 69–70

enthalpy change 64–65

metals 51, 65

periodicity 145–146, 154–155, 160–161

states of matter 64

metal halides 165–166

metals

boiling points 51

bonding and structures 44, 50–51, 65, 67–69

electrical conductivity 50–51, 68, 70

melting points 51, 65

Periodic Table 50, 142–146, 155–159

properties 50–51

methane 47, 61

methanol 113

molar mass 23–26, 30, 74

molecular covalent bonding 47–49, 52–53, 65–67

molecular crystals 65–66

molecular formulae 30, 32–34, 177–178, 180

molecular ions 12

molecular shape 54–55, 60–63, 68–70

moles 10, 23–29

concentrations 25–26, 38

formulae determination 30–34
gases 28–29
mass relationship 23–26, 30
reacting quantities 39–41, 73
solutions 25–26
volume relationships 25–29
monomers 224–226

neutralisation of acids 72–73, 79–81
neutrons 4–5, 6–7
noble gas compounds 44
nomenclature of organic molecules 181–191, 206, 216, 218, 230–231
non-metals 142–143
nucleons 4–5
nucleophiles 207
nucleophilic substitution 208–210, 212
nucleus 4

octahedral molecules 61
OIL RIG 125, 130, 162
orbitals 14–16, 47, 150–151
organic chemistry 174–253
alcohols 230–241, 242
aldehydes 182, 236–239, 242
alkanes 181, 182, 190–215
alkenes 181, 182, 196, 216–229, 242
analytical methods 242–249
bonding 176–189, 211, 216–226
carbon compounds 176–187
carboxylic acids 182, 236–238, 242
double bonds 178, 181, 211, 216–226
formulae 177–191, 206, 216, 230–231
functional groups 182–184, 186, 242
halogenoalkanes 182, 201–203, 206–215
homologous series 185
IUPAC rules 181, 217
IR spectroscopy 245–249
isomers/isomerism 176, 183–191, 211–213, 217–218, 235–236
ketones 182, 236, 238–239
mass spectrometry 243–244
nomenclature 181–191, 206, 216, 218, 230–231

prefixes 181–182, 206, 218, 230–231
suffixes 181–182, 206, 218, 230–231
test tube reactions 242
oxidation 124–137, 156, 162
alcohols 236–239
alkaline earth metals 156
halogens 162
oxidation states 127–133
oxidising agents 124, 126
ozone layer 102–103, 199, 209

partial pressure 121
particles, atomic structure 4–5
Pauling scale 52
p-block 143–144
percentage error 26
percentage yield 39, 40–41
periodicity/Periodic Table 142–153
alkaline earth metals 155–159
atomic structure 147–148, 154, 160–161
blocks 143–144, 154
boiling points 145–146, 155, 160–161
electron arrangement 143–151, 154–155, 160–161
electronegativity 52–53
electronic structures 143–151, 154–155, 160–161
elements 143–149
Group 2 elements 154–159
Group 7 elements 160–167
groups 144, 148–151, 154–171
halogens 161–167
history 142
inorganic chemistry 142–171
ionisation energy 18–19, 148–155
melting points 147–148, 154–155, 160–161
metals 50, 142–146, 155–159
names and numbers 142–144
trends 145–151
periods 144
pH 156
physical properties
alcohols 232
alkanes 192, 206–207
alkenes 219
bonding 64–70
Group 2 elements 154–155

Group 7 elements 160–161
halogenoalkanes 206–207
planar molecules 60
plastics 226
see also polymers
polar bonds 52–55, 206–207
polarity 52–55, 93, 192, 206–207
pollution 198–120, 246
polymers 224–227
p-orbitals 67
positional isomers 183–184, 186, 217
positive inductive effect 221
potassium hydroxides 209, 211–212
prefixes 181–182, 206, 218, 230–231, 261
pressure
enthalpy changes 75–95
equilibria 109, 121–123
ideal gas equation 27–29
primary alcohols 232, 236
production of ethanol 233–234
propagation of chain reactions 201, 202
proton number 6–7
protons 4–5

quantum mechanics 14
quantum theory 8

radiocarbon dating 7
reactions
alcohols 235–239
alkanes 182, 192, 201–202, 206–215
alkenes 219–226
amount of substances 35–41
atom economy 39–40, 176
balanced equations/quantities 35–41
carbons 180, 218–209
chlorine 162, 169, 201–203
endothermic reactions 72–95, 97, 109–110
enthalpy change 64–65, 73–95
equilibria 106–123
exothermic reactions 72–95, 97, 109–110
free radicals 180, 201–203
Group 2 elements 156–157
halogens 162–167
mechanisms of carbons 180
oxidation 124–137, 156, 162

percentage yields 39, 40–41
   redox reactions 3, 124–126, 130–137
   reductions 124–137, 162, 165–166
reactivity trends 144, 192, 207, 219, 231
recycling polymers 227
redox equations 130–133
redox reactions 3, 124–126, 130–137, 164–166
reducing agents 124, 126
reduction reactions 124–137, 162, 164–165
relative atomic mass 10–12, 22–24, 30–33
relative formula mass 23–24
relative masses 4–5, 10–12, 22–24, 28–29
relative molecular mass 10, 22–24, 28–29, 177
repulsion of electron pairs 60–63
rotation of alkene bonds 216–217

salts *see* halogens
saturated hydrocarbons 181, 182, 190, 216
s-block 143–144
secondary alcohols 232, 236, 238
second ionisation energy 154–155
shapes 54–55, 60–63, 68–70
shared paired electrons 60
shells of electrons 9, 14–16, 60
silver ions 165–166
silver mirror test 239
silver nitrate 165–166
simple molecular structures 60–61, 68–70
SI units 27
skeletal formulae 179–180, 191
sodium halide reactions 164–165
sodium hydroxide 209, 211–212
solids
   bonding 64–66, 70
   electrical conductivity 70
   enthalpy change 76
   heating 64–65
solubility
   alkanes 192, 206–207
   alkenes 219
   Group 2 elements 154, 157

solutions
   amount of substances 25–26, 38
   concentrations 25–26, 38
   electrical conductivity 70
   enthalpy change 75–77
   moles in 25–26
   redox equations 131–132
space, mass spectrometers 12–13
specific heat capacity 78–79
spectator ions 36–37, 125
spectrometry 10–13, 243–244
spin, electrons 15
square planar ions 63
standard molar enthalpy change of reaction 77
states of matter 64–67, 70
state symbols 35–36
stereoisomerism 187
stoichiometry 34
straight chain alkanes 190
structural formulae 178–179, 185, 186
structural isomerism 186
structure
   alcohols 231
   alkanes 191
   alkenes 216–217
   ice 57
   Periodic table 142–143
sub-atomic particles 4–5
sub-shells of electrons 14
substitution reactions 208–210, 212
successive ionisation energies 17–18, 150–151, 154–155
suffixes 181–182, 206, 218, 230–231, 259
sulfates 157
sulfuric acid 166–168, 223
symmetry of alkenes 220–222, 223
systematic names 181, 184

temperature
   equilibria 109–110, 119, 122–123
   ideal gas equation 27–29
   Maxwell–Boltzmann distribution 99
   see also heat
termination of chain reactions 201, 202
tertiary alcohols 232, 236

test-tube reactions 242
tetrahedral ions 60–61, 62
thermal cracking 196–197, 216
thermochemical cycles 85–90, 93
thermochemistry 72–95
three-dimensional structural formulae 179–180
time of flight (TOF) mass spectrometers 10–11, 243
titrations 38, 114–115
Tollens' test 239
trans-isomerism 187, 218
transition state 97
trends in periodicity/Periodic Table 145–151, 154–157, 164
triangular pyramids 62
trigonal structures 60
triple bonds 178

ultraviolet (UV) light 202
unbranched chains 190
unsaturated hydrocarbons 216
   see also alkenes
uses of nucleophilic substitution 210

van der Waals forces 54, 55–56, 66–67
vaporisation 65
vibration 64
volumes
   equilibrium constants 115
   gases 27–29
   moles 25–29

water
   alkene reactions 223
   boiling points 57
   bromine extraction 163
   chlorine reactions 167
   Group 2 element reactions 156
   halogen reactions 163
   hydrogen bonding 48, 56–58
   ice 57
   mass spectrometry 244
   molecular shape/bond angles 62
   sampling 244
wavenumbers 245–249

zeolites 102

The authors would like to acknowledge Colin Chamber and Lawrie Ryan, as well as the contribution of their editors – Sophie Ladden, Alison Schrecker, Sadie Ann Garratt, and Sarah Ryan.

The authors and publisher are grateful to the following for permission to reproduce photographs and other copyright material in this book.

Acknowledgements

p2-3(b/g): Rynio Productions/Shutterstock; **p4**: IBM; **p10**: John Mclean/Science Photo Library; **p12**: Jpl-Caltech/Msss/NASA; **p23**: Martyn Chillmaid; **p27**: Hulton-Deutsch Collection/Corbis; **p31**: Martyn Chillmaid; **p62**: Nicolas/iStockphoto; **p66**: AptTone/Shutterstock; **p72**: Martyn F. Chillmaid/Science Photo Library; **p80**: Martyn F. Chillmaid; **p102**(T): Foxterrier2005/Shutterstock; **p102**(B): Constantine Pankin/Shutterstock; **p108**: Science Photo Library; **p114**: Science Photo Library; **p128**: Martyn F. Chillmaid/Science Photo Library; **p131**: Charles D. Winters/Science Photo Library; **p140-141**(b/g): Kuttelvaserova Stuchelova/Shutterstock; **p142**: SCIENCE PHOTO LIBRARY; **p155**: Andrew Fletcher/Shutterstock; **p156**(B): Russ Munn/Agstockusa/Science Photo Library; **p156**(T): Green Gate Publishing Services, Tonbridge, Kent; **p160**: Sciencephotos/Alamy; **p166**: Andrew Lambert Photography/Science Photo Library; **p174-175** (b/g): Sergio Stakhnyk/Shutterstock; **p184**(T): Molekuul.Be/Shutterstock; **p184**(B): Meunierd/Shutterstock; **p192**(T): Martyn F. Chillmaid; **p192**(B): Martyn F. Chillmaid/Science Photo Library; **p194**: Terry Poche/Fotolia; **p196**: Green Gate Publishing Services, Tonbridge, Kent; **p198**: E.R.Degginger/Science Photo Library; **p199**(T): Kzenon/Shutterstock; **p199**(B): Science Photo Library; **p201**: Fotostorm/iStockphoto; **p206**: Martyn F. Chillmaid/Science Photo Library; **p217**: Martyn F. Chillmaid; **p224**: Corel; **p225**: Martyn F. Chillmaid/Science Photo Library; **p226**(T): Andrew Lambert Photography/Science Photo Library; **p226**(B): Woraput/iStockphoto; **p230**: Martyn F. Chillmaid/Science Photo Library; **p231**: urbanbuzz/Shutterstock; **p235**: Martyn F. Chillmaid; **p237**: Isopix/Rex Features; **p239**(T): Andrew Lambert Photography/Science Photo Library; **p239**(B): Martin Shields/Alamy; **p243**: Corel; **p244**: Colin Cuthbert/Newcastle University/ Science Photo Library; p256: Science Photo Library;

Artwork by Q2A Media